The Best ACT® Math Books Ever
Book 2: Numbers, Stats, Trig, and Geometry

Created by Brooke P. Hanson

The Best ACT® Math Books Ever
Book 2: Numbers, Stats, Trig, and Geometry

Created by Brooke P. Hanson

Additional Contributions by SupertutorTV

Layout by Jennifer Wang

Cover design by Skyler Thiot

First Edition

Published by Supertutor Media, Inc.

ACT® is a registered trademark of ACT®, Inc., and was not involved in the making of this book.

Copyright © 2019 by Supertutor Media, Inc.
All rights reserved. No portion of this book may be reproduced, reused, or transmitted in any form.

ISBN-13: 978-1-7322320-1-3
ISBN-10: 1-7322320-1-6

Any questions, comments, or errors can be submitted to info@supertutortv.com.

WANT MORE AWESOME PREP?

THE BEST ACT PREP COURSE EVER

Head to SUPERTUTORTV.COM

Video-Based Online ACT Prep by a Perfect Scoring Tutor*!

- A comprehensive ACT course with over 100 hours of video content
- Know-how for every section of the ACT: English, Math, Reading, Science and Essay
- Our math videos cover EVERY lesson in this book series. It's like having a private tutor walk you through each chapter!
- Includes a FREE copy of the Official ACT Prep Guide***
- Downloadable English Content Drills and Practice Essay Topics, plus digital versions of the math problems and lessons in this book series
- Explanations for 8+ Official ACT practice tests***
- Private tutoring experience & results at a fraction of the cost!

Students have raised their ACT score by **up to 12 points** using our course, with an average score increase of 5 points**.

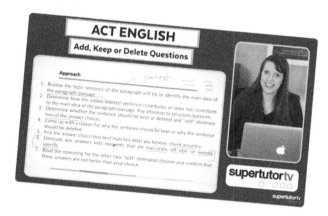

supertutortv

Or get BUNDLE SAVINGS with our SAT & ACT Course Bundle!

*As an adult, Brooke has scored a perfect 36 composite and 12/12 essay on the ACT. **As self reported by course users *** Some tests may require additional book not included. See course description page for more information: https://supertutortv.com/the-best-act-prep-course-ever; ACT is a registered trademark of ACT Inc. This course is not endorsed by or associated with ACT Inc.

Access Your Free Online Resources*

FREE ACT calculator programs for download, book errata, our pdf format ACT math formula "cheat sheet" and more. Find them at:

supertutortv.com/BookOwners

*Creation of an account may be required

Need Private Tutoring?

Brooke offers tutoring to a limited number of one-on-one private students, both online via Skype and in her Los Angeles office for the ACT, SAT, ISEE/SSAT and writing. She also offers college admissions consulting and essay coaching. For more information, visit:

supertutortv.com/tutoring-information

Join our Mailing List!

Join our mailing list for FREE to keep up to date on our YouTube video releases, new products, tips, reminders and exclusive offers! Sign up at:

supertutortv.com/subscribe

Follow us!

YouTube.com/SupertutorTV
Instagram @SupertutorTV
Facebook.com/SupertutorTV
Twitter @SupertutorTV

TABLE OF CONTENTS

Introduction vii

Part One: Numbers 1

1. Scientific Notation 3
   Problem Set & Answers 7

2. Properties of Numbers 9
   Problem Set & Answers 16

3. LCM & GCF 22
   Problem Set & Answers 27

4. Fractions 31
   Problem Set & Answers 39

5. Percents 44
   Problem Set & Answers 51

6. Sequences and Series 56
   Problem Set & Answers 66

Part Two: Probability and Statistics 73

7. Averages 75
   Problem Set & Answers 84

8. Data Analysis 94
   Problem Set & Answers 101

9. Counting and Arrangements 108
   Problem Set & Answers 124

10. Probability 132
    Problem Set & Answers 143

Part Three: Geometry 153

11. Angles and Lines 155
Problem Set & Answers 163

12. Triangles 175
Problem Set & Answers 188

13. Circles 195
Problem Set & Answers 206

14. Polygons 216
Problem Set & Answers 229

15. Similar Shapes 239
Problem Set & Answers 247

16. Solids 251
Problem Set & Answers 265

17. Vectors 271
Problem Set & Answers 278

Part Four: Trigonometry 281

18. SOHCAHTOA 283
Problem Set & Answers 290

19. Trigonometry 296
Problem Set & Answers 309

20. Law of Sines and Cosines 323
Problem Set & Answers 326

21. Trig Graphs 331
Problem Set & Answers 342

FOREWORD

As a private tutor, I always try to ensure my students have total mastery of the content necessary to crush any exam. On the ACT math section, that means exposure to every piece of math knowledge that might be on the test. But in my 15+ years of tutoring, I could never find all that math knowledge in one place. I tried book after book, but while one had nice lessons, it offered a measly 3 practice problems per section. While another seemed to have drills, it left out major sections covered by the exam. As a result, I was constantly assigning my students a mishmash of pdf worksheets, chapters from their sister's Algebra book, and portions of multiple ACT prep books. True, I had some students score perfectly on the ACT math exam. But I always wanted to have a single source for all the formulas, skills, and drills my students needed to master the math section once and for all. Thus this book series was born.

This series is the culmination of 3+ years of research and work. Every problem in the series is inspired by an authentic ACT problem. With the help of my team, I've reviewed dozens of released official ACT exams, sorting through thousands of problems, to ensure we cover what you need to know. Each chapter is organized by topic area. I list the skills you need to know, introduce each topic, and highlight any key terms or formulas you'll need. Then, I show you at least one step-by-step example of how to approach each type of problem. Each lesson is followed by a problem set with full explanatory answers, typically with ~20 or so problems, enough that you still have more to drill with if you need to peek at the answers until you get the hang of things. If you're wanting more explanation of these topics, I walk through every lesson in this book in our online ACT video based course, The Best ACT Prep Course Ever, at SupertutorTV.com.

To accommodate the 1000+ practice problems in the series, I've split the material into two volumes. This second volume covers numbers, statistics, geometry and statistics. If you're looking for help in Algebra, check out the 1st volume in this series: The Best ACT Math Books Ever, Book 1: Algebra.

It's taken a village to finish these books. From cutting out and sorting questions by type to proofreading questions, spooling answer choices, drafting explanations, and drawing geometry figures, over a dozen people contributed to the problem sets in this book, sharing their problem solving know how. Thanks so much to all of you!

Contributors:
Estefania Lahera, Elijah Spiegel, Jennifer Cook, Heidi Chiu, Gabriel Schneider, Lianne Huang, Matthew Morawiec, Nathan Wenger, Megan Roudebush, Daniel Lee, Thierno Diallo, Liane Beatriz Capiral, Leslie Kim, Celestine Seo, Arjun Verma

We've proofread, but there still may be errors. If you find any, email us at info@supertutortv.com. If any explanations confuse you, don't hesitate to reach out!

HOW TO USE THIS BOOK

Method 1: Do EVERYTHING

If you have 6+ months to prep, at least a couple of hours per week, and want to get the most insanely high score possible, you can do this book and the other book in this series start to finish. Feel free to skip over problems that are crazy easy for you as you go. On the other hand, if you're not aiming for a top score, feel free to skip the toughest problems so you can get an all around baseline on the easier ones.

Method 2: Target & Drill

For everyone else, I recommend taking an official ACT practice test and grading it. You can download one for free from our resources page at **SupertutorTV.com/resources** if you need one. Then figure out what kinds of questions you missed, identify the corresponding chapters, and "assign" these chapters to yourself. If you're an online course subscriber, I typically will indicate which problem type each question is (i.e. suggest what book chapters to review per the question you missed) in my corresponding explanation video. We have explanations for about 30 math questions from the 2017-2018 Official ACT practice test available free on our YouTube Channel as well: **youtube.com/SupertutorTV**. Go to our page and scroll down to look for the playlist.

Armed with your study list, read or skim each relevant chapter before the problem set, depending on your level of comfort with the particular problem type. (If you are a subscriber to our online course, you can also follow along with our lesson video). Then attempt the problem set. If you like, you can first start with the odds on the first pass, grade and review with the explanations, and then return again to the evens to solidify your learning and see if you can get the rest correct without peeking at any more explanations. Once you're done, take another practice test and repeat!

The chapters in this book are not entirely in order of difficulty, though they're approximately in sequential learning order and grouped with similar content.

If you miss a particular type of question more than once, start with these. I also recommend starting with problems that correspond to lower numbered test questions first and leaving the hard stuff (questions 50-60) for last.

Remember, you don't have to do every page or problem in this book to succeed on the test. Prep smart. Just because I write something doesn't mean YOU need it. Figure out your weak areas and do what you need the most help with.

PART ONE: NUMBERS

CHAPTER

1 SCIENTIFIC NOTATION

> ### SKILLS TO KNOW
> - How to complete basic problems that ask for answers in scientific notation
> - How to complete word problems involving scientific notation

THE BASICS

Most of you have probably learned scientific notation back in elementary or middle school—it's one of those skills that you may have known once, but may be a bit rusty now, so let's review some of the basics:

1. What Is It?

For a number in the form $a \times 10^n$ to be in scientific notation, it must contain a value (a) greater than or equal to 1 and less than 10 $(1 \leq a < 10)$ multiplied by ten to power (n), where n is an integer.

Some ACT® problems may be easier to solve when you can eliminate answers NOT officially in scientific notation off that technicality. Both numbers below have the same value, but only one is in scientific notation.

This **IS** in scientific notation:
$$5.5 \times 10^{-7}$$

This is **NOT** in scientific notation:
$$55 \times 10^{-6}$$

2. Positive Exponents

When a number is multiplied by ten to a positive exponent of degree n, move the decimal point n places to the RIGHT. (NOTE: The ACT will not likely have a question this easy).

> Expand the following number:
> $$4.3 \times 10^9$$

Since the exponent is 9, move the decimal points nine places to the right:

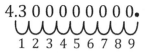

4,300,000,000

As you can see, I draw a little loop for each hop, and count all the loops.

Answer: 4.3 billion.

To change the number back to scientific notation, move the decimal back to the left, and count. When you are to the point that only one number remains to the left, you're done. Count the number of moves—that is your exponent.

3. Negative Exponents

When a number is multiplied by ten to a negative exponent of degree n, move the decimal point n places to the LEFT.

$$5.854 \times 10^{-5} + 3.56 \times 10^{-3} = ?$$

First we'll put the first number in regular form, then the second. Then we'll add the numbers and convert back to scientific notation.

$$.00005.854 \quad + \quad .003.56$$
$$5\,4\,3\,2\,1 3\,2\,1$$

Then line up the decimal points to add the numbers (don't forget to carry the "1"):

$$\begin{array}{r} 1 \\ 0.00005854 \\ +0.00356000 \\ \hline 0.00361854 \end{array}$$

Now we convert back to scientific notation:

$$0.003.61854$$
$$1\,2\,3$$

As you can see, we need to move the decimal point three places to the right—so we add an exponent of negative 3 to our 10:

$$3.61854 \times 10^{-3}$$

That's our answer!

WORD PROBLEMS

We'll also need to know how to solve word problems that involve scientific notation. In general, for these problems, first convert the scientific notation numbers into standard form. Then, solve the problem. Finally, convert your answer back to scientific notation.

The area of four oceans is given in the following chart in square miles. What is the total area that these four oceans cover?

Indian	2.84×10^7
Pacific	6.25×10^7
Atlantic	4.11×10^7
Arctic	5.43×10^6

For the problem, because three of the oceans have the same power of 10, we can add those integers first, factoring out the 10^7 term.

$$2.84 \times 10^7 + 6.25 \times 10^7 + 4.11 \times 10^7 = 10^7 \times (2.84 + 6.25 + 4.11)$$

Using your calculator, you can find that:

$$2.84 + 6.25 + 4.11 = 13.2$$

Now we can convert that to standard form:

$$13.2\,0\,0\,0\,0\,0\,0.$$

That's 132 million: $132{,}000{,}000$.

Now we convert the Arctic Ocean to standard form:

$$5.4\,3\,0\,0\,0\,0.$$

That's $5{,}430{,}000$ or 5.43 million.

Now we add 132 million plus 5.43 million:

$$\begin{array}{r} 132{,}000{,}000 \\ +5{,}430{,}000 \\ \hline 137{,}430{,}000 \end{array}$$

Finally, we convert back to scientific notation:

$$1.3\,7\,0\,0\,0\,0\,0\,0.$$

Answer: 1.3743×10^8.

TRICKY PROBLEMS

> Whenever x and y are both integers, what is the approximate value of $(4.56 \times 10^x)(5.22 \times 10^y)$ expressed in scientific notation?
>
> A. $2.38 \times 10^{yx+1}$ B. $2.38 \times 10^{x+y}$ C. $2.38 \times 10^{x+y+1}$
>
> D. $2.38 \times 10^{x+y-1}$ E. $2.38 \times 10^{x-y-1}$

Occasionally, you may have scientific notation problems that involve variables in the answer choices. For these problems, you could make up numbers, find the pattern and plug in what you made up into each answer choice to discover the answer. You can also attempt algebraically.

Algebraic Method:

Here you can first rearrange to get:

$$(4.56 \times 5.22)(10^x \times 10^y)$$

Applying our exponent rule:
$$\left(a^x \times a^y = a^{x+y}\right)$$

That's approximately:
$$(23.8)\left(10^{x+y}\right)$$

Now I put the 23.8 in scientific notation:
$$\left(2.38 \times 10^1\right)\left(10^{x+y}\right)$$

And finally, I combine the 10 with the other 10 powers:
$$(2.38)\left(10^1 \times 10^{x+y}\right)$$

Again I apply the power rule: $\left(a^x \times a^y = a^{x+y}\right)$
$$(2.38)\left(10^{x+y+1}\right)$$

Answer: **C**.

 NOTE: For more help with exponents, see the chapter Exponents and Radicals in Book 1. For more help on units and ratios check out Chapter 9 in Book 1 on Ratio, Rates, & Units. Many word problems here overlap with these skills.

1. The platelet count of a healthy adult is about 2.5×10^{-5} parts per 1 million parts blood. In a patient with thrombocytopenia, a disorder in which the body produces fewer platelets, the platelet level is 100 times lower than this average. What is the approximate patient's platelet count in parts per million?

    A. 2.5×10^{-7}
    B. 2.5×10^{-3}
    C. 2.5×10^{-105}
    D. -9.9×10
    E. 2.5×10^{7}

2. The average diameter of a monocyte, a type of white blood cell, is 2.00×10^{-5} meters. A monocyte's diameter is about how many times the diameter of a red blood cell, whose average diameter is 6.50×10^{-6}?

    A. 3.25×10
    B. 3.08×10
    C. 3.08
    D. 3.25
    E. 3.08×10^{-11}

3. A hummingbird beats its wings approximately 14 times per second. After 5 hours and 20 minutes, how many times has the bird beat its wings?

    A. 7.47×10^{1}
    B. 2.62×10^{5}
    C. 4.36×10^{3}
    D. 2.69×10^{5}
    E. 4.48×10^{2}

4. At sea level, the speed of sound is $3.40 * 10^{2}$ meters per second. After 3 hours and 30 minutes, about how many kilometers could a sound wave travel?

    A. 4.28×10^{3} km
    B. 7.14×10^{3} km
    C. 7.14×10^{4} km
    D. 6.73×10^{4} km
    E. 4.28×10^{5} km

5. According to the EPA, water is no longer safe to drink once the level of lead in the water reaches 15,000,000 parts per billion. What is the corresponding ratio of lead to water in a sample at this level of contamination written in scientific notation?

    A. $1.5 \times 10^{-1} : 1$
    B. $1.5 \times 10^{-2} : 1$
    C. $1.5 \times 10^{-3} : 1$
    D. $1.5 \times 10 : 1$
    E. $1.5 \times 10^{2} : 1$

6. Which is the closest to 6.5×10^{-4}?

    A. 0.065
    B. 0.0065
    C. 0.00065
    D. 0.000065
    E. 0.0000065

7. Whenever x and y are both integers, what is $\dfrac{3.5 \times 10^{y}}{6.2 \times 10^{x}}$ expressed in scientific notation?

    A. $5.6 \times 10^{y-x-1}$
    B. $5.6 \times 10^{y-x}$
    C. $5.6 \times 10^{\frac{y}{x}-1}$
    D. $5.6 \times 10^{y+x}$
    E. $5.6 \times 10^{x-y-1}$

SCIENTIFIC NOTATION — ANSWERS

ANSWER KEY
1. A   2. C   3. D   4. A   5. B   6. C   7. A

ANSWER EXPLANATIONS

1. **A.** If the platelet count we need is 100 times lower than the healthy amount of 2.5×10^{-5} ppm, then we divide to find the lower platelet count: $\dfrac{2.5 \times 10^{-5}}{100} = \dfrac{2.5 \times 10^{-5}}{10^2} \to 2.5 \times 10^{-7}$ parts per million.

2. **C.** 2.00×10^{-5} is x times 6.50×10^{-6}. Solving for x, we get:
$$2.00 \times 10^{-5} = x(6.50 \times 10^{-6}) \to x = \dfrac{2.00 \times 10^{-5}}{6.50 \times 10^{-6}} = \dfrac{2}{6.50} \times 10^{-5-(-6)} = 0.308 \times 10^{1} = 3.08.$$

3. **D.** We'll use dimensional analysis (covered in Ch 9 in book 1) to solve. First figure out what we have. First, we're given a rate: 14 beats PER second. Remember, per means divide. I can write that as $\dfrac{14 \text{ beats}}{1 \text{ sec}}$ and next we're given an amount of time 5 hours and 20 minutes (or 5 and 1/3 hours). Essentially, we will use the rate to "convert" this time into total number of beats to find the answer. To set it up, I put the time first, then convert it to seconds, then apply the given rate. I put what I want to cancel on the bottom. When it comes to my rate, I put what I "want" (beats) on top. All units cancel except the beats: $5\dfrac{1}{3} \text{ hours} \left(\dfrac{60 \text{ min}}{1 \text{ hr}}\right)\left(\dfrac{60 \text{ sec}}{1 \text{ min}}\right)\left(\dfrac{14 \text{ beats}}{\text{sec}}\right) = (320)(60)(14) \text{ beats} \to 268{,}800 \text{ beat}$. Writing this in scientific notation, we have 2.69×10^{5}.

4. **A.** One way to solve is to convert the rate we have to "minutes" and the time we have to "minutes" so that we can then use the rate to convert the time to distance, setting up fractions that allow units to cancel. Additionally, we must convert the distance to kilometers. We could do this in one big line as in question 3, or we can do it in steps: At sea level, the speed of sound is $\dfrac{3.4 \times 10^{2} \text{ meters}}{\text{sec}} \times \dfrac{60 \text{ sec}}{1 \text{ min}} = 60(3.4) \times 10^{2} = 204 \times 10^{2} = 2.04 \times 10^{4}$ meters per minute. We calculate 3 hours and 30 minutes is equal to $3(60) + 30 = 210$ min. So, the distance travelled is approximately $\dfrac{2.04 \times 10^{4} \text{ meters}}{\text{min}} \times \dfrac{1 \text{ kilometer}}{1000 \text{ meters}} \times 210 \text{ min} = 428.4 \times 10 = 4.28 \times 10^{3}$ kilometers. The method for this, called dimensional analysis, is covered in depth in Ch 9 of Book 1.

5. **B.** We need to convert what is in parts per billion to parts per part, where "1" is the part on the bottom (as all answer choices are a ratio compared to 1). Per means divide, so 15,000,000 parts per billion is equal to $\dfrac{15{,}000{,}000}{1{,}000{,}000{,}000}$. Canceling out the zeros on the numerator and denominator, we get $\dfrac{15}{1000} = 0.015 = 1.5 \times 10^{-2}$. Look for the answer with this to the left of the 1. For every 1 unit of water, this is how much lead is present.

6. **C.** $6.5 \times 10^{-4} = 0.00065$. So, the value closest to that, with the same number of zeros preceding the first non-zero integer, is 0.00065.

7. **A.** $\dfrac{3.5 \times 10^{y}}{6.2 \times 10^{x}}$ is $\dfrac{3.5}{6.2} \times 10^{y-x} = 0.56 \times 10^{y-x} = 5.6 \times 10^{-1} \times 10^{y-x} = 5.6 \times 10^{y-x-1}$.

CHAPTER

2 PROPERTIES OF NUMBERS

> ## SKILLS TO KNOW
> - Definitions of key math terms & how to solve problems reliant on these terms
> - The "divisibility" rules and how to apply them
> - How to solve "digits" problems
> - How to solve problems involving integers, odd/even numbers, or basic counting
> - How to approach "could be true" or "must be true" questions

NOTE: Factoring is covered more in depth in **LCM & GCF, Chapter 3** in this book. Skip ahead if you need to so you can handle problems involving prime factorization, etc. Counting is covered more in depth in **Chapter 9, Counting & Arrangements**.

KEY MATH TERMS

To answer this style of problem correctly, you'll need to know several basic math terms and how to apply them.

INTEGER: A number that is a whole number and includes no fraction or decimal parts. Integers can be negative, zero or positive. (Examples: -400, -7, 0, 1, 3, 56, 230).

ZERO: A number that is **even** and **neither positive nor negative**. Also, an integer. Numbers cannot be divided by zero, and zero cannot be the value of the denominator (bottom) of a fraction. Sometimes, this word is also used to indicate an x-intercept of a polynomial (when y is zero and x is a number).

IRRATIONAL NUMBERS: Decimals that are neither terminating nor repeating are called irrational because they cannot be written as fractions (Examples: $\sqrt{2}$, π).

RATIONAL NUMBERS: A number that can be written as a fraction. Rational numbers include fractions and decimals (repeating or terminal) as well as natural numbers, whole numbers, and integers (Examples: $-\dfrac{1}{2}$, 0.444, $0.\overline{5}$, 0.56, -4).

NATURAL NUMBERS: All positive integers, plus zero (Examples: 0, 1, 2, 3, 4, etc.).

REAL NUMBERS: Numbers that do not have an imaginary component. The set of rational numbers and the set of irrational numbers together make up the set of real numbers.

COMPLEX NUMBERS: A complex number is a number that can be expressed in the form $a+bi$, where a and b are real numbers (note: b *can* equal 0) and i is the imaginary unit which satisfies the equation $i=\sqrt{-1}$. Complex numbers include all imaginary numbers AND all real numbers—the coefficient of "i" in a non-imaginary (AKA real) number is simply zero. For example, 5 could be expressed as $5+0i$. In other words, all numbers in the world are complex numbers, but not all numbers are "imaginary" (Examples: $5i$, $\sqrt{-6}$, 67, 0, $0+0i$).

PROPERTIES OF NUMBERS — SKILLS

FACTOR: A factor of a number is an integer which can be multiplied by another integer to produce that number. (Example: 5 is a factor of 10 because 5 multiplied by 2 is equal to 10. This means that 2 is also a factor of 10). Prime numbers have no factors other than 1 and themselves. Remember, FACTORS are like pieces of a number that multiply together—don't confuse them with MULTIPLES which are numbers that any given number is a factor of (i.e., 8 is a multiple of 2, but 2 is a factor of 8).

PRIME FACTORS: Prime factors are factors (not including 1!) that are prime (have no factors other than 1 and themselves) and when multiplied together form a given number (Example: the prime factors of 63 are 7, 3 and 3 because 7, 3, and 3 are all prime, and when you multiply $(7)(3)(3)$ you get 63). We only list positive factors when listing the prime factorization of a number.

PRIME NUMBER: A number for which the only factors are itself and 1.
<u>IMPORTANT</u>: 1 (one) is **NOT** a prime number!

PERFECT SQUARE: A perfect square is an integer that has an integer square root (Example: 4 is a perfect square, because its square root is 2, an integer).

DIVISIBILITY RULES

A number is divisible by:	Example	Reasoning
2 if the ones digit is divisible by 2 (if the number is even).	42	"2" is the ones digit.
3 if the sum of its digits is divisible by 3.	39	$3+9=12$ 12 is divisible by 3. 39 is divisible by 3.
4 if the last two digits are divisible by 4.	124	24 is divisible by 4. 124 is divisible by 4.
5 if the ones digit is 0 or 5.	40	"0" is the ones digit. 40 is divisible by 5.
6 if the number is divisible by 2 and 3.	66	$6+6=12$; 12 is divisible by 3. 6 (ones digit) is divisible by 2. 66 is divisible by 6.
9 if the sum of its digits is divisible by 9.	423	$4+2+3=9$ 9 is divisible by 9. 423 is divisible by 9.
10 if the ones digit is 0.	450	"0" is the ones digit.

> How many prime numbers are there between 1 and 50?

This question is a hybrid of a question that requires your knowledge of prime numbers and what I'll call a "counting" problem—one in which you must write down a list of numbers of sorts and count up how many are in that list.

First, let's review what prime means: a number that is not divisible by anything except itself and one. Let's also remember, **<u>1 is not prime.</u>**

Start by listing off prime numbers by counting up through each number—you can also write down all the numbers, and slash off any that are not prime or circle those that are. Remember, the only even number that is prime is 2.

$$2, 3, 5, 7, 11, 13, 17, 19, 23, 29, 31, 37, 41, 43, 47$$

Now if there are any numbers you're not sure about, apply the divisibility rules:

If divisible by 2: Even number ending in 0, 2, 4, 6, 8

If divisible by 3: Add digits—is sum divisible by 3?

Thus, we can say that there are 15 primes numbers between 1 and 50.

FINDING A NUMBER(S) OR PARTICULAR DIGIT(S) BASED ON ITS PROPERTIES

The next thing you need to know is how to break down problems into digits.

TIP: Remember that you can always express a number's value based on its digits as follows:

Let's say abc is a three-digit number, where a represents the hundreds place, b the tens place, and c the ones place. $100a + 10b + c =$ the numeric value of a number whose digits are $\underline{a}$ $\underline{b}$ $\underline{c}$. Note: each place value is multiplied by a power of 10, e.g. $10^2 a \times 10^1 b \times 10^0 c$.

> Let a, b and c stand for digits between 0 and 9, and suppose that:
>
> $$\begin{array}{r} 38a \\ +b37 \\ \hline 11c5 \end{array}$$
>
> What does the expression $\dfrac{ab}{c}$ equal?

First, we can solve for the value of a. When we add the ones places in the addition problem, we see that the result of adding a and 7 is equal to 5, but we are not sure what the value of the tens place is because that value is carried into the next column. However, because a must be between 0 and 9, we can determine that $a = 8$, because no other value between 0 and 9 would result in a 5 in the ones place. Finally, because $7 + 8 = 15$, we must carry a 1 into the tens column.

Now that we know what will be carried into the next column, we can solve for the value of c. We know the value of $8 + 3 + 1$ (including the carry value from the addition in the ones place). $8 + 3 + 1 = 12$, but since the 1 is carried into the next column, the value of c is 2.

Next, we can solve for the value of b. Because this is the last column, we don't have to worry about carrying the value of our addition; we know that $3 + b + 1$ (the carry value from the addition in the tens place) $= 11$. Simplifying using algebra, we find $b = 7$.

CHAPTER 2

PROPERTIES OF NUMBERS   SKILLS

Finally, we can solve for the value of the expression $\frac{ab}{c}$. Plugging in our values of 8, 7, and 2 and then simplifying gives an answer of 28.

Answer: 28.

INTEGER QUESTIONS

What is the largest integer value that satisfies the inequality $\frac{x}{12} < \frac{17}{33}$?

Don't be intimidated by the less than sign! Most of the algebraic rules that you are familiar with using are still valid here. Just as in an equation, you can add or subtract whatever you want, and can multiply or divide by any positive number. Multiplying both sides by 12 gives:

$$x < 12 \times \frac{17}{33}$$

Plugging in the expression $12 \times 17 \div 33$ to your calculator shows a result of ≈ 6.181. Because the problem asks for the largest integer that satisfies this expression, we simply find the largest integer less than 6.181, which is 6.

Answer: **6**

ODD AND EVEN NUMBERS

 TIP: **KNOW THE FOLLOWING RULES OF HOW ODD & EVEN NUMBERS WORK**:

        An ODD times an ODD is an ODD: $(3)(3) = 9$
        An EVEN times ANYTHING is EVEN: $2(17) = 34$; $4(6) = 24$
        An ODD plus an ODD is EVEN: $7 + 5 = 12$
        An EVEN plus an EVEN is EVEN: $4 + 8 = 12$
        An EVEN plus an ODD is ODD: $2 + 5 = 7$
        An EVEN exponent creates a positive solution: $(-2)^4 = 16$
        An ODD exponent keeps the sign (+ or –) the same: $(-3)^3 = -27$

Let a be a negative, even integer. The expression $\frac{a^2}{b}$ is a negative, odd integer when b is what type of number?

    A. Positive even integer
    B. Positive odd integer
    C. Negative odd integer
    D. Negative even integer less than -3
    E. Negative even integer greater than -3

In many problems, you'll need to know how certain numbers behave:

Even exponents, for example, always create non-negative solutions. Even numbers multiplied by even numbers create more even numbers.

We thus know a^2 must be an even positive integer, as it is both the product of two even integers, and a negative number squared (a negative times a negative equals a positive). Furthermore, we know it must be divisible by four (two even numbers must each have 2 as a factor—when you multiply them together these two "2's" make 4).

In order to get an odd quotient when starting with an even number (a^2) in the numerator, we must eliminate all the even factors, or divide out all the even "parts" of the number—i.e. all the 2's. If we must divide by a number that has the twos in it, then that number must be even, so we can eliminate answers (B) and (C).

We also know that we need our quotient to be negative. Remember—the numerator, because it has an even exponent, must be a non-negative number. The only way to make this quotient negative would be to divide the numerator by a negative denominator. We can eliminate (A).

Now left with (D) and (E), we have to be careful—when we deal with negative numbers "greater than" means to the right on the number line, or towards zero. So answer (E) is basically the same as saying -2, but -2 won't work. Because the top of the fraction has at least two even factors, it must have at least two "2's" in its factorization and be divisible by at least 4—or some number whose absolute value is at least 4. I know it's confusing because of the negative sign—but a single negative two won't divide out both even pieces hidden in a^2.

Thus, (D) is the answer.

Now if that whole explanation confuses you, then the best thing to do is make up numbers—often this will work just as quickly as reasoning—but sometimes it won't. In this particular instance, making up numbers works fine.

Let $a = -2$:
$$\frac{a^2}{b} = \frac{(-2)^2}{b} = \frac{4}{b}$$

Now we can read through the choices and try to make up a number that works for b, such as -4, which would create a quotient of -1. Clearly, (D) is still the answer.

However, being able to understand the logic behind why (D) works can really help on tougher versions of this problem. If you're aiming for a score above a 32, I recommend you try to understand the logic behind questions like this, too.

Answer: **D.**

PROPERTIES OF NUMBERS — SKILLS

COUNTING PROBLEMS INVOLVING NUMBER PROPERTIES

More counting problems are covered in **Chapter 9: Counting & Arrangements.**

> How many numbers between 1 and 100 have a tens digits that is the square of the ones digit?

I know what you're thinking: this is going to take forever!

Don't freak out, these aren't as bad as they look. Still, counting problems, on average, can take a bit more time than some other types of questions. If you have a lot of trouble with time, you can skip these and come back to them later.

The best way to do the problem is simply to start figuring it out—start counting up from one and ask which numbers work:

Does 1 work? No—it doesn't even have a tens digit—let's skip to 10.
Does 10 work? No—1 is not a perfect square of 0.
But 11 does work—1 (tens digit) is a perfect square of 1 (ones digit).

Now let's start to think—what are we doing here? I want to get to the point that I have a "2" in the ones digit and the tens digit equals "2" squared. 2 squared is 4—so that number is 42.

Now we can go to 3 in the ones digit—three squared is 9, so that number is 93.
I could also think of the set up like this—

$$\_1$$
$$\_2$$
$$\_3$$
$$\_4$$
$$\ldots$$

and I can fill in the tens digits by squaring the ones digit.

By the time I get to four, however, four squared is 16—that's not a single digit.

As you can see, there are only three numbers that work for this parameter—not nearly as many as you may have originally thought! You might also notice that I tend to "dive in" before I overthink the problem. Sure, staring with "1" may seem a waste of time to some. True, I could try to think out all the ideas behind the problem, but often simply jumping into the "sandbox" and starting to play around with numbers helps you come to an understanding of how the problem works more quickly. That way, even if you don't see the secret reasoning right away, you are at least making progress. Once you see the problem's logic, speed up and apply what you know. Your first guess doesn't have to be right, but it does need to get your brain cells moving so that you have hope of finding the pattern or way that the problem works.

Answer: **3**.

SKILLS   PROPERTIES OF NUMBERS

COULD BE TRUE / MUST BE TRUE

These are questions that ask "which of the following," "could be true," "must be true," or something similar.

For these problems, you can reason algebraically or plug in numbers. If you can't see an algebraic solution, I typically recommend plugging in a variety of numbers from the following list:

- Tiny number ($-10{,}000$)
- A negative number (-15, -4, -6)
- Negative one (-1)
- Negative fraction ($-\frac{1}{2}$, $-\frac{1}{9}$)
- Zero (0)
- Fraction ($\frac{1}{4}$, $\frac{1}{7}$)
- One (1)
- Two (if "odd" / "even" problem)
- Positive number (7)
- Giant number ($10{,}000$)

If you've tried all of these, chances are your conclusions will hold. Still, you'll likely be able to figure out a solution with just a few wisely chosen test values.

> Two numbers are reciprocals if their product is equal to 1. If a and b are reciprocals and $a > 1$, then b must be:
>
> **A.** Greater than 1
> **B.** Between 0 and 1
> **C.** Equal to 0
> **D.** Between 0 and -1
> **E.** Less than -1

Plug in some numbers and see what happens!

Let's say $a = 4$. Then we take a's reciprocal and get $b = \frac{1}{4}$.

Because our answer choices on this problem are "mutually exclusive" (i.e. they don't overlap), we know only choice B works. However, if more than one answer choice included the value 1/4, we would need to continue making up numbers until we could narrow our options. Thus, the answer is **B**.

Answer: **B.**

TIP: Pay attention to the wording of your question. "Must" be true is not the same as "could" be true. Analyze the situation accordingly. You may need to test more than one example you make up depending on this wording and the available answer choices.

PROPERTIES OF NUMBERS   QUESTIONS

1. If a is a positive integer, then the sum of $9a$ and $4a$ is always divisible by which of the following?

   A. 13
   B. 9
   C. 4
   D. 5
   E. 36

2. Which of the following could be the last 3 digits of an integer that is a perfect square?

   A. 232
   B. 753
   C. 689
   D. 597
   E. 168

3. What percent (to the nearest tenth) of numbers from 10 to 99 have a tens digit that is 2 times as large as the units digit?

   A. 4.1%
   B. 4.2%
   C. 4.3%
   D. 4.4%
   E. 4.7%

4. For every 2-digit integer number, a, with tens digit x and units digit y, let b be the 2-digit number reversing the digits of a. Which expression is equivalent to $a+b$?

   A. $2xy$
   B. $10(x+y)$
   C. $10x+10y$
   D. $11(x+y)$
   E. $x+y$

5. Let a be a positive, odd integer. The expression ab^2 is a positive, odd integer when b is part of what set of integers?

   A. Positive Even Integers
   B. Negative Even Integers
   C. Positive Odd Integers
   D. All Integers
   E. All Odd Integers

6. x and y are consecutive integers such that $x > y$. All of the following are true EXCEPT?

   A. $x-y$ is odd
   B. $x-y$ is even
   C. $x+y$ is odd
   D. $y-x<0$
   E. x^2+y^2 is odd

7. Which set gives the 3 largest prime numbers that are less than 100?

   A. $\{37, 47, 59\}$
   B. $\{83, 89, 97\}$
   C. $\{59, 61, 67\}$
   D. $\{97, 98, 99\}$
   E. $\{95, 97, 99\}$

8. The sum of any two prime numbers except the number 2 must be which of the following?

   A. Prime Number
   B. Perfect Square
   C. Even Number
   D. Odd Number
   E. A multiple of 3

9. How many positive prime numbers are factors of the number 60?

   A. 1
   B. 2
   C. 3
   D. 4
   E. 5

10. On the real number line, -0.1395 is between $\dfrac{a}{1000}$ and $\dfrac{a+1}{1000}$ for some integer a. What is a?

    A. -138
    B. -139
    C. -140
    D. -141
    E. -142

11. The variables x, y, and z are all integers and $x+y+z=200$. If $45<x<75$ and $3<y<18$ what is the maximum possible value of z?

    A. 107
    B. 152
    C. 151
    D. 150
    E. 149

12. What is the correct order from greatest to smallest for the following values: $6\frac{4}{9}$, $\frac{44}{7}$, 6.374?

    A. $\frac{44}{7}$, $6\frac{4}{9}$, 6.374
    B. $6\frac{4}{9}$, 6.374, $\frac{44}{7}$
    C. $6\frac{4}{9}$, $\frac{44}{7}$, 6.374
    D. 6.374, $\frac{44}{7}$, $6\frac{4}{9}$
    E. 6.374, $6\frac{4}{9}$, $\frac{44}{7}$

13. What is the correct order from least to greatest of the following values: $\sqrt{5}$, 2.1, $\frac{9}{4}$, $2\frac{2}{3}$?

    A. $\sqrt{5}$, $\frac{9}{4}$, $2\frac{2}{3}$, 2.1
    B. $\sqrt{5}$, $\frac{9}{4}$, 2.1, $2\frac{2}{3}$
    C. 2.1, $\sqrt{5}$, $\frac{9}{4}$, $2\frac{2}{3}$
    D. 2.1, $2\frac{2}{3}$, $\sqrt{5}$, $\frac{9}{4}$
    E. $\frac{9}{4}$, 2.1, $2\frac{2}{3}$, $\sqrt{5}$

14. x, y, and z are positive real numbers such that $3x=4z$ and $\frac{1}{3}x=\frac{1}{5}y$. Which of the following inequalities is true?

    A. $y>z>x$
    B. $x>y>z$
    C. $z>x>y$
    D. $x>z>y$
    E. $y>x>z$

15. For every pair of integers n and m, to which of the following sets must $\frac{m}{n}$ belong?

    I. The natural numbers
    II. The integers
    III. The rational numbers
    IV. The real numbers
    V. The complex numbers

    A. I, II, and III only
    B. II, III, and IV only
    C. III and IV only
    D. III, IV, and V only
    E. IV and V only

16. What is the lowest possible value for the product of two integers that differ by ten?

    A. −36
    B. −16
    C. −25
    D. −24
    E. −4

17. x is a factor of 12 and y is a factor of 39. Which of the following cannot be the value of xy?

    A. 18
    B. 78
    C. 72
    D. 36
    E. 468

18. The integer x is 5 more than positive integer y. The integer z is 3 less than b. The product of x and z is 36. What one of the answer choices could be b?

    A. 15
    B. 21
    C. 7
    D. 12
    E. 8

19. There are 3 integers A, B, and C. A is a prime factor of 18. B is a factor of 36 such that $6 < B < 12$. C is a perfect cube such that $10 < C < 30$. What is one possible value of $\frac{AB}{C}$?

   A. 3
   B. 1
   C. $\frac{4}{9}$
   D. $\frac{8}{9}$
   E. $\frac{4}{3}$

20. Two numbers are reciprocals if their product is equal to 1. If a and b are reciprocals and $a < -1$, then b must be:

   A. Greater than 1
   B. Between 0 and 1
   C. Equal to 0
   D. Between 0 and -1
   E. Less than -1

21. If m and n are integers such that $6 \leq m \leq 18$, and $-3 \leq n \leq 9$, then the minimum value for $\frac{m}{n}$ is:

   A. -18
   B. -9
   C. -6
   D. 3
   E. 2

22. If A, B, and C are real numbers, and if $ABC = -1$, which of the following conditions must be true?

   A. $AB = \frac{1}{C}$
   B. A, B, and C must all be negative
   C. Either A, B, or C must equal -1
   D. Either A, B, or C is between -1 and 0
   E. $BC = -\frac{1}{A}$

23. Suppose $z < -1$. Which of the following has the greatest value?

   A. $\frac{1}{z}$
   B. z^2
   C. z^3
   D. $\sqrt{-z}$
   E. $-\frac{z}{2}$

24. a is a positive real number and $\frac{9a^2}{16a}$ is a rational number. Which statement about a must be true?

   A. a is rational
   B. a is irrational
   C. $a = \frac{16}{9}$
   D. $a = \frac{4}{3}$
   E. $a = \frac{3}{4}$

25. Which of the following is irrational?

   A. $\sqrt{\frac{4}{9}}$
   B. $\sqrt{\frac{16}{25}}$
   C. $\sqrt{\frac{100}{36}}$
   D. $\sqrt{\frac{3}{4}}$
   E. $\sqrt{\frac{64}{81}}$

26. Which of the following statements is false about rational and/or irrational numbers?

   A. The sum of any 2 irrational numbers is irrational
   B. The product of any 2 irrational numbers is always rational
   C. The quotient of any 2 rational numbers is always rational
   D. The product of a rational and irrational number is irrational
   E. The quotient of any 2 irrational numbers could be irrational

27. The square root of some positive integer is n. What expression gives the square root of next integer greater than the first integer?

   A. $\sqrt{(n^2 + 1)}$
   B. $\sqrt{n+1}$
   C. $n^2 + 1$
   D. $n + 1$
   E. $(n+1)^2$

ANSWER KEY

1. A 2. C 3. D 4. D 5. E 6. B 7. B 8. C 9. C 10. C 11. D 12. B 13. C 14. E
15. D 16. C 17. C 18. C 19. B 20. D 21. A 22. E 23. B 24. A 25. D 26. B 27. A

ANSWER EXPLANATIONS

1. **A.** $9a + 4a = 13a$. So, $13a$ is always divisible by 13 and a.

2. **C.** The squares of the first 10 digits are $1^2 = 1, 2^2 = 4, 3^2 = 9, 4^2 = 16, 5^2 = 25, 6^2 = 36, 7^2 = 49, 8^2 = 64, 7^2 = 49, 8^2 = 64, 9^2 = 81$, and $10^2 = 100$. These only end with the digits $1, 4, 9, 5, 6$ and 0. In the answer choices, only answer choice (C) ends with a 9. So, only answer choice (C) could be the last 3 digits of a square number.

3. **D.** From 10 to 99 there are 90 numbers inclusive. You can think of this as eliminating the nine numbers 1-9 from the 99 numbers included in 1-99 to get 90. Of these 90 numbers, the numbers that have tens digits that are twice as large as the units digits include $21, 42, 63$ and 84. So, 4 out of 90 numbers $= \frac{4}{90}(100\%) \to 4.4\%$ of the numbers from 10 to 99 satisfy the description.

4. **D.** The 2-digit number a can be written as $10x + y$. When we reverse x and y, we get $b = 10y + x$. So, $a + b = 10x + y + 10y + x \to 11x + 11y \to 11(x+y)$.

5. **E.** To get an odd answer to a multiplication problem, ALL of its factors must be odd. So, in order for ab^2 to be odd, b^2 must be odd and b must be odd. It does not matter if b is positive or negative because all integers are positive when squared. So, b belongs to the set of all odd integers.

6. **B.** If x and y are consecutive integers with $x > y$, that means $x = y + 1$. So, evaluating the expression $x - y$, we can substitute in $x = y + 1$ for x to get $y + 1 - y = 1$. Since 1 is odd, answer choice (A) is true. The statement "$x - y$ is even" in answer (B) is the only false statement. Answer (C) is true because $x + y = y + 1 + y = 2y + 1$ is an even number plus an odd number $=$ odd. Answer (D) is true because $y - x = y - (y+1) \to -1 < 0$. Answer (E) is true because $x^2 + y^2 = (y+1)^2 + y^2 = y^2 + 2y + 1 + y^2 = 2y^2 + 2y + 1 = 2(y^2 + y) + 1$ as an even $+$ an odd $=$ odd.

7. **B.** To determine if a number is prime, we can divide the number by all integers less than its square root, and if none of the integers divide the number, then it is prime. So, we start counting down from 99 until we find three numbers that are prime. Use divisibility rules to speed things up. To make the process faster, we also only need to evaluate the numbers given to us in the answer choices, and we can start with the largest available. 99 is clearly divisible by 9, so it is not prime. You can also add its digits to verify this $9 + 9 = 18$, which is divisible by 9. Thus Choices D and E are incorrect. 97 might be prime. Its square root (given that its less than 100) will be between 9 and 10, so we only need to test factors less than 10. We know it's not divisible by 3 or 9 as its digits sum to 16. We also know it's not even so need not test any even factors. It doesn't end in five, so it's not divisible by 5. Finally we test 7. $\frac{97}{7} = 13.86$. Because 97 is not divisible by any integer less than or equal to 9, it is prime. Now, eliminate choices A and C, which do not include 97, and find the answer is (B). At this point most students should move on. If you have time, test if 89 and 83 are prime. Remember we only need to test up to the square roots, which I can estimate are ~9. Neither end in 5, so 5 is not a factor. Neither are even, so we don't need to test even factors. Neither are divisible by 3 (their digits' sums are not divisible by 3), so 3 and 9 are not factors. Thus the only number remaining to test is 7. $\frac{89}{7} = 12.714$. 89 is not divisible by any integer less than 9. $\sqrt{83} = 9.11$ and $\frac{83}{7} = 11.857$ so 83 is not divisible by any integer less than 9 either. Thus both 89 and 83 are prime.

8. **C.** 2 is the only even prime number, so if we sum all the prime numbers other than two, we are summing odd numbers only. The sum of any two odd numbers is always an even number.

9. **C.** We find the prime factorization of 60. $60 = 6(10) = 3(2)(5)(2) = 2^2(3)(5)$. So, 60 has three prime factors: $2, 3,$ and 5. For more help with factoring, see Chapter 3 in this book.

10. **C.** The problem tells us that $\frac{a}{1000} < -0.1395 < \frac{a+1}{1000}$. So, multiplying both sides by 1000, we get $a < -139.5 < a + 1$. We can tell that the number -139.5 is in between the numbers -140 and -139. So, $a = -140$ and $a + 1 = -139$.

PROPERTIES OF NUMBERS   ANSWERS

11. **D.** Since $x+y+z=200$, to find the maximum possible value of z we want to find the smallest possible values of x and y Since $45<x<75$ and $3<y<18$, the smallest integer values for x and y are 46 and 4. So, plugging in $x=46$ and $y=4$, we have $46+4+z=200$. Subtracting $(46+4)$ on both sides, we get $z=150$.

12. **B.** Plugging these values into the calculator, we get $6\frac{4}{9}=6.444$, $\frac{44}{7}=6.286$, and $6.374=6.374$. So, putting these in order from greatest to smallest, we have $6.444>6.374>6.286$. Putting these values in their original forms, we get $6\frac{4}{9}>6.374>\frac{44}{7}$.

13. **C.** Plugging these values into the calculator, we get $\sqrt{5}=2.236$, $2.1=2.1$, $\frac{9}{4}=2.25$, and $2\frac{2}{3}=2.667$. So, putting these in order from least to greatest, we have $2.1<2.236<2.25<2.667$. Putting these values in their original forms, we get $2.1<\sqrt{5}<\frac{9}{4}<2\frac{2}{3}$.

14. **E.** Multiplying both sides of $\frac{1}{3}x=\frac{1}{5}y$ by 15, we get $5x=3y$. So, we know that $y>x$ because it has a smaller coefficient. Similarly, we know $3x=4z$ means that $x>z$ because x has the smaller coefficient. So, we can write $y>x>z$.

15. **D.** Recognize that every element in the list "falls under" the item after it. For example, all natural numbers are integers (but just the positive ones), all integers are rational numbers (anything that can be expressed as a fraction), and all real numbers are complex numbers (anything that can be put in the form a+bi). Many students don't realize complex numbers INCLUDE both real AND imaginary numbers. (I know!!) Thus, we only need to prove the most specific case, and the rest will follow. $\frac{m}{n}$ does not have to be a natural number, since it could be negative. $\frac{m}{n}$ doesn't have to be an integer, since it could fall between the integers, for example, as $\frac{3}{2}$. However, $\frac{m}{n}$ must be a rational number, since a rational number can be expressed as a fraction of rational numbers, and m and n are both rational (since they are integers.) So III is true, and as we pointed out before, all the rest follow, so III, IV, and V are true.

16. **C.** If we take two negative numbers, their product will be positive. If we take two positive numbers, their product will also be positive. So, we are looking for one positive and one negative integer that differ by 10. This narrows the options down to 9 pairs of numbers: $(-1,9),(-2,8),(-3,7),(-4,6),(-5,5),(-6,4),(-7,3),(-8,2),(-9,1)$. Now imagine the product of each pair to find the one with the largest negative value: $(-5,5)\to(-5)5=-25$. Alternatively, solve algebraically, creating an equation to represent the product of x and a number ten less than x: $y=x(x-10)$. The lowest value for y in this upward facing parabola is at the vertex; we can find it by graphing or averaging the zeros (i.e. when y=0, x=10 or 0) to find the x-value of the vertex (5) and then plugging that into $y=x(x-10)$ to get the y-value of the vertex (minimum) $5(5-10)=-25$. See Quadratics and Polynomials in Book 1 for more ways to find a vertex of a parabola.

17. **C.** Since x is a factor of 12, $x=\frac{12}{a}$ for some integer a. Likewise, since y is a factor of 39, $y=\frac{39}{b}$ for some integer b. This means that $xy=\frac{12}{a}\left(\frac{39}{b}\right)\to\frac{(3)(4)(3)(13)}{ab}\to\frac{2^2 3^2 13}{ab}$. All answer choices are a combination of the factors in the prime factorization in the numerator of this fraction except for 72 whose prime factorization is $72=3^3 2$. The factor 3^3 makes $\frac{2^2 3^2 13}{ab}=3^3 2$ because it has too many 3's. Alternatively, list the prime factors of 12 and 39 using a factor tree or rainbow (see Ch 3). Then go through the answers and factor each. Then try to make up x and y values to equal each answer choice using the factors of 12 and 39.

18. **C.** y is a positive integer (i.e. at least 1) and $x=5+y$ so $x\geq6$. Z is 3 less than b, so $z=b-3$ or $b=z+3$ The product of x and z is 36 so $x\times z=36$. Since we know that 36 is positive and $x\geq6$, we can narrow down the values of x and z. We can represent the values of x and z as ordered factor pairs of 36, where x is the bigger (or equal) number: (6,6), (9,4), (12,3), and (36,1). So, the possible values of z are 6, 4, 3, and 1. Thus the values of $b=z+3$ could be $6+3=9$, $4+3=7$, $3+3=6$, or $1+3=4$. The only value that matches a possible value for b is answer choice C.

19. **B.** $18 = 9(2) = 3^2(2)$ so the prime factors of 18 are 2 and 3, thus A must be either 2 or 3. The factors of 36 are 1, 2, 3, 4, 6, 9, 12, 18, 36, and the only factor that satisfies the condition $6 < B < 12$ is 9, so $B = 9$. The only perfect cube between 10 and 30 is $27 = 3^3$, so $C = 27$. Plugging in $A = 2$, $B = 9$, and $C = 27$ to the expression $\frac{AB}{C}$, we get $\frac{2(9)}{27} = \frac{18}{27} \to \frac{2}{3}$. Plugging $A = 3$ $B = 9$ and $C = 27$ we get $\frac{3(9)}{27} = \frac{27}{27} \to 1$. $\frac{2}{3}$ is not one of the answer options, but 1 is.

20. **D.** Because $a < -1$, we know that a is a negative number that isn't -1 or a fraction between -1 and zero. In order for ab to equal 1, b must be $\frac{1}{a}$. We can quickly make up a possible a value such as -5 and then calculate its reciprocal b, $-\frac{1}{5}$. Thus b is a negative fraction or decimal between -1 and 0, and answer D is correct.

21. **A.** To get the minimum value for $\frac{m}{n}$, since negative numbers are an option, we want the number to be as far away from zero to the left, i.e. as "negative," as possible. Remember that negative numbers with "large" magnitudes (i.e. absolute values) are actually SMALL. To get a negative quotient, we'll need one of the two chosen values to be negative. Because we can chose from negative numbers only for the denominator, $\frac{m}{n}$ would be smallest when the positive number with the largest magnitude (absolute value) is in the numerator and the negative number with the smallest magnitude (absolute value) is in the denominator. First, let's find the highest positive number for the numerator: $m = 18$. Now, let's find the smallest magnitude negative number for the denominator; we want it to be negative and small in absolute value, or very close to zero, but not zero. $n = -1$ is our best bet. $\frac{m}{n} = \frac{18}{-1} = -18$. If you're not sure on questions like this, or you find the explanation confusing, just move fast and plug in as many options as you can to find the trend.

22. **E.** If $ABC = -1$, by dividing by A on both sides, we get $BC = -\frac{1}{A}$.

23. **B.** In a question like this, we can start by considering which answers will be positive and which will be negative, and eliminating the negative ones. Since z is negative, answer choice A, $\frac{1}{z}$ will also be negative. Answer choice B, z^2 will be positive. Answer choice C, z^3, will be negative. Answer choice D, $\sqrt{-z}$, will be positive. Answer choice E, $-\frac{z}{2}$, will be positive. The positive answers are thus B, D, and E. To find the largest one, consider that answer C, because z is less than -1 and therefore *not* a fraction, z^2 will produce a number larger than $|z|$, while D and E produce numbers smaller than $|z|$, taking the square root of the absolute value and dividing by 2, respectively. Note that if the problem said only $z < 0$, this answer could not be determined. But because we are told that $z < -1$, answer choice C works.

24. **A.** If $\frac{9a^2}{16a}$ is a rational number, that means that it could be expressed as a fraction with integers in the numerator and denominator. Simplifying the fraction to get $\frac{9a^2}{16a} = \frac{9a}{16}$ we know that a must be rational in order for $\frac{9a}{16}$ to be rational.

25. **D.** Every answer choice except for answer choice D can be simplified to be a rational number because they are fractions with perfect squares in the numerator and denominator.

26. **B.** Irrational numbers cannot be expressed as a fraction. Multiplying any 2 irrational numbers by each other will not make them rational. For example, π is irrational, and multiplying it by π produces π^2, which is also irrational. Thus, (B) is false.

27. **A.** Let the positive integer be represented as x. Then, $n = \sqrt{x}$. So, $x = n^2$. The next integer greater than x is then equal to $n^2 + 1$ and the square root of that is $\sqrt{(n^2 + 1)}$. If this is confusing, make up numbers to help you.

CHAPTER 3: LCM & GCF

> ## SKILLS TO KNOW
> - How to use a Factor Tree
> - How to use a Factor Rainbow
> - How to find Least Common Multiples
> - How to find Greatest Common Factors
> - Solving word problems involving LCM and/or GCF

FACTOR TREES

One skill every ACT® math student should know is how to use a "factor tree." Factor trees are tools used to determine the prime factorization of a number.

Remember that prime factorization finds all the prime numbers that when multiplied together equal the given number.

Let's say we want to factor a number like 72:

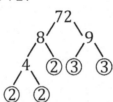

We branch off pairs of factors until we cannot branch off anymore. At this point, we get to the prime factorization (circled above). It doesn't matter which pair of factors you choose to start with, so long as you keep going until no more factors can be divided out. For example, I could have started with 2 and 36 instead of 8 and 9. If you're not sure whether a number is prime, use the divisibility rules (see **Chapter 2,** page 10).

FACTOR RAINBOWS

Another valuable tool to know is how to break down a number with a factor rainbow. For this tool, we start by drawing a "rainbow" from the number 1 to the number we are trying to factor. Below we'll use 36 as an example.

Then you count up one number at a time, adding a "rainbow" arch for each pair of factors. For instance, I count from 1 to 2, and ask, does 2 go into 36? Yes, it does, 18 times. So I add 2 and 18 to my rainbow.

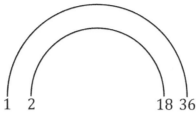

Then I count to 3 and ask, does 3 go into 36? If so how many times, and so forth. I continue, adding pairs 3 and 12, 4 and 9, then I get to 5, but 5 does not go into 36, so I write it down below the rainbow and slash it out. Then I get to 6, the square root of 36, and I'm done.

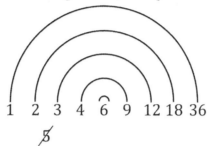

What the factor rainbow gives us is not our PRIME factors, but ALL of the factors of a number. It also gives us all the factor pairs. If you are required by a problem to count the total number of factors, rather than to find the prime factors, this is your best tool.

LEAST COMMON MULTIPLE

The least common multiple of two or more numbers, often abbreviated as LCM, is the smallest whole number that has those two or more numbers among its factors. For example, the LCM of 2 and 3 is 6, since it is the smallest whole number that has both 2 and 3 among its factors.

The elementary school method for finding the Least Common Multiple is to create lists of multiples. Skip count and make a list for each number you want to find the LCM of. Then find the first number that appears on all your lists. Below I've done so for 2 and 3. I have circled the answer, six.

$$2: 2, 4, ⑥, 8, 10$$
$$3: 3, ⑥, 9, 12$$

That's fine and dandy, but there's a more efficient way to find LCM: using factors.

> What is the least common multiple of 24, 9, and 60?

Step 1: List all numbers' prime factorizations:
$$24 = 2 \times 2 \times 2 \times 3$$
$$9 = 3 \times 3$$
$$60 = 2 \times 2 \times 3 \times 5$$

If you can't see these numbers easily, use a factor tree to find these factors:

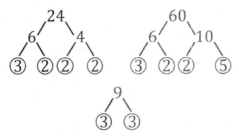

Step 2: Identify each unique prime factor and the greatest number of times it occurs in any of the numbers.

2 occurs a maximum of 3 times (in 24).
3 occurs a maximum of 2 times (in 9).
5 occurs a maximum of 1 time (in 60).

Step 3: Multiply the unique prime factors by the greatest number of times each occurs, and then multiply all of the numbers together to form your LCM:

$$\underbrace{2\times 2\times 2}_{3\ times} \times \underbrace{3\times 3}_{2\ times} \times \underbrace{5}_{1\ time} = 8\times 9\times 5 = 360$$

Answer: **360 is the LCM of 24, 9, and 60.**

Many of my students, though, struggle in memorizing this process, while other confuse LCM and GCF. To remember this method, I like to think of LCM as building a "Lego" kit (easily enough, it starts with "L"! and GCF does not start with "L"). Do you know how Lego kits sometimes have a certain object you can build, say a pirate ship, but then when you open the box there are instructions for two other pirate ships that you could also build? But you can't build all three ships at once? Lego only includes enough pieces to build one ship at a time, but some pieces may only be used on one ship or another. That's similar to the way LCM works. Each LCM is like an entire pirate ship Lego Kit, each number that you're trying to find the LCM of is like a pirate ship, and the numbers that are prime factors of these "pirate ship" numbers are like Lego pieces—the pieces that multiply together to build each number.

For example, if I wanted the LCM of 18 and 24, the pirate ships are 18 and 24, and the Lego pieces are the prime numbers that make up these "pirate ships:" 3, 3, 2 for 18 and 3, 2, 2, and 2 for 24. To build both "pirate ships" I'll need two 3's (because 18 needs two 3's but 24 only needs one) and three 2's (because 24 needs three 2's but 18 only needs 1). In your LCM or Lego kit you need to have the elements necessary to build all two numbers (or pirate ships) but you can reuse the pieces (or prime factors) that are needed to build each. To then find the LCM, you multiply all the "pieces" you have together, so I'd multiply to get $3\times 3\times 2\times 2\times 2$ to get 72.

> What is the least common denominator for the expression below?
> $$\frac{1}{17\times 29\times 31} + \frac{1}{17^2\times 31^2} + \frac{1}{17\times 23\times 29}$$

We find the LCM of all the denominators to find the least common denominator (LCD). In other words, use the same process to determine LCM and LCD. Here, our denominators are already factored into prime numbers. Let's find the greatest number of times each factor occurs:

$$17^1 \times \boxed{29^1} \times 31^1$$
$$17^2 \times \boxed{31^2}$$
$$\boxed{17^1} \times \boxed{23^1} \times 29^1$$

Circled are instances of the highest powers of the prime factors. The least common denominator will be the product of the highest power of each respective prime.

Answer: $17^2 \times 23 \times 29 \times 31^2$.

GREATEST COMMON FACTOR

The greatest common factor of two or more numbers, often abbreviated as GCF, is the largest whole number that is a factor of all the numbers. For example, the GCF of 18 and 24 is 6 because 6 is the largest whole number that is a factor of 18 and 24. We're looking for what prime factors (or even just factors) overlap, and then we simply multiply these numbers together to get the GCF.

How to find the GCF

What is the greatest common factor of 4620 and 1575?

Step 1: List both numbers' prime factorizations (use a factor tree if necessary):

$$4620: 2 \times 2 \times 3 \times 5 \times 7 \times 11$$
$$1575: 3 \times 3 \times 5 \times 5 \times 7$$

Step 2: Match all the numbers that appear in both prime factorizations:

$$3, 5, \text{ and } 7$$

(Note: we only list 5 once; if 5 appeared twice in both prime factorizations, we would list it twice.)

Step 3: Multiply all the common prime factors:

$$3 \times 5 \times 7 = 105$$

Answer: **105 is the GCF of 4620 and 1575.**

TIP: If there are no matches between the prime factorizations, the GCF is 1. The ACT® is notorious for "trick" questions, so be sure to know this fact!

GCF AND LCM WITH VARIABLES

The same concept can be applied to variables in monomials.

What is the greatest common factor of $21x^3y$ and $63x^2y^2$?

Even though there are variables now, we can go through the same steps as before.

$$21x^3y : 7 \times 3 \times x \times x \times x \times y$$

$$63x^2y^2 : 3 \times 3 \times 7 \times x \times x \times y \times y$$

Matching the common numbers and variables, we find the product of the overlapping numbers:

$$3 \times 7 \times x \times x \times y, \text{ which is } 21x^2y.$$

Answer: $21x^2y$.

SPEED TIP: To save time, don't factor down the variables; simply look for the lowest powered version of the variables that overlap. Above, x^2 and y are the lowest powered overlapping variables. Also, if you spot common factors earlier, don't factor the numbers down all the way. I.e. above, 21 goes into 63, so it will be the GCF of those two numbers. If this confuses you, stick to what I wrote above.

SOLVING WORD PROBLEMS WITH GCF/LCM

It can be tricky to figure out when GCF or LCM apply in word problems. Here are some tips.

USE GCF:
- To break apart two or more larger amounts into same size smaller pieces/groups and to maximize the size of the smaller pieces/groups

For example, if you have two ribbons, 6ft and 9ft, and want to cut into equal pieces and want to maximize the size of those equal pieces, find the GCF.

USE LCM:
- To find the overlap in a pattern that repeats over time (clocks that chime so many minutes apart, lights that blink at certain intervals, etc.) that is described as a "first," "least," or "minimum."

> One light blinks every 8 seconds. Another light blinks every 6 seconds. When you see them blink at the same time, how long will it be until the first time the lights blink at the same time again?

Because the problem is asking for the least number of seconds that it will take for the lights to blink together, finding the LCM is the correct technique to apply.

Using the method we wrote about previously, let's list all the prime factorizations:

$$8 : 2 \times 2 \times 2$$

$$6 : 2 \times 3$$

Then, we list each unique prime factorization that appears the most amount of times. So, for 8, we have 2:3 times. For 6, we have 3:1 time.

Then, we multiply them:

$$2 \times 2 \times 2 \times 3 = 24$$

Answer: **24 seconds.**

QUESTIONS — LCM & GCF

1. What is the largest 3-digit integer that is divisible by 8 and is a multiple of 5?

   A. 800
   B. 850
   C. 910
   D. 960
   E. 990

2. What is the lowest positive common multiple of 14 and 50?

   A. 350
   B. 150
   C. 700
   D. 1050
   E. 400

3. Two numbers have a greatest common factor of 6 and a least common multiple of 36. Which of the following pairs could they be?

   A. 18 and 36
   B. 6 and 12
   C. 6 and 18
   D. 12 and 18
   E. 12 and 36

4. One bell rings every 5 seconds. Another bell rings every 9 seconds. If they ring together and you start counting seconds, how many seconds will pass before the two bells ring together again?

   A. 15
   B. 30
   C. 45
   D. 60
   E. 90

5. At the hardware store, screws come in packs of 15 and nuts come in packs of 18. If Alexa uses the same number of screws as nuts and uses all of the screws and nuts she buys, what is the minimum number of nuts she can use?

   A. 3
   B. 6
   C. 30
   D. 90
   E. 270

6. Which of the following is not a factor of 2002?

   A. 17
   B. 14
   C. 13
   D. 11
   E. 2

7. Cib sells pens in packs of 6, and Kulmus sells pens in bundles of 8. If both sold the same number of pens last week, what is the smallest number of pens each could have sold?

   A. 48
   B. 24
   C. 14
   D. 12
   E. 9

8. For all positive integers x, what is the greatest common factor of the two numbers $90x$ and $300x$?

   A. 30
   B. x
   C. $10x$
   D. $30x$
   E. $900x$

9. The least common multiple (LCM) of 2 numbers is 1,300. The lesser of the 2 numbers is 50. What is the minimum value of the other number?

   A. 2
   B. 13
   C. 52
   D. 26
   E. 260

10. What is the greatest common factor of 500; 1,000; and 12,500?

    A. 5
    B. 100
    C. 125
    D. 250
    E. 500

11. The sum of 0.12 and 0.06 can be written as a fraction where the numerator and the denominator are both positive integers. When the numerator and the denominator are both divided by their greatest common factor, what is the sum of the numerator and the denominator of the resulting fraction?

    A. 18
    B. 59
    C. 61
    D. 118
    E. 236

CHAPTER 3

LCM & GCF — QUESTIONS

12. If the least common multiple of 7, 10, 14, and v is 420, which of the following could be v?

- A. 14
- B. 18
- C. 84
- D. 5
- E. 35

13. What is the least common multiple of 25, 18, and 45?

- A. 90
- B. 180
- C. 360
- D. 450
- E. 18,000

14. If a is a factor of 36 and b is a factor of 45, the product of a and b could NOT be which of the following?

- A. 1620
- B. 324
- C. 90
- D. 21
- E. 1

15. What is the least common multiple of $8a^2$, 2, $3a$, $6b$, and $4ab$?

- A. $16a^3 b$
- B. $24a^2 b$
- C. $24a^3 b$
- D. $54a^2 b$
- E. $60a^3 b$

16. Which is the greatest term that must be a factor of both a and b?

- A. 0
- B. 1
- C. a
- D. b
- E. ab

17. What is the least common denominator of $\dfrac{4}{35}$, $\dfrac{1}{28}$, and $\dfrac{3}{8}$?

- A. 6
- B. 35
- C. 140
- D. 280
- E. 560

18. The least common multiple (LCM) of 2 numbers is 144. The greater of the 2 numbers is 72. What is the maximum value of the other number?

- A. 2
- B. 8
- C. 24
- D. 36
- E. 48

19. What is the least common denominator of the fractions $\dfrac{3}{8}, \dfrac{4}{9}$, and $\dfrac{7}{30}$?

- A. 47
- B. 220
- C. 360
- D. 500
- E. 2,160

20. What is the least common denominator for adding fractions $\dfrac{2}{45}, \dfrac{4}{63}$, and $\dfrac{3}{14}$?

- A. 126
- B. 315
- C. 630
- D. 2,835
- E. 39,690

21. Which of the following is the least common denominator for the expression:

$$\dfrac{1}{29^2 \times 89 \times 907^2} + \dfrac{1}{29 \times 89^3} + \dfrac{1}{29^3 \times 907}?$$

- A. 29×89
- B. $29 \times 89 \times 907$
- C. $29 \times 89 \times 907^2$
- D. $29^3 \times 89^3 \times 907^2$
- E. $29^6 \times 89^4 \times 907^3$

ANSWERS — LCM & GCF

ANSWER KEY

1. D   2. A   3. D   4. C   5. D   6. A   7. B   8. D   9. C   10. E   11. B   12. C   13. D   14. D
15. B   16. B   17. D   18. E   19. C   20. C   21. D

ANSWER EXPLANATIONS

1. **D.** The 3-digit integer is divisible by 8 and a multiple of 5, so it is divisible by both 8 and 5. Thus it can be written in the form $x = 40y$ because 40 is the LCM of 8 and 5. So, we want to find the largest multiple of 40 that is 3 digits (i.e. less than 999). We do this by taking $\frac{999}{40} = 24.975$, rounding this number down to 24, and then multiplying $40(24) = 960$.

2. **A.** We can compute the least common multiple of 14 and 50 by writing out the prime factorization of 14 and 50 and then multiplying each prime factor to the greatest power that appears throughout the factorizations. We get $14 = 2 \times 7$ and $50 = 5^2 \times 2$. The greatest powered version of each factor appearing in the terms are $5^2, 7,$ and 2. So, the product of those factors is $5^2 \times 7 \times 2 = 25 \times 14 = 350$.

3. **D.** Use the answers. To find the greatest common factor and least common multiple for each pair, first reduce each term to its prime factors. For choice A, this produces $18 = 2 \times 3 \times 3$ and $36 = 2 \times 2 \times 3 \times 3$. Using the regular method for finding the GCF gives $18 = 2 \times 2 \times 3$, and using the regular method for finding the LCM gives $36 = 2 \times 2 \times 3 \times 3$. Applying this method to all the solutions reveals that D is the correct answer choice.

4. **C.** The bells will ring together when the time reaches a number of seconds that is a multiple of both 5 and 9. The smallest number that satisfies this is the least common multiple of 5 and 9, which is $5 \times 9 = 45$.

5. **D.** The minimum number of nuts that Alexa can use is the least common multiple of 15 and 18. The prime factorization of 15 is 3×5, and the prime factorization of 18 is $2 \times 3 \times 3$. 2 appears a maximum of 1 time, 3 appears a maximum of 2 times, and 5 appears a maximum of 1 time. Therefore the least common multiple is $2^1 3^2 5^1 = 2 \times 9 \times 5 = 90$.

6. **A.** Get out your calculator and divide 2002 by each answer choice. 2002 is evenly divisible by each answer except 17.

7. **B.** If the two companies sell the same number of pens in different quantities, the minimum number of pens each sells will be the least common multiple of the different quantities. The least common multiple of 6 and 8 is the product of each distinct factor to the greatest power to which it occurs in the prime factorizations of the numbers. The prime factorization of 6 is 2×3 and the prime factorization of 8 is $2 \times 2 \times 2$. We thus need three 2's and one 3: $2 \times 2 \times 2 \times 3 = 24$.

8. **D.** The greatest common factor of two numbers is the product of all of the prime numbers common to both numbers' prime factorizations. The prime factorization of $90x$ is $2 \times 3 \times 3 \times 5 \times x$. The prime factorization of $300x$ is $2 \times 2 \times 3 \times 5 \times 5 \times x$. The product of the numbers in common to both is $2 \times 3 \times 5 \times x = 30x$. Note that, while x is included in the prime factorizations, it is not necessarily prime. However, whichever prime numbers make it up would be in common with each other, and thus are included altogether in the final answer as x.

9. **C.** LCM, remember, is like a Lego kit. This "kit" has to contain all the pieces necessary to "build" out each number that it is the LCM of. We know 1300 factors into $13(100) = 13(2)(2)(5)(5)$. At this point I can see that 50 is $(5)(5)(2)$, so those "pieces" are necessary to build out 50. For 1300 to be the LCM of 50 and another number, it is also going to contain "pieces" the other number needs to build itself out. The smallest "other" number would be the product of the pieces not included in the factors of 50 <u>and</u> the highest powered factors of 50, if the "power" happens to be higher than that of 50's factors. Because the LCM has two 2's but 50 only has one 2, our other number must include 2^2 in its factorization. It also must contain 13, as 13 is not a factor of 50. Thus, we multiply 2^2 times 13 to get 52.

10. **E.** The greatest common factor of the three numbers is the product of all of the prime numbers common to their prime factorizations. The prime factorization of 500 is $2 \times 2 \times 5 \times 5 \times 5$. The prime factorization of 1,000 is $2 \times 2 \times 2 \times 5 \times 5 \times 5$. The prime factorization of 12,500 is $2 \times 2 \times 5 \times 5 \times 5 \times 5 \times 5$. The product of the numbers common to all factorizations is $2 \times 2 \times 5 \times 5 \times 5 = 500$.

11. **B.** The sum of 0.12 and 0.06 is 0.18. It can be expressed as $\frac{18}{100}$. The rest of the problem is essentially a round about

way to say put the fraction in lowest terms and add the numerator and denominator. The greatest common factor of 18 and 100 is 2. Thus, dividing both by 2 gives us $\frac{9}{50}$. The sum of the numerator and denominator is 59.

12. **C.** Notice that 420 isn't a multiple of 18, as its digits do not sum to a multiple of 9, and 18 is divisible by 9, so rule out Choice (B). Now find the least common multiple (LCM) of 7, 10, and 14. The LCM of 7 and 10 is 70. The number 70 is also a multiple of 14, so the LCM of 7, 10, and 14 is 70. Choice (A), 14 is already part of the original set, and would still divide evenly into the LCM 70 so it doesn't affect the LCM and is something of a trick answer that we can ignore. Now, use a calculator to divide 70 by 5, and then again to divide 70 by 35. We find that 70 also is a multiple of 5, and 35. Because all these numbers divide evenly into 70, if any of these numbers were v, the LCM of 7, 10, 14, and v would be 70. As a result, you can rule out Choices (D), and (E), leaving Choice (C) as the only answer.

13. **D.** Factor each number to its prime factorization and find the product of the greatest powered unique factors among the factors you find. 18: $3^2(2)$, 25: 5^2, 45: $3^2(5)$. We take the highest powered version of each unique factor. We need two 5's, two 3's, and one 2: $(5^2)(3^2)(2) = 450$. If you selected choice (E), be careful—this is a common multiple of all three numbers, but it is not the *least* common multiple.

14. **D.** Factor 36 and 45. 36 has factors 36, 18, 9, 3, and 1. 45 has factors 45, 15, 9, 5, 3, and 1. Answer choice (A) is 36×45, (B) is 36×9, (C) is 18×5, (E) is 1×1, and (D) cannot be made by multiplying 1 number from the set of factors of 36 by 1 number in the set of factors of 45.

15. **B.** First, factor each term. In this problem, the given numbers are all products of 2, 3, a, and b. To find the lowest common multiple of the given values, figure out the maximum number of times each component (2, 3, a, and b) appears in any one of our given values. $8a^2 = 2 \times 2 \times 2 \times a \times a$, so the least common multiple must have $2 \times 2 \times 2 \times a \times a$ as a factor. No value has more than one factor of 3, so our number is only required to have one factor of 3. Finally, our least common multiple must have one b (two a's are already included). Multiply the necessary factors together to get $24a^2b$.

16. **B.** The factors of a are a and 1, while the factors of b are b and 1. Therefore, the greatest factor they must share is 1.

17. **D.** First, write the prime factorization of the numbers. $35 = 7 \times 5$, $28 = 2 \times 2 \times 7$, $8 = 2^3$. Multiply each number raised to the highest power then multiply. $7 \times 5 \times 2^3 = 280$.

18. **E.** 72 is evenly divisible by 2, 8, 24, and 36, which you may be able to just see or you can divide quickly using your calculator to determine. Thus these cannot be the answers, as then the LCM would be 72; the only possible answer is 48.

19. **C.** The least common denominator is the least common multiple of the denominators: 8, 9, and 30. First, find the prime factorizations of 8, 9, and 30: $2 \times 2 \times 2$, 3×3, and $2 \times 3 \times 5$. The maximum number of 2's in the prime factorization of any of the numbers is three, so 2^3 must be part of the prime factorization of the LCD. The maximum number of 3's in the prime factorization of any of the numbers is two, so 3^2 is also part of our LCD. Finally, the maximum number of 5's in the prime factorization of any of the numbers is 1, giving us 5^1. Our least common multiple is the product of all of these: $2^3 3^2 5^1 = 8 \times 9 \times 5 = 360$.

20. **C.** The least common denominator is the least common multiple of the denominators: 45, 63, and 14. We first find their prime factorizations: $3 \times 3 \times 5$, $3 \times 3 \times 7$, and 2×7, respectively. Every single distinct number in these prime factorizations must appear in our LCD, and the number must at least appear the maximum number of times it appears in any one factorization. The max number of 2's in the prime factorization of any of the numbers is 1, so 2^1 is in our LCD. The maximum number of 3's in the prime factorization of any of the numbers is two, giving us 3^2 for our LCD. The maximum number of 5's in the prime factorization of any of the numbers is one, which tells us 5^1 is in our LCD. Finally, the maximum number of 7's in the prime factorization of any of the numbers is one, giving us 7^1. The least common multiple is the product of these: $2^1 3^2 5^1 7^1 = 2 \times 9 \times 5 \times 7 = 630$.

21. **D.** The least common denominator is the least common multiple of the denominators: $29^2 \times 89 \times 907^2$, 29×89^3, and $29^3 \times 907$. The number 29 appears three times at most in any one of the denominators, giving us 29^3. The number 89 also appears three times at most in any one of the denominators, giving us 89^3. The number 907 appears two times at most in any single one of the denominators, giving us 907^2. Our least common multiple is the product of this: $29^3 89^3 907^2$.

CHAPTER

4

FRACTIONS

SKILLS TO KNOW
- How to use the "FRAC" function on your calculator
- Comparing fractions, decimals, and percents
- How to multiply, divide, add, and subtract fractions
- Setting up and solving fraction word problems

 NOTE: **Chapter 3, LCM & GCF,** covers least common denominator (LCD) fraction problems. Fraction problems also arise in **Chapter 2, Properties of Numbers,** and throughout the book series.

THE "FRAC" FUNCTION

TI-84 calculators (and many other calculators) have the amazing ability to turn decimal answers into fractions. On the ACT®, this is a very valuable tool. To use this function on a TI-84, click the MATH key on the upper left on your calculator, and then hit ENTER to select FRAC on the menu and ENTER again to turn the last number on your calculator into a fraction. This series of steps will turn whatever decimal is in your calculator into a fraction.

We can't cover every calculator in this book, so if you have a different one, get on the internet and look up how to convert decimals to fractions. It's a must know trick for tackling ACT® fraction problems.

Knowing this function allows you to bypass the perils of fraction problems and treat them as you would any problem with integers or decimals, using your calculator for any calculations.

For examples of how to incorporate this idea, see the word problems examples at the end of this chapter, and look for the calculator icon.

FRACTIONS VS DECIMALS VS PERCENTS

Remember, a fraction bar means DIVIDE. All you need to do to compare fractions and decimals is simply use your handy calculator and divide the top number by the bottom number to convert fractions to decimals. Percent values can be converted to decimals by moving the decimal point over two to the left. 25% is .25 or $\frac{25}{100} = \frac{1}{4}$. Decimal form is usually the easiest form in which to compare numbers.

Which of the following gives the fractions $-\frac{4}{9}$, $-\frac{3}{5}$, and $-\frac{5}{8}$ in order from least to greatest?

A. $-\frac{5}{8} < -\frac{3}{5} < -\frac{4}{9}$ B. $-\frac{4}{9} < -\frac{3}{5} < -\frac{5}{8}$ C. $-\frac{4}{9} < -\frac{5}{8} < -\frac{3}{5}$

D. $-\frac{3}{5} < -\frac{4}{9} < -\frac{5}{8}$ E. $-\frac{3}{5} < -\frac{5}{8} < -\frac{4}{9}$

To solve, we just divide the top number (numerator) by the bottom number (denominator) in our calculator—i.e. for $-\dfrac{4}{9}$, punch in $-4 \div 9$ into your calculator:

$$-\frac{4}{9} = -0.\overline{4}$$

$$-\frac{3}{5} = -0.6$$

$$-\frac{5}{8} = -0.625$$

Now remember that a negative number which has a BIG absolute value is actually less than a negative number with a smaller absolute value. Don't get confused! Create the proper order based on your decimals, then match your inequality back up with the original fractions:

$$-0.625 < -0.6 < -0.\overline{4}$$

$$-\frac{5}{8} < -\frac{3}{5} < -\frac{4}{9}$$

Answer: **A**.

OPERATIONS WITH FRACTIONS

First, remember that on the ACT® you can use a calculator. Oftentimes that means you can make your calculator do all the work, and won't need to do the work of finding common denominators, etc. However, because some problems have variables or other challenges, it's important to know the basics, too. Below is a quick summary of basic fraction operations.

Multiplication

To multiply fractions, multiply the numerators and denominators separately. It doesn't matter whether the denominators are the same number or not. Here's an example:

$$\frac{3}{5} \times \frac{6}{7} = \frac{3 \times 6}{5 \times 7} \rightarrow \frac{18}{35}$$

Division

To divide by a fraction, multiply by the reciprocal. When you divide by a fraction, simply multiply by the reciprocal (the fraction turned upside-down), i.e.:

$$5 \div \frac{1}{12} = 5 \times \frac{12}{1}$$

$$5 \times 12 = 60$$

Just flip the fraction AFTER the division sign upside down and multiply—remember the order MUST stay the same:

$$\frac{2}{3} \div -\frac{4}{5} = \frac{2}{3} \times -\frac{5}{4}$$

$$\frac{2}{2+\dfrac{1}{2-\dfrac{1}{2}}}=?$$

With these types of problems, always start with the most buried fraction and work your way up. Solve out each piece and substitute back in to the expression. Here we'll start with the $2-\dfrac{1}{2}$ at the bottom:

On the left we have the original expression: On the right we solve out each small piece:

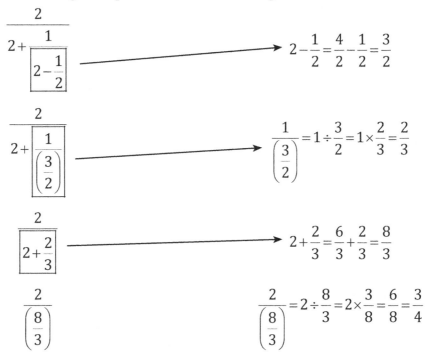

Addition and Subtraction

For us to add or subtract fractions, they need to have the same denominator. This can be achieved by multiplying both fractions by another fraction equal to one, "de-simplifying" them so as to change the way that they are numerically represented but not changing their values.

$$\frac{4}{7}+\frac{1}{3}=?$$

First, we multiply the fractions by a fraction equal to one which has the opposite denominator so the two fractions have a common denominator. For more information on finding the least common denominators, check out Chapter 3 on LCM/GCF.

For this problem, the common denominator between 7 and 3 is 21.

We multiply by the version of "one" that can convert each denominator to 21:

$$\frac{4}{7}\times\frac{3}{3} \text{ and } \frac{1}{3}\times\frac{7}{7}$$

This is mathematically valid because both fractions are multiplied by 1, which does not change their numerical values.

This gives us:
$$\frac{12}{21}+\frac{7}{21}=\frac{19}{21}$$

Answer: $\frac{19}{21}$.

$$\frac{\frac{5}{3}-\frac{3}{4}}{\frac{5}{3}+\frac{3}{4}}=?$$

NOTE: You can do this problem accurately and rapidly by using your calculator. To enter a fraction, divide the top number by the bottom number within a set of parentheses. At the end, use the FRAC function as described at the beginning of the chapter to convert everything back into a fraction.

I'll also show you how to do this process manually. We need to add the two fractions in the denominator and subtract the two in the numerator (remember the top of a fraction or bottom is essentially in "parentheses" so we add and subtract our elements within the numerator and denominator first before further simplifying).

$$\frac{\frac{5}{3}-\frac{3}{4}}{\frac{5}{3}+\frac{3}{4}}$$

The least common denominator of 3 and 4 is 12, so we multiply to convert all the fractions to equivalent fractions with a denominator of 12:

$$\frac{\frac{5}{3}\left(\frac{4}{4}\right)-\frac{3}{4}\left(\frac{3}{3}\right)}{\frac{5}{3}\left(\frac{4}{4}\right)+\frac{3}{4}\left(\frac{3}{3}\right)}=\frac{\frac{20}{12}-\frac{9}{12}}{\frac{20}{12}+\frac{9}{12}}=\frac{\frac{11}{12}}{\frac{29}{12}}$$

If the fraction format is confusing, we can also write it like this:

$$=\frac{11}{12}\div\frac{29}{12}$$

Now, we are trying to divide this fraction. As mentioned before, we do this by multiplying the top or first fraction by the reciprocal of the bottom or second one.

$$=\frac{11}{12}\times\frac{12}{29}$$

We can cross out the 12's because they cancel out, which leads us to the answer:

$$\frac{11}{29}$$

Answer: $\frac{11}{29}$.

> For all real numbers x such that $x \neq 0$, $\dfrac{3}{4} + \dfrac{7}{x} = ?$
>
> A. $\dfrac{10}{4+x}$ B. $\dfrac{21}{4x}$ C. $\dfrac{3x+28}{4x}$ D. $\dfrac{10}{4x}$ E. $\dfrac{7x+12}{4+x}$

For this problem we need to simply apply the same rules of fractions we use in fraction problems with numbers only to the variable. We first find our least common denominator:

$$4x$$

We'll add these fractions by converting each fraction into an equivalent fraction that shares this denominator:

$$\left(\frac{3}{4}\right)\left(\frac{x}{x}\right) + \left(\frac{7}{x}\right)\left(\frac{4}{4}\right)$$

$$\left(\frac{3x}{4x}\right) + \left(\frac{28}{4x}\right)$$

Now that the denominators match, we can add the top of the fractions:

$$\frac{3x+28}{4x}$$

Answer: **C**.

WORD PROBLEMS WITH FRACTIONS

For word problems with fractions, sometimes we need to translate the words into an equation and then solve. These are very similar to problems with **Percents (Chapter 5)**. Just as with percent problems, you'll also sometimes need to distinguish between what you're taking the fraction "of" so you don't calculate the wrong amount. If you "translate" the problem correctly, you should be fine.

> What fraction of an empty 750 ml boiling flask can be filled by a 1250 ml volumetric flask filled halfway?

Translate this one step at a time, just as we did in the percents chapter.

"What fraction" → write n

(We simply know that n will be a fraction, but it's just like any other unknown number, don't let the word "fraction" throw you off. Unlike percents, fractions don't require you to put the number over 100 or anything. "What fraction" just means make up a variable).

Of → write the sign for multiply

750 ml flask → write **750**

Can be filled by → write =

A 1250 ml Flask → write **1250**

Filled half-way → multiply by $\dfrac{1}{2}$

FRACTIONS   SKILLS

What fraction of an empty 750 ml flask can be filled by a 1250 ml flask filled halfway?

$$n \times 750 = 1250 \times \frac{1}{2}$$

$$750n = \frac{1}{2}(1250)$$

$$750n = 625$$

$$n = \frac{625}{750}$$

 To reduce this fraction on a TI-84, use the FRAC function on your calculator: type in $625 \div 750$ then hit ENTER, MATH, be sure FRAC is selected, and hit ENTER again. This process will show you the reduced fraction form:

$$n = \frac{5}{6}$$

Answer: $\frac{5}{6}$.

 TIP: Some word problems that include fractions, like the question below, are rates problems, so if you struggle with those, check out **Chapter 9: Rates, Ratios and Units in Book 1**.

 For English Literature class, Amelia must read Ethan Frome in 7 days. She reads $\frac{1}{10}$ of the book each of the first three days. For the remaining four days, what fraction of the book must Amelia read per day, on average?

This question is both a fraction and a rates problem. Remember we want fraction per day—that's the total fraction of the book left, divided by the number of days left (remember PER means DIVIDE).

First, let's figure out what fraction of the book she must read in the last four days.

She's already read $\frac{1}{10}$ of the book three days in a row, so we can add those three tenths together or multiply that fraction times 3 to get:

$$\frac{1}{10}(3) = \frac{3}{10}$$

If she has read $\frac{3}{10}$ then she has $\frac{7}{10}$ to go. We get $\frac{7}{10}$ by subtracting $\frac{3}{10}$ from 1. Remember in word problems, the whole amount or total is always 1 as the fractions add up to that whole.

Now she must read $\frac{7}{10}$ of the book over the course of 4 days. To figure out the fraction per day we take the fraction she has left to read and divide it by 4. You can do this in your calculator (0.7×0.25, then ENTER, then hit MATH, select FRAC, and hit ENTER), or multiply by $\frac{1}{4}$ to the same effect:

$$\frac{7}{10} \times \frac{1}{4} = \frac{7}{40}$$

Answer: $\frac{7}{40}$.

$\frac{1}{5}$ of what number is $\frac{1}{7}$ of 28?

This is another translation problem, also similar to percent problems. Translate it one piece at at time:

$\frac{1}{5}$	of	what number	is	$\frac{1}{7}$	of	28?
$\frac{1}{5}$	×	n	=	$\frac{1}{7}$	×	28
		$\frac{1}{5}n$	=	4		
		$n = 20$				

Answer: 20.

As you can see, if you simply translate each word and number, the problem is straightforward.

What fraction of a 9-inch-diameter pizza contains the same amount of pizza as a single slice taken from a 16-inch-diameter pizza cut into 12 slices, assuming both pizzas are of uniform, equal thickness?

For this question, we can apply the same logic as the last problem.

Because they are of equal depth or thickness, we can assume this is just an area problem and calculate everything based on surface area of the pizzas.

"What fraction" indicates that we need a variable (x will do!). We can do an initial set up using words to help us know what we need:

What fraction	of	a 9" diameter pizza	is the same	as a single slice	of	a 16" diameter pizza
x	×	total area of a 9" diameter pizza	=	$\frac{1}{12}$	×	total area of a 16" diameter pizza

Now we can solve for total areas.

First let's find the total area of the 9" diameter pizza:

Area of a circle: $A = \pi r^2$

$r = 4.5$ because radius is half of the diameter (9)

Always be careful not to confuse your radius and diameter!

$$A = \pi(4.5)^2$$
$$A = 20.25\pi$$

I'm going to leave the π in because I think it might cancel out later. Now I just need to figure out the area of the other larger pizza:

$$x \times 20.25\pi = \frac{1}{12} \times \text{total area of a } 16\text{" diameter pizza}$$

Area of a circle: $A = \pi r^2$
$r = 8$ because radius is half of the diameter (16)
Area of larger pizza: $A = \pi(8^2) = 64\pi$

We can now plug this into the above (initial) equation:

$$x \times 20.25\pi = \frac{1}{12} \times (64\pi)$$

The π cancels on each side and we get:

$$x \times 20.25 = \frac{1}{12} \times (64)$$

Use your calculator to divide 64 by 12 (you can use the FRAC function described earlier in this chapter or know that $0.3333\ldots$ is $\frac{1}{3}$).

$$20.25x = 5\frac{1}{3}$$

Now I use my calculator to divide $5\frac{1}{3}$ by 20.25 and get:

$$0.2633744\ldots$$

To turn this into a fraction, I hit MATH, then make sure FRAC is selected and press ENTER to get the answer:

$$\frac{64}{243}$$

Answer: $\frac{64}{243}$.

TIP: The size of the denominator has an INVERSE relationship with the value of a fraction. When dealing with positive fractions, to "maximize" a fraction's value, make its numerator larger and/or its denominator smaller. For example, 3/5 is smaller than 3/4, which is smaller than 3/3. **As the denominator shrinks, the value rises.** To "minimize" a fraction's value, make its numerator smaller and/or its denominator bigger. 4/5 is larger than 4/6, which is larger than 4/7. **As the denominator grows, the value shrinks.**

1. Which of the following is less than $\frac{2}{3}$?

   A. $\frac{3}{4}$

   B. $\frac{3}{5}$

   C. $\frac{6}{9}$

   D. $\frac{7}{10}$

   E. $\frac{9}{13}$

2. Seta has a long list of numbers that she first must divide by $\frac{2}{3}$ and then multiply by $\frac{8}{9}$ for a school assignment. She could achieve the same result in one step by dividing each number on the list by which of the following numbers?

   A. $\frac{3}{4}$

   B. $\frac{1}{2}$

   C. $\frac{1}{4}$

   D. 3

   E. 4

3. Mitch is making trail mix that needs $3\frac{1}{3}$ cups of raisins. However, he only has $\frac{1}{2}$ cup of raisins at home. The amount of raisins Mitch has at home is what fraction of the amount he needs for the recipe?

   A. $\frac{2}{3}$

   B. $\frac{1}{2}$

   C. $\frac{1}{4}$

   D. $\frac{2}{15}$

   E. $\frac{3}{20}$

4. Sidra is building a birdhouse. She drills a hole with a drill bit and tries to find a corresponding screw size. The $\frac{1}{4}$ inch screw is too big, and the $\frac{1}{8}$ inch screw is too small. Which of the following could be the size of the drill bit Sidra used?

   A. $\frac{1}{2}$-inch

   B. $\frac{3}{16}$-inch

   C. $\frac{5}{16}$-inch

   D. $\frac{4}{32}$-inch

   E. $\frac{8}{32}$-inch

5. Sugar Rush Candy Company has x chocolate bars and places $25 gift certificates for their stores in y of them as a way to attract more customers. Which of the following is a general expression for the fraction of the chocolate bars that do NOT have gift certificates?

   A. $\frac{y}{x}$

   B. $\frac{x}{y}$

   C. $\frac{x-y}{x}$

   D. $\frac{x-y}{y}$

   E. $\frac{y-x}{x}$

6. Of the 907 juniors in Walker High School, approximately $\frac{4}{5}$ of them are taking a foreign language, and approximately $\frac{3}{4}$ of those are taking Spanish. Which of the following is the closest estimate for how many juniors are taking Spanish?

   A. 180
   B. 430
   C. 550
   D. 750
   E. 825

FRACTIONS QUESTIONS

7. The expression $\dfrac{6-\dfrac{1}{8}}{2+\dfrac{1}{16}}$ is equal to:

   A. $\dfrac{94}{33}$

   B. $\dfrac{1551}{128}$

   C. 3

   D. 6

   E. 10

8. Which of the following is equivalent to $\dfrac{28x}{9} + \dfrac{4x}{3}$?

   A. $2\dfrac{2}{9}$

   B. $4x$

   C. $4\dfrac{2}{3}x$

   D. $4\dfrac{4}{9}x$

   E. $13\dfrac{1}{3}x$

9. The expression $\dfrac{\left(\dfrac{\dfrac{1}{4}}{\dfrac{1}{5}+\dfrac{1}{6}}\right)}{\dfrac{3}{4}-\dfrac{1}{3}}$ equals:

   A. $\dfrac{11}{144}$

   B. $\dfrac{9}{2}$

   C. $\dfrac{11}{20}$

   D. $\dfrac{18}{11}$

   E. $\dfrac{54}{11}$

10. $\dfrac{1}{4} \times \dfrac{2}{5} \times \dfrac{3}{6} \times \dfrac{4}{7} \times \dfrac{5}{8} \times \dfrac{6}{9} \times \dfrac{7}{10} \times \dfrac{8}{11} = ?$

    A. $\dfrac{1}{82}$

    B. $\dfrac{3}{55}$

    C. $\dfrac{1}{165}$

    D. 1

    E. $\dfrac{1}{110}$

11. In what order should $\dfrac{5}{2}, \dfrac{9}{3}, \dfrac{5}{4}$ and $\dfrac{7}{5}$ be listed to be arranged by increasing size?

    A. $\dfrac{5}{4}<\dfrac{7}{5}<\dfrac{5}{2}<\dfrac{9}{3}$

    B. $\dfrac{5}{4}<\dfrac{7}{5}<\dfrac{9}{3}<\dfrac{5}{2}$

    C. $\dfrac{9}{3}<\dfrac{5}{4}<\dfrac{9}{3}<\dfrac{5}{2}$

    D. $\dfrac{5}{2}<\dfrac{5}{4}<\dfrac{9}{3}<\dfrac{7}{5}$

    E. $\dfrac{7}{5}<\dfrac{5}{4}<\dfrac{5}{2}<\dfrac{9}{3}$

12. What fraction of $3\dfrac{1}{2}$ is $1\dfrac{1}{6}$?

    A. $\dfrac{1}{2}$

    B. $\dfrac{1}{3}$

    C. $\dfrac{3}{1}$

    D. $\dfrac{1}{6}$

    E. $\dfrac{5}{12}$

13. What fraction is halfway between $\frac{1}{7}$ and $\frac{1}{5}$?

    A. $\frac{1}{3}$

    B. $\frac{2}{35}$

    C. $\frac{1}{6}$

    D. $\frac{12}{35}$

    E. $\frac{6}{35}$

14. The value of x that solves $\frac{x}{3}+1=\frac{5}{6}$ lies between which of the following numbers?

    A. -3 and -1
    B. -1 and 0
    C. 0 and 1
    D. 1 and 3
    E. 3 and 6

15. Which of the following gives the range of numbers that are within $\frac{4}{3}$ of the number $\frac{3}{4}$?

    A. $-\frac{25}{12}$ to $\frac{7}{12}$

    B. $-\frac{25}{12}$ to $\frac{25}{12}$

    C. $-\frac{7}{12}$ to $\frac{7}{12}$

    D. $-\frac{7}{12}$ to $\frac{25}{12}$

    E. -1 to 1

16. When $1 \leq x \leq 6$ and $18 \leq y \leq 24$, the largest possible value for $\frac{4}{y-x}$ is:

    A. $\frac{4}{23}$

    B. $\frac{2}{9}$

    C. $\frac{4}{17}$

    D. $\frac{1}{3}$

    E. 3

17. For American Literature class, Liz must read *As I Lay Dying* in 11 days. She reads $\frac{1}{16}$ of the book each day for the first 4 days. For the remaining 7 days, on average, what fraction of the book must Liz read per day?

    A. $\frac{15}{112}$

    B. $\frac{3}{28}$

    C. $\frac{3}{4}$

    D. $\frac{1}{4}$

    E. $\frac{1}{16}$

FRACTIONS   ANSWERS

ANSWER KEY
1. B 2. A 3. E 4. B 5. C 6. C 7. A 8. D 9. D 10. C 11. A 12. B 13. E 14. B
15. D 16. D 17. B

ANSWER EXPLANATIONS

1. **B.** We can use decimal notation as an easy way to compare values. Remember that $\frac{2}{3} \approx 0.66$. Answer choice A, $\frac{3}{4} = 0.75$, is greater, so it is not the answer. Answer choice B, $\frac{3}{5} = \frac{6}{10} = 0.6$, is less than 0.66, so B is correct.

2. **A.** Dividing by a fraction is the same as multiplying by the reciprocal of that fraction. Thus, Seta's action is the same as multiplying by the reciprocal of $\frac{2}{3}$, which is $\frac{3}{2}$, and multiplying again by $\frac{8}{9}$. We can simplify these two steps into a signle step by multiplying $\frac{3}{2} \times \frac{8}{9} = \frac{4}{3}$. Thus Seta's series of actions is the equivalent of multiplying by $\frac{4}{3}$. However, the question asks for what we can divide by to equal her actions, so our answer is the reciprocal, $\frac{3}{4}$.

3. **E.** Translate: the amount Mitch has $\left(\frac{1}{2}\right)$ is (=) what fraction (x) of (multiply) the amount he needs $\left(3\frac{1}{3}\right)$, i.e. $\frac{1}{2} = x\left(3\frac{1}{3}\right)$. Now divide both sides by $3\frac{1}{3}$ and simplify to solve for x: $\frac{\frac{1}{2}}{3\frac{1}{3}} = \frac{\frac{1}{2}}{\frac{10}{3}} = \frac{1}{2} \times \frac{3}{10} = \frac{3}{20}$.

4. **B.** We can solve this problem by creating a common denominator among all of the fractions; Let's use the LCD, 16. In this case, we have $\frac{1}{4} = \frac{4}{16}$, the size that is too big, and $\frac{1}{8} = \frac{2}{16}$, the size that is too small. Answer A, $\frac{1}{2} = \frac{8}{16}$, is too big. Answer B, $\frac{3}{16}$, falls between $\frac{4}{16}$ and $\frac{2}{16}$, so it is correct. We could alternatively put everything in decimal form and compare.

5. **C.** Since x is the total number of bars and y is the number of bars with gift certificates, $x - y$ must be the number of bars that do not have certificates. To express the number of bars without certificates as a fraction of the whole, we divide $x - y$ by the whole number of chocolate bars, x. Thus, our final answer is $\frac{x-y}{x}$.

6. **C.** To find the fraction of a population as a whole number, we multiply the whole number by the fraction. First find the number of students taking a foreign language: $907 \times \frac{4}{5} = 725.6 \approx 726$. Next, find the fraction of these students who are taking Spanish: $726 \times \frac{3}{4} = 544.5 \approx 545$. Per the wording of the question, we only need the closest answer choice, which is 550, C.

7. **A.** First, simplify the numerator and denominator into single fractions. For the top, $6 - \frac{1}{8}$ is equal to $\frac{48}{8} - \frac{1}{8}$. For the bottom, $2 + \frac{1}{16}$ is equal to $\frac{32}{16} + \frac{1}{16}$. Substituting these in for the top and bottom gives us $\frac{\frac{48-1}{8}}{\frac{32+1}{16}} = \frac{\frac{47}{8}}{\frac{33}{16}}$. Since dividing by a fraction is the same as multiplying by its reciprocal, this is $\frac{47}{8} \times \frac{16}{33} = \frac{47}{1} \times \frac{2}{33} = \frac{94}{33}$.

8. **D.** First, multiply $\frac{4x}{3}$ by $\frac{3}{3}$ to express the fractions in terms of their least common denominator, 9. We get $\frac{28x}{9} + \frac{12x}{9} = \frac{40x}{9}$. None of the answer choices has an improper fraction, so we turn this into a mixed number: $\frac{40x}{9} = \left(\frac{36}{9} + \frac{4}{9}\right)x = \left(4 + \frac{4}{9}\right)x = 4\frac{4}{9}x$.

42   CHAPTER 4

9. **D.** First, simplify each "horizontal" level of the fraction. $\frac{1}{5}+\frac{1}{6}$ becomes $\frac{6}{30}+\frac{5}{30}=\frac{11}{30}$. $\frac{3}{4}-\frac{1}{3}$ becomes $\frac{9}{12}-\frac{4}{12}=\frac{5}{12}$. Now our expression equals $\dfrac{\frac{1}{4}}{\dfrac{\frac{11}{30}}{\frac{5}{12}}}$. We will divide by the fractions, starting with the numerator $\left(\dfrac{\frac{1}{4}}{\frac{11}{30}}\right)$, multiplying by the reciprocal of its denominator. $\dfrac{\frac{1}{4}\times\frac{30}{11}}{\frac{5}{12}}=\dfrac{\frac{30}{44}}{\frac{5}{12}}=\frac{30}{44}\times\frac{12}{5}=\frac{6}{44}\times\frac{12}{1}=\frac{6}{11}\times\frac{3}{1}=\frac{18}{11}$.

10. **C.** We can solve this problem by plugging it into a calculator, but it is actually easier to start by canceling out equal terms on the top and bottom. $4,5,6,7,$ and 8 on the top and the bottom cancel out, $\frac{1}{4}\times\frac{2}{5}\times\frac{3}{6}\times\frac{\cancel{4}}{\cancel{7}}\times\frac{\cancel{5}}{\cancel{8}}\times\frac{\cancel{6}}{9}\times\frac{\cancel{7}}{10}\times\frac{\cancel{8}}{11}$ leaving us with $\frac{1\times2\times3}{9\times10\times11}$. We can further simplify by canceling out the 2 and 10 to get 1 and 5, and by canceling out the 3 and 9 to get 1 and 3. We are now left with $\frac{1}{3\times5\times11}$. This equals $\frac{1}{165}$.

11. **A.** This problem is simpler to solve when we express the fractions as decimals. $\frac{5}{2}$ becomes 2.5, $\frac{9}{3}$ becomes 3. $\frac{5}{4}$ becomes 1.25. Use your calculator to find that $\frac{7}{5}$ is 1.4. Then, order the decimals in increasing size: $1.25, 1.4, 2.5, 3$. Now, by matching the decimals with the corresponding fractions from the beginning, we get $\frac{5}{4},\frac{7}{5},\frac{5}{2},\frac{9}{3}$.

12. **B.** If $1\frac{1}{6}$ is a certain fraction of $3\frac{1}{2}$, that means that $3\frac{1}{2}$ multiplied by that fraction will equal $1\frac{1}{6}$. We set up our equation as $3\frac{1}{2}n=1\frac{1}{6}$. Convert from mixed fractions to improper fractions: $\frac{7}{2}n=\frac{7}{6}$. Next, isolate n by multiplying both sides by the conjugate of n's coefficient: $\frac{2}{7}\left(\frac{7}{2}n\right)=\left(\frac{7}{6}\right)\frac{2}{7}$. This becomes $n=\frac{2}{6}=\frac{1}{3}$.

13. **E.** To find the number halfway between the two numbers, we add them together and halve the sum. $\frac{1}{2}\left(\frac{1}{7}+\frac{1}{5}\right)=\frac{1}{2}\left(\frac{5}{35}+\frac{7}{35}\right)=\frac{1}{2}\left(\frac{12}{35}\right)=\frac{6}{35}$.

14. **B.** Solve for x. $\frac{x}{3}+1=\frac{5}{6}$; $\frac{x}{3}=-\frac{1}{6}$; $x=-\frac{1}{2}$. $-\frac{1}{2}$ lies between -1 and 0.

15. **D.** We are looking for the range of numbers 4/3 less than 3/4 to 4/3 greater than 3/4. To find the left bound of the range of numbers, find $\frac{3}{4}-\frac{4}{3}=\frac{9}{12}-\frac{16}{12}=-\frac{7}{12}$. To find the right bound of the range of numbers, find $\frac{3}{4}+\frac{4}{3}=\frac{9}{12}+\frac{16}{12}=\frac{25}{12}$. Thus, the bounds of the range are $-\frac{7}{12}$ and $\frac{25}{12}$.

16. **D.** The largest possible value will have the smallest possible denominator. Thus, y will be its smallest possible value, and x will be its largest possible value. $\frac{4}{18-6}=\frac{4}{12}=\frac{1}{3}$.

17. **B.** After reading $\frac{1}{16}$ of the book for 4 days, she has completed $\frac{4}{16}=\frac{1}{4}$ of the book. Now, she has 7 days left to finish the remaining $1-\frac{1}{4}=\frac{3}{4}$ of the book. Her average rate she must read per day to finish on time equals the total fraction of the book left over total days she has to read, or $\dfrac{\frac{3}{4}}{7}=\frac{3}{28}$.

CHAPTER 4

CHAPTER 5

PERCENTS

> ### SKILLS TO KNOW
> - Basic Percent Translation Problems
> - Percent Increase and Decrease
> - Percent Short Cuts
> - Percent Word Problems
>   - The "Macy's" Problem (Multiple percent manipulations)
>   - Start with Percents / End with Percents/Ratios problems
>   - The Percent Proportion

PERCENT BASICS

Percents represent **PART over WHOLE**. A percent is numerically expressed as itself divided by 100. 25%, for example, is equal to the fraction 25/100 or the decimal 0.25.

To find a percent, we can use **the percent proportion.** Take the PART divide it by the WHOLE and set this quantity equal to the PERCENT over 100 (below, x represents our percent).

$$\frac{Part}{Whole}=\frac{x}{100}$$

We'll use this proportion idea in word problems. For now let's dive into some percent-speak:

TRANSLATING PERCENT PROBLEMS

The basic type of percent problem involves translating English expressions involved in percent calculations into "math" language. Many percent problems involve words like "what," "is," "percent," and "of." These words can be translated to math:

- **"WHAT"** means an unknown. Put a question mark or a new* variable (i.e. x or n).
- **"PERCENT"** means divide by 100. Always divide **the number mentioned before the word percent** by 100 (i.e. 5 percent $=\frac{5}{100}$).
- **"WHAT PERCENT"** means divide a new* variable by 100 (i.e. "what percent" means you should write $\frac{x}{100}$).
- **"x PERCENT"** whenever you see a variable next to "percent" write the variable over 100 (i.e. $x\%=\frac{x}{100}$).
- **"IS"** means equals (i.e. =).
- **"OF"** means multiply (i.e. write * or ×).

*If you see the word "WHAT" DO NOT reuse a variable already used in the problem somewhere. Make up a new variable!

When you see these words, simply translate them one at a time to form equations.

> If 10 is x percent of 50, what is x percent of 30?

$$\begin{array}{ccccc} 10 & \text{is} & x \text{ percent} & \text{of} & 50 \\ \downarrow & \downarrow & \downarrow & \downarrow & \downarrow \\ 10 & = & \dfrac{x}{100} & \times & 50 \end{array}$$

First Equation: $10 = \dfrac{x}{100}(50)$

$$\begin{array}{ccccc} \text{what} & \text{is} & x \text{ percent} & \text{of} & 30 \\ \downarrow & \downarrow & \downarrow & \downarrow & \downarrow \\ ? & = & \dfrac{x}{100} & \times & 30 \end{array}$$

Second Equation: $? = \dfrac{x}{100}(30)$

To solve, we first simplify the first equation and get:

$$10 = \dfrac{x}{100}(50)$$
$$10 = \dfrac{50x}{100}$$
$$10 = \dfrac{x}{2}$$
$$20 = x$$

Now we plug in $20 = x$:

$$? = \dfrac{20}{100}(30)$$
$$? = \dfrac{1}{5}(30)$$
$$? = 6$$

Answer: **6**.

PERCENT INCREASE AND DECREASE

PERCENT INCREASE OR DECREASE FORMULA

$$\dfrac{New\ Number - Original\ Number}{Original\ Number} \times 100$$

A quick way to think of it: **NEW minus OLD over OLD**. If the answer is **POSITIVE** it is a percent **increase**. If the answer is **NEGATIVE** it is a percent **decrease**.

PERCENTS   SKILLS

> Ashley worked a total of 80 hours last week and 56 hours this week. By what percentage did her work hours change?

Because her number of hours went down, we know this question is asking for a percent decrease. We can plug in the numbers given (80 is the original number, 56 is the new number) into our percent decrease equation and solve:

$$\frac{56-80}{80} \times 100 = -30\%$$

Answer: **30% decrease**.

Another way to think of percent changes is: $\dfrac{Numeric\ Change}{Original}$

> If the number of shirts increased by 5 and there were originally 10 shirts, then what is the percent increase in the number of shirts?

$$\frac{Numeric\ Change}{Original} = \frac{5}{10} \text{ or } 50\%$$

Answer: **50%**.

PERCENT SHORT CUTS

Most students learn the traditional set up for percent equations and handle a typical word problem in the following fashion:

> Marley bought four bandannas for $3.00 each, and two shorts for $15.00 each. If she pays 8.5% sales tax on her purchase, what is her total cost?

Step 1: Calculate the cost of the items before the tax:
Four bandannas times $3.00 each $=\$4\times3=\$12$ total for bandannas
Two shorts for $15.00 each $=\$15\times2=\$30$ total for shorts
$\$12+\$30=\$42$ total for all items.

Step 2: Calculate the sales tax:
Sales tax equals 8.5% of the total, $42:

$$\frac{8.5}{100} \times \$42 = \$3.57 \text{ or } 0.085 \times \$42 = \$3.57$$

Step 3: Add the sales tax to the pre-tax total:
$$\$42+\$3.57=\$45.57$$

BUT THERE IS A FASTER WAY!

The above problem utilizes the following formula (albeit in steps):

$$\text{Original Amount} + \frac{\text{Percent Increase}}{100}(\text{Original Amount}) = \text{Total Amount}$$

Let's imagine that our original pre-tax amount is "x":

$$x + \frac{\text{Percent Increase}}{100}(x) = \text{Total}$$

Once we know that percent increase, let's fill it in:

$$x + \frac{8.5}{100}(x) = \text{Total} \text{ or } x + 0.085x = \text{Total}$$

AHA! Do you see that we can simplify that expression on the right?

$$x + 0.085x = 1.085x$$

The thing is, this idea holds for EVERY percent increase (and decrease) problem. Thus, we can add "1" to the decimal form of the percent, put that in front of "x" and do this problem not in three steps, but in two. The same principle holds for percent decrease—but in this case we subtract from 1.

SPEED TIP: To increase a number by x percent, multiply it by $(1+\frac{x}{100})$. To decrease a number by x percent, multiply it by $(1-\frac{x}{100})$. Remember $\frac{x}{100}$ is the same as the percent in decimal form (i.e. $25\% = \frac{25}{100} = 0.25$.).

At Step 2, we simple calculate $(1+0.085)(\$42)$ or $1.085 \times \$42$ and get our answer with one calculator entry!

Answer: $\$45.57$.

> Marni buys 4 shirts originally priced at n dollars each, but has a 20% off discount coupon that is applied at checkout. Which of the following represents her total cost before tax?
>
> A. $4.8n$ B. $4n$ C. $3.2n$ D. $0.8n$ E. $0.2n$

Step 1: Find her pre-discount cost: the number of shirts times the cost per shirt: $4n$

Step 2: Multiply by $1-d$ where d is the discount expressed as the numeric percent value (i.e. the percent in decimal form or divided by 100 already). Because the discount is 20%, d=0.2.

$$(1-0.2)(4n) \text{ or } 0.8(4n)$$

Step 3: Simplify:

$$0.8(4n) = 3.2n$$

Answer: **C**.

WORD PROBLEMS

The "Macy's" Problem

Many word problems involve multiple percent discounts. At first these might seem most easily solved by adding together all the discount percents and then reducing the original number by that amount. However, that strategy will often backfire: you must reduce according to how the problem is worded.

I call these "Macy's" problems, because they reflect a common advertising trick that Macy's (and other stores) often use. Rather than give customers a flat 50% off, Macy's will advertise 30% and then an additional 10% off with coupon and 10% off for using your Macy's card! Sounds like a great deal, right? Like 50% off, right? Not so fast...

With these kinds of incentives, the additional **discounts come off of the intermediate, already reduced prices, NOT off of the original.** As a result, you need to do these problems in steps.

> Sweta buys a sweater originally priced at $120. The sweater is subject to a 30% off sale, and Sweta has a coupon for an additional 10% off any sale item. She additionally gets a 10% off discount off the coupon discounted price as part of an early bird promotion. What does she pay, before tax, for her sweater?

First take the 30% off—I'll use my shortcut method described above:
$$0.7(\$120)$$

Then take the additional 10% off:
$$0.9(0.7(\$120))$$

Then take another 10% off:
$$0.9\Big(0.9\big(0.7(\$120)\big)\Big) = \$68.04$$

Answer: $68.04.

As you can see, she didn't pay $60 (or 50% of the original price)—the store eked another $8.04 out of her!

> On June 1, a used car was put up for sale. On June 15, the price was reduced by 15% After a month, the price was reduced by 10% of the June 15 price and then sold. The final sale price was $8415. What was the price of the car on June 1st?
>
> A. $6437   B. $6732   C. $10,519   D. $10,645   E. $11,000

MISTAKE ALERT: A common mistake in percent problems is to apply the percent to the wrong amount. For example, the above problem gives us the FINAL sales price and asks for the ORIGINAL price. Remember, percents are always applied to the STARTING number, so you can't apply the percents to the number we are given, $8415, and work backwards. **Percents don't work in reverse!** We must make up a variable, applying the percents to the original, starting number.

Let x represent our original price on June 1st. Now use the shortcut to discount x by 15%, or keep 85%. Again we must apply the percents in steps and cannot add the percents together first.
$$0.85x$$
Now use the shortcut again to discount this value by another 10% (i.e. keep 90%) and we set this total equal to our final price.

$$(0.9(0.85x)) = 8415$$

Now simplify using algebra:

$$0.765x = 8415$$
$$x = 11,000$$

Answer: **E**.

Starts with Percents (or Ratios)/ Ends with Percents (or Ratios)

Whenever you have a percent or ratio problem that has NO REAL VALUES—i.e. you have no idea regarding the actual amounts involved—make up a number to make the problem easier to solve! **I prefer to make up 100 or 10 or some multiple thereof to make the calculations easier.** The example below shows this idea with percents, but this method also works with ratio problems.

> A class has an activities fund; 20% of the fund is spent in September, and 10% of the remaining monies are spent in October. What percent of the original fund is left by the end of October?
>
> A. 30%   B. 42%   C. 67%   D. 70%   E. 72%

As you can see, we have no idea how much is in this fund. All we have is percents. So let's make up a number: $100.

If the fund has $100, then they spent $20 in September:

$$0.2(\$100) = \$20$$

And have $80 left at the start of October:

$$\$100 - (0.2(\$100)) = \$80 \text{ or } (0.8)(\$100) = \$80$$

Now we must be careful. The problem referenced 10% "of remaining monies." The most common error, again, that students make is to take this percent off the original money amount. During October, they spent 10% of the remaining $80, which is $8:

$$0.10(\$80) = \$8$$

Thus, $80 − $8 = $72 of the original fund remains. We could also find this by thinking that we need 0.9 or 90% of the $80 to remain after 10% was spent:

$$0.9(\$80) = \$72$$

$72 is then what percent of the original amount, $100? That's easy because we choose 100 as our starting number! It's the final number over the original:

$$\frac{72}{100} = 72\%$$

Answer: **E**.

Percent Proportions

Sometimes you need to think about what percent means to solve a problem: part over whole.

> Julie has won 52 out of 80 tennis matches. If she wins every game from now, what is the least number of games she must win to improve her win percentage to at least 70%?

For this problem we start with part over whole:

$$\frac{52}{80} = \frac{Matches\ Won}{Total\ Matches} = Her\ Current\ Percentage = 65\%$$

We want to get that up to $70\%+$—to do that she'll need to play more games and win them. We'll call the number of games she plays and wins "g." We'll add g to the top of the fraction because it adds to the number won—but also to the bottom of the fraction—as it adds to the total number of matches. We'll now figure out how many wins will get her to 70.

$$\frac{52+g}{80+g} = 0.7$$
$$52+g = 0.7(80+g)$$
$$52+g = 56+0.7g$$
$$0.3g = 56-52$$
$$0.3g = 4$$
$$g \approx 13.3333$$

We now round—she needs at least 13.3333 matches, so that means at 14 matches she'll finally be above 70%. If you're not sure, plug in 13 and you'll find that it creates an answer below 70%!

Be careful when rounding in real world situation word problems. You don't always simply round to the nearest whole number! When in doubt, double check your work and test your answer.

Answer: **14**.

1. 60% of a number is 72. What is 130% of that number?

   A. 56.16
   B. 92.31
   C. 120
   D. 156
   E. 1560

2. x is 175% of y. 35% of y is what percent of x?

   A. 5%
   B. 20%
   C. 25%
   D. 110%
   E. 500%

3. There is a basket with 20 apples. 40% of the apples are taken by Jimmy. 25% of the remaining apples are taken away by Joe. How many apples remain in the basket?

   A. 9
   B. 11
   C. 12
   D. 17
   E. 18

4. In a group of people, 30% of the people leave, and 10% of the remaining people leave a few hours later. What percent of the original group is left?

   A. 37%
   B. 40%
   C. 60%
   D. 63%
   E. 71%

5. The temperature at the beginning of the day is x. By noon the temperature increased 35%, by the end of the day, the temperature decreased 25% from the temperature at noon. The temperature at the end of the day is what percent of x?

   A. 10%
   B. 33.75%
   C. 60%
   D. 101.25%
   E. 110%

6. James' commute to work has been increasing every year due to increasing traffic. From 2012 to 2013 his commute time increased by 25%. From 2013 to 2014 his commute time increased by 15%. From 2012 to 2014, by what percent did his commute time increase?

   A. 3.75%
   B. 18.75%
   C. 28.75%
   D. 40%
   E. 43.75%

7. In January, Gary could run a mile in 6 minutes. In February, he decreased his time by 15%. In March, Gary decreased his time by another 12%. His mile time in March is what percent of his original mile time?

   A. 74.8%
   B. 73%
   C. 25.2%
   D. 10.2%
   E. 1.8%

8. At a used car dealer, 55% of the cars have 7 or more seats. Of the remaining cars, 40% have 5 or 6 seats. What percent of cars have fewer than 5 seats?

   A. 5%
   B. 22%
   C. 23%
   D. 27%
   E. 45%

9. A tennis player plays her first 30 matches and wins 18 of them. Then she loses the next 5 matches in a row. What is the minimum number of matches she would have to win to match or surpass her original winning percentage?

   A. 7
   B. 8
   C. 2
   D. 6
   E. 4

PERCENTS   QUESTIONS

10. Matt is playing an online game and has won 30 out of 100 times. If he wins every game from now, what is the least number of games he must win to improve his win percentage to 40%?

   A. 10
   B. 13
   C. 15
   D. 17
   E. 18

11. Ben and Charlotte went to a restaurant and their bill was $36.21. They decided to leave a tip that would be 18% of their bill. Which of the following is closest to that amount?

   A. $7.23
   B. $7.00
   C. $6.50
   D. $6.07
   E. $5.50

12. What is 20% of 10% of 5?

   A. .001
   B. 1
   C. .1
   D. .2
   E. .5

13. What is 3% of 9.81×10^5?

   A. 32700
   B. 327000
   C. 2943000
   D. 294300
   E. 29430

14. What percent of $\frac{4}{5}$ is $\frac{1}{5}$?

   A. 25%
   B. 20%
   C. 80%
   D. 16%
   E. 400%

15. Due to inflation, prices have risen by 30%. What does a calculator that previously cost $80.00 now cost?

   A. $104.00
   B. $110.00
   C. $80.30
   D. $83.00
   E. $90.00

16. Every year a car depreciates (loses value). A particular car that was worth x dollars last year now sells for 20% less. Which of the following calculations gives the current cost, in dollars, of the car?

   A. $x - .2x$
   B. $.2x$
   C. $x - .02x$
   D. $x - .8x$
   E. $1.2x$

17. During a sale at a department store, all ties are discounted by 75%. A customer buys a tie that was originally priced at $135. If an 8% sales tax was added after the discount was taken, about how much money does the customer have to pay?

   A. $109.35
   B. $60.75
   C. $36.45
   D. $33.75
   E. $68.00

18. Ms. Murphy gives her class a quiz with 25 questions. Assuming no partial credit is given for any questions, only one of the following percents is possible for a student to score on this quiz of 25 questions. Which one is it?

   A. 85%
   B. 88%
   C. 90%
   D. 94%
   E. 97%

19. The school took a survey of the student body's favorite flavor of ice cream. 20% of students said they prefer chocolate, 15% said they prefer vanilla, 20% said they prefer strawberry, 30% said they prefer cookie dough, and the rest said they prefer sorbet. If each student could only pick one flavor, and n students said they preferred sorbet, how many students in terms of n were surveyed?

A. $100n$

B. $\dfrac{100-n}{15}$

C. $\dfrac{15n}{100}$

D. $\dfrac{100n}{15}$

E. $.85n$

20. A chair that was originally sold for $150 is discounted to $135. What is the percent discount on this chair?

A. 90%
B. 15%
C. 11%
D. 10%
E. 1%

21. The table below shows the age distribution of a student body at a large university.

Age, in years	17	18	19	20	21	22
Percent of students	4%	24%	21%	26%	18%	7%

What percent of students are younger than 21?

A. 49%
B. 74%
C. 75%
D. 51%
E. 26%

PERCENTS   ANSWERS

ANSWER KEY

1. D 2. B 3. A 4. D 5. D 6. E 7. A 8. D 9. B 10. D 11. C 12. C 13. E 14. A
15. A 16. A 17. C 18. B 19. D 20. D 21. C

ANSWER EXPLANATIONS

1. **D.** Translate each piece: 60% $\left(\frac{60}{100}\right)$ ... of (multiply) ...a number (x) ... is $(=)$ ...72, i.e. $0.6x = 72$. So, $x = \frac{72}{0.6} = 120$. 130% of 120 is $1.3 \times 120 = 156$.

2. **B.** Translate each piece; let "n" equal the missing percent we need: x ... is $(=)$ ...175% $\left(\frac{175}{100}\right)$ ... of (multiply) ... y means $x = 1.75y$. So $y = \frac{x}{1.75}$. 35% $\left(\frac{35}{100}\right)$ of (multiply) y ... is $(=)$ ... what percent $\left(\frac{n}{100}\right)$ ... of (multiply) ... x means $0.35y = \left(\frac{n}{100}\right)(x)$. Now we can substitute in for y, placing its value from the first equation into the second, and then solve for our missing percent, n: $0.35y = \frac{n}{100}x \rightarrow .35(\frac{x}{1.75}) = \frac{n}{100}x \rightarrow .2x = \frac{n}{100}x \rightarrow .2 = \frac{n}{100} \rightarrow n = 20$

3. **A.** 40% of the 20 apples translates to $0.4 \times 20 = 8$ apples taken by Jimmy. This means there are $20 - 8 = 12$ apples remaining. Of these 12 remaining apples, Joe takes 25% or $0.25 \times 12 = 3$ of them. This means there are $12 - 3 = 9$ apples left in the basket.

4. **D.** Let x represent the number of people the group started with. So, after 30% of the people leave, we have $x - 0.3x = 0.7x$ people left. A few hours later, 10% of the $0.7x$ people leave, leaving $0.7x - (0.1(0.7x)) = 0.7 - 0.07x = 0.63x$ people left. This is 63% of the original x people.

5. **D.** After the temperature increases 35%, it is $1.35x$. When $1.35x$ decreases 25%, we "keep" 75% of this number, so we multiply by $1 - 0.25$ or $.75$: $0.75 \times 1.35x \rightarrow 1.0125x$. This is 101.25% of the original temperature x.

6. **E.** Let x represent the original commute time. From 2012 to 2013, his commute increased by 25%, which means it was $1.25x$. Then, from 2013 to 2014, his then commute time of $1.25x$ increased by 15%, making it $1.15 \times 1.25x = 1.4375x$. So, in total from 2012 to 2014, his commute time is 143.75% of the original commute time x, which means that it increased by 43.75%.

7. **A.** In February, Gary's original time of 6 minutes decreased by 15%, which means that his time was then $6 - 0.15 \times 6 \rightarrow 0.85 \times 6 \rightarrow 5.1$. Then, in March, he decreased his time of 5.1 minutes by another 12%, which means that his time in March was $5.1 - 0.12 \times 5.1 \rightarrow 0.88 \times 5.1 \rightarrow 4.488$ minutes. This is $\frac{4.488}{6} = 0.748 = 74.8\%$ of his original time.

8. **D.** Let x equal the total number of cars. 55% of the cars have 7 or more seats, which means that $0.55x$ cars have 7 or more seats and $x - 0.55x = 0.45x$ cars, or 45%, do not have 7 or more seats. Of those $0.45x$ cars, 40% have 5 or 6 seats. This means that 100-40 or 60% of the remaining vehicles have fewer than 5 seats. We can use the numeric version of 60% "(60/100 or .6) and multiply this times 45% of x to find the percent of the total with fewer than 5 seats: $0.6 \times 0.45x \rightarrow 0.27x$ cars have fewer than 5 seats. Convert the decimal .27 to a percent: 27% of the original x cars.

9. **B.** The tennis player wins 18 out of 30 matches, which is $\frac{18}{30} \times 100\% = 60\%$ of her matches. She loses the next 5 in a row, which means that she now has won 18 out of 35 matches. If she wins x number of games in a row, she can match or surpass her original percentage of 60%. We can calculate her percentage after winning x games in a row as $\frac{18+x}{35+x} = \frac{\%\ wins}{100}$, adding the wins both to the denominator, which represents the number of wins, and the numer-

ator, which represents the number of total matches. We want a winning percentage of 60%, i.e. 60/100 or .6, so we have $\frac{18+x}{35+x}=0.6$. Multiplying both sides by $35+x$, we get $18+x=0.6(35+x) \rightarrow 18+x=21+0.6x$. Subtracting $0.6x+18$ on both sides, we get $0.4x=3$. So, $x=7.5$. Since she must win at least 7.5 matches in a row, but can't win half a game; winning 8 matches would enable her to surpass her original winning percentage.

10. **D.** We can calculate Matt's percentage after winning x games in a row as $\frac{30+x}{100+x}=\frac{\% \text{ wins}}{100}$. We want his percent wins to be 40%, so we have $\frac{30+x}{100+x}=0.4$. Multiplying both sides by $100+x$, we get $30+x=0.4\times(100+x) \rightarrow 30+x=40+0.4x$. Subtracting $0.4x+30$ on both sides, we get $0.6x=10$. So, $x=16.667$. Since he must win at least 16.667 games in a row, winning 17 games would enable him to improve his winning percentage to 40%.

11. **C.** Translate: we need 18% ($\frac{18}{100}$ or .18) of (multiply) their bill (insert the total amount, 36.41). Solving: $(.18)(\$36.21)=6.5178 \approx 6.50$, answer (C).

12. **C.** $\left(\frac{20}{100}\right)\left(\frac{10}{100}\right)(5)=\left(\frac{1}{5}\right)\left(\frac{1}{10}\right)(5)=.1$, answer (C).

13. **E.** $9.81\times 10^5(.03)=29430$, answer (E).

14. **A.** Translate: What percent (x/100) of (multiply) 4/5 is (=) 1/5 becomes $\frac{x}{100}(\frac{4}{5})=\frac{1}{5} \rightarrow \frac{x}{100}=(\frac{1}{5})(\frac{5}{4}) \rightarrow \frac{x}{100}=\frac{1}{4} \rightarrow x=\frac{1}{4}(100) \rightarrow x=25$, answer (A).

15. **A.** To increase by 30% ($\frac{30}{100}$ or .3) we multiply by $1+0.3$: $\$80(1.3)=\$104$, answer (A).

16. **A.** The current value is the past year's values minus 20% (or 0.2) of the past year's value. Since the past year's value was x, the current value is $x-.2x$, answer (A).

17. **C.** A 75% discount means the ties cost 25% of the original price. The sales tax will increase the price by 8%, so a customer will pay 100% plus 8%, or 1.08 times the final price of the item. Since both these operations are multiplication, and multiplication is commutative, they can be done in one step. $\$135(.25)(1.08)=\$36.45$, which is answer (C).

18. **B.** We can represent this situation as part over whole with the percent proportion: $\frac{n}{25}=\frac{score}{100}$ where n is an integer value representing the number of questions correct on the test (we assume a student can either get full credit or no credit on each question) and 25 is the total number of test questions. We need a percent score that when divided by 100 can reduce to a fraction with a denominator of 25. From looking at this fraction, I know I divide 100 by 4 to get to 25. Thus I will divide the score by 4 to get to n. In other words, I need a score divisible by 4. Looking at the answers provided, the only answer choice divisible by 4 is 88%, answer (B).

19. **D.** Subtracting the known percents from the total percentage amount, we find $100\%-20\%-15\%-20\%-30\%=15\%$ of students prefer sorbet. Let n equal the number of people who prefer sorbet. These n people represent 15% of the total number of students, which we can call S, as modeled by the percent proportion: $\frac{n}{S}=\frac{15}{100}$. Solve, and we see that $S=\frac{100n}{15}$. All the other answers are expressions that use n and numbers in the correct expression, but position them incorrectly.

20. **D.** Remember percent decrease is numeric change ($15) over the original ($150): 15/150=.1 or 10%. Alternatively, set up an equation $135=150x$, where x is a decimal form of a percent. $\frac{135}{150}=.9$, which means that $135 is 90% of the original price, or that there is a 10% (100%−90%) discount, answer (D).

21. **C.** Sum the percent of students at age twenty and below. $4\%+24\%+21\%+26\%=75\%$, answer (C). Be careful with the details. Make sure all the percents add up to 100% and that there aren't "missing" elements on the chart.

CHAPTER 6
SEQUENCES AND SERIES

SKILLS TO KNOW

- Sequence vs. Series
- Arithmetic Sequences
- Geometric Sequences
- The Sum of a finite Arithmetic Sequence (Arithmetic Series)
- Miscellaneous Patterns

A **sequence** is a string of numbers, possibly going on infinitely, that follows a strict pattern. The first number in the sequence is the first term, the second is the second term, and so on.

A **series** is the sum of all of the numbers in a sequence.

One kind of sequence is the **arithmetic sequence**. In this type of sequence, every term increases by a constant amount. In other words, to move up one value in the sequence, you add a certain number. Then you add that number again to reach the next value in the sequence.

Let's look at a few arithmetic sequences:
$$0, 2, 4, 6$$

Here's another:
$$\{6, 12, 18, ...\}$$

And one more:
$$\{5, 8, 11, 14, ...\}$$

As you can see, in an arithmetic sequence every term increases by a constant amount. The first sequence started with 0 and increased in increments of 2. The second sequence started with 6 and increased in increments of 6. The third sequence started with 5 and increased in increments of 3. We call these increments the **common difference**. Arithmetic sequences can also subtract an equal increment or use increments that are fractions or decimals.

$$1.5, 1, .5, 0, -.5, -1...$$

In the arithmetic sequence above, for example, the common difference is -0.5: we subtract 0.5 to find each subsequent term.

Sometimes, arithmetic sequence problems are easy enough to do by simply working forward with the understanding of the pattern at play.

> What is the fifth term in the arithmetic sequence $\{3, 8, 13, ...\}$?

Here, all we need to do is figure out how much we increase each term by, and then work forward two more numbers in the sequence. To get from 3 to 8 we add 5 (since $8-3=5$), so 5 is our common difference. We can now calculate the 4th and 5th terms:

$$4\text{th term: } 13 + 5 = 18$$
$$5\text{th term: } 18 + 5 = 23$$

Answer: 23.

However, if you are looking for, say, the 450th term, it would be absurd to add the same number over and over 449 times. For cases such as this, you'll need to either know the formula or extrapolate the pattern of the sequence and apply that pattern.

Remember that the sum of a single number (let's use d for **common difference**) added together x times $\overbrace{(d + d + d + d + ... + d)}^{x \text{ times}}$ equals dx. Since we are adding the same number (our **common difference**) over and over, we can use **multiplication** to speed up the process. If the first term is a_1, and we increase by increments of d, we can express the nth **term in the sequence** (a_n) as:

FORMULA FOR THE nth TERM OF AN ARITHMETIC SEQUENCE

$$a_n = a_1 + (n-1)d$$

Where a_n is the nth term in the sequence, a_1 is the first term in the sequence, n is the number of the term in the sequence, and d is the common difference.

We use $n-1$ because when n equals one, we don't add d yet. We only add d for the first time in reaching the second term. We'll add a 2nd "d" when we get to the third term, a 3rd "d" when we get to the fourth term, etc. Therefore, we must subtract one from the number term we are on to figure out how many d's we have added.

To see why this works, let's look at an example.

> In the arithmetic sequence, $3, 7, 11, 15, ...$, what would be the 21st term?
>
> **A.** 86   **B.** 83   **C.** 80   **D.** 4   **E.** 19

First, I'll say that I really am NOT a huge fan of formulas. I find them difficult to remember and am constantly confusing them. On questions like these, I find the best way to learn the formula is to understand where it comes from so that I can derive it at will. After deriving it enough times, I "see" the formula because of how it works and then apply it correctly. Some of the time, in fact, I use the ideas of the formula without actually using the formula itself.

For this question, I'll start by punching out a chart that goes through each term and tracks the pattern. Remember n is used to simply keep track of which number in the sequence I'm on. I'll break each piece down into what I'm adding:

n	Value of the term	Calculation of the term	What's happening
1	3	3	First term
2	7	$3+4$	First term plus 1×4
3	11	$3+4+4$	First term plus 2×4
4	15	$3+4+4+4$	First term plus 3×4

At this point, I see the pattern: I add the first term to 4 times one less than n.

When $n=2$, for example, I have only one 4 to add. When $n=3$, I add two 4's. So for the 21st term, I'll add 20 4's:

$$3 + 20(4) = 83$$

I can also solve this directly using the formula. Again, the more I write out a chart like this, the better I get at remembering that formula, too. If you're aiming for a 32+ on the math, mastering the formula after understanding why it works is a good idea.

We are looking for the **21st** term, so $n = 21$. Our **first term** is 3, so $a_1 = 3$. The **common difference** between each term is 4, so $d = 4$. Plugging all these in:

$$a_n = a_1 + (n-1)d$$
$$3 + (21-1)4 = 3 + (20)4$$
$$= 3 + 80$$
$$= 83$$

Answer: **B**.

GEOMETRIC SEQUENCE

A geometric sequence is to the arithmetic sequence what multiplication is to addition. That is, instead of adding by something every term, we multiply by something every term. We call this multiplication increment the **common ratio**, often denoted by the variable r.

Here's an example of a simple geometric sequence:

$$\{1, 2, 4, 8, 16, ...\}$$

The first term is 1 and every subsequent term is multiplied by 2. 2 is our **common ratio**.

Let's take a look at another example, and try to find a relationship between our original number and those that follow. If we can find that pattern, we'll see how to create a formula to solve for a term in a geometric sequence.

$$2, 6, 18, 54, 162, ...$$

Let's list out each term in a chart of sorts and try to deduce what is going on at each step:

n	Value of the term (a_n)	Calculation of the term	What's happening
1	2	2	First term
2	6	$2(3)$	When $n=2$, multiply first term by 3 once
3	18	$2(3)(3)$	When $n=3$, multiply first term by 3 twice
4	54	$2(3)(3)(3)$	When $n=4$, multiply first term by 3^3
5	162	$2(3)(3)(3)(3)$	When $n=5$, multiply first term by 3^4
n	a_n	$a_1(3^{n-1})$	Multiply first term, a_1, by 3^{n-1}

Over time, I realize that the exponent of 3 is always one less than my term n. That's where the $(n-1)$ comes from. Now I can see why the following formula works.

FORMULA FOR THE nth TERM OF A GEOMETRIC SEQUENCE

To find the nth term in a geometric sequence, where a_n is the nth term in the sequence, a_1 is the first term in the sequence, and r is the common ratio (what we multiply each subsequent term by):

$$a_n = a_1 r^{n-1}$$

What is the 8th term of the geometric sequence $-1, 3, -9, 27, \ldots$?

A. -81 B. 243 C. -729 D. 2187 E. -6561

Using the formula above, we can plug in numbers and find the answer. We know we're looking for the **8th term**, so $n=8$, and that the **first term** is -1, so $a_1 = -1$. The **common ratio**, or amount we multiply by each time is -3, so $r = -3$.

$$a_n = a_1 r^{n-1}$$
$$a_n = -1 \times (-3)^{8-1}$$
$$= -1 \times (-3)^7$$
$$= -1 \times -2187$$
$$= 2187$$

Answer: **D**.

Alternatively, the numbers are small enough in this problem that I could have figured out the common ratio -3 and then multiplied by it several times in my calculator until I reached the 8th term. If you use this method, just move as fast as you can and get the problem done! No shame in the easy method if it works.

SEQUENCES/SERIES — SKILLS

ARITHMETIC SERIES (OR THE SUM OF AN ARITHMETIC SEQUENCE)

NOTE: **These are RARE on the ACT.** Skip this section if you're not aiming for a 30+ or are short on time.

Sometimes, a question or word problem will ask you to find the sum of a certain number of terms in an arithmetic sequence. This sum is called an **arithmetic series**.

This, too, has a formula, but it's actually one of the easiest ones to remember, because essentially, it's the same as the **average formula**. (See **Chapter 7: Averages** for more on this formula.)

> Amelia is saving for a bicycle. She saves $5 the first day. She saves an additional $6 the second day, an additional $7 the third day, and so on, such that each day she increases her daily addition to her savings by $1. If she saves money in this fashion for 25 days, saving an additional $29 on the 25th and final day once she reaches her savings goal, how much money does Amelia save, in total, for her bicycle?

To answer this question, we could try to count out every day, but that would be time consuming:

$$\$5 + \$6 + \$7 + \ldots + \$29$$

I would have to type in 25 different numbers into my calculator!

Another way we could do this would be by stacking our list and finding a common sum:

$$\begin{array}{cccccc} 5 & 6 & 7 & 8 & 16 & 17 \\ +29 & +28 & +27 & +26 & \ldots +18 & +(0) \\ \hline 34 & 34 & 34 & 34 & 34 & 17 \end{array}$$

As you can see, I've stacked the highest term with the lowest term, and counted down on my bottom list as I count up on my top list. I have to figure out then how I "turn the corner," which I do by dividing 34 by 2, getting 17. Because 17 is a whole number, I know I'll turn the corner with the pair 16/18, have 17 by "itself" and then continue the pattern. If I had a midpoint that was 0.5 (such as 17.5) I know I'd be splitting 17 & 18 as I "turn" the corner and move from top to bottom as I count.

Seeing this pattern, I know there will be 12 pairs (to find the term from the $$ amount, I simply subtract 4 given the original pattern, or I cut 25 in half, and know there are 12 pairs and 1 outlier at the corner turn), plus the outlier, 17.

So I calculate as follows:

$$34(12) + 17 = 425$$

From this, I have our answer.

But we can move much faster if we understand that the principle behind this stacking: averages.

Each "pair" we chose was equidistant from our mean/median of the data set. Because the two values are the same "distance" apart (remember we add the common difference to each term, so the spacing is always identical), when we add them together, they create a common sum. They also create a common average, that also happens to be the average of the whole set: if we divide by two with any of the pairs in the above we get 17: that's the mean and the median of our data. We can see that as 17 is the number we hit when we "round the corner" from the top list to the bottom list.

 FUN FACT: In an **arithmetic sequence**, the **mean** and the **median** are the same amount.

Essentially, we know that the average of the first and the last term is also the average of the set as a whole. Armed with that information, we can find the sum much more quickly. All we need is the first term and the last term. Then we can average those two to find the average of the entire set. From there, we can use the average equation.

$$Average = \frac{Sum}{Number\ of\ items}$$

Now we can solve for the sum, clearing the fraction to get what we want in terms of the sequence:

$$Average(Number\ of\ items) = Sum$$

The average is the first term plus the last term divided by 2, so we can substitute that in:

$$\left(\overbrace{\frac{first\ term + last\ term}{2}}^{Average\ of\ first\ and\ last\ terms}\right)(Number\ of\ items) = Sum$$

We can rearrange this to form the formal equation (though memorizing the above is probably easier and works just fine!)

THE SUM OF THE FIRST n TERMS OF AN ARITHMETIC SEQUENCE

$$S_n = \frac{n(a_1 + a_n)}{2}$$

Where S_n = sum of the first n terms of arithmetic sequence, n = number of terms in the sequence, a_1 = first term in the sequence, and a_n equals the nth term in the sequence.

And now we have a formula for the sum of an arithmetic sequence! We plug in the first and last terms, 5 and 29 and the number of items in the list, 25:

$$\left(\frac{5+29}{2}\right)(25) = Sum$$
$$(17)(25) = 425$$

As you can see, this is the same answer we got above.

Answer: 425.

SEQUENCES/SERIES   SKILLS

> The second term in an arithmetic sequence is 63 and the seventh term is 78. What is the common difference of the sequence?

For this problem, we first need to know vocabulary, including arithmetic sequence and common difference. You should know these terms as we reviewed them above. The common difference (we'll call this d) is the amount we add each time to move up one term in the sequence. For each additional term, we add d. We can think of this as follows:

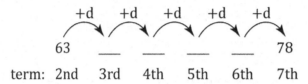

As you can see, we'll add d five times to get to the 7th term from the 2nd one.

Common mistakes people make here:

- They look at the four blanks and assume that means they add d four times (you need to add it FIVE times though, look at the "hops" on top, not the blanks).
- They assume the pattern starts at the first and not the 2nd term.
- They don't draw it out, and in doing so, get confused or are off by one term (again, adding d four times or perhaps six times). Drawing the problem out helps prevent carelessness.

From the picture, we create an algebraic formula and solve:

$$63 + 5d = 78$$
$$5d = 15$$
$$d = 3$$

Answer: 3.

Again, we could have also used the formula for this, but I am not a huge fan of formulas unless I need them. In this case, the formula method actually takes a bit longer. Also, formulas are too easy to remember incorrectly! Still, if you prefer to use the formula, the solution that way looks something like this:

$$a_n = a_1 + (n-1)d$$

We know that when $n = 2$, $a_n = 63$. Plugging in we get:

$$63 = a_1 + (2-1)d$$
$$63 = a_1 + d$$

We also know that when $n=7$, $a_n=78$. Plugging in we get:

$$78 = a_1 + (7-1)d$$
$$78 = a_1 + 6d$$

Now we have two equations and two unknowns (a_1, d). We can solve this system of equations by elimination or substitution. Because I know I want d, I will solve for a_1 to eliminate it (remember, isolate the variable you want to eliminate.)

$$63 = a_1 + d$$
$$63 - d = a_1$$

Now I substitute this value in for a_1 in my other equation:

$$78 = a_1 + 6d$$
$$78 = (63 - d) + 6d$$
$$78 = 63 + 5d$$
$$15 = 5d$$
$$3 = d$$

Again, I get the answer of 3 (but hopefully you're convinced that this is the long way!).

REMAINDER PROBLEMS

NOTE: For similar problems, see the **Complex Numbers** Chapter in Book 1.

Another type of problem that occurs is a sometimes called a remainder problem. For these problems, a pattern will be established and you'll need to identify which member of the pattern occurs at a particular point. For example, you might be asked if today is Tuesday, what day of the week was it 201 days ago?* Or a group of 122 campers counts off by 5 (assigning 1 to the first camper, 2 to the second...5 to the fifth, 1 to the sixth and so forth) to break into groups. If 63 campers have not yet received their numbers, what is the next number that will be called?* *Answers on next page.

> Jemal is making a friendship bracelet that follows the pattern of one white bead, one red bead, one yellow bead, one green bead, one orange bead, and then one blue bead. If he follows this pattern in this order, what will be the 35th bead in the sequence?

Here we have a pattern that repeats every six terms. I can write out the pattern using variables for each bead:

```
1  2  3  4  5  6
W  R  Y  G  O  B
```

When I get to the 7th term I go back to white:

```
7  8  9  10  11  12
W  R  Y  G   O   B
```

As you can see, I am always going to end my row of a pattern with a number that is divisible by 6, because these are in groups of 6. So we call this a "Remainder" problem, because I can solve by dividing the number of the term by 6 and finding the remainder.

35 divided by 6 is 5 remainder 5. What this means is that we go through 5 complete WRYGOB patterns and then walk through the pattern 5 more members in after completing those 5 patterns.

I can imagine a multiple of 6 and thus "B" in the pattern at the number 30:

$$\begin{matrix} & & & & & 30 \\ W & R & Y & G & O & B \end{matrix}$$

At this point I can extend my pattern and write out the remaining elements from there, or count in according the remainder (5 elements) from the first element, W. Here we see O, or orange, is the color of the 35th bead.

$$\begin{matrix} 31 & 32 & 33 & 34 & 35 & \\ W & R & Y & G & O & B \end{matrix}$$

Answer: **Orange**.

WARNING: Be careful with tricky remainder problems. **I always recommend writing out the pattern** as I do above at least once rather than only using the formula. This carefulness can be vital when counting backwards or dealing with atypical patterns.

*Answers to two example problems on previous page:

Question 1: **Thursday** (201/7 has a remainder of 5, so backwards five days from Tuesday is Thursday. Be careful not to count days FORWARD! Write out the pattern to be extra careful.)

Question 2: **Five** (If 63 campers don't have numbers, 122-63=59 do. The next camper called is number 60, which is a multiple of 5, so her number is five.)

MISCELLANEOUS PATTERNS

NOTE: Spatial arrangements and patterns are covered at the end of **Chapter 9: Counting and Arrangements**.

Sometimes you're given a sequence that doesn't follow the above patterns. In this case, you can't rely on memorized formulas. You'll need to trace out the pattern, figure out the rule at play and apply that rule to find the value you need. Often you'll need to step these out one piece at a time, by hand.

Sometimes these are called **recursive sequences**, because you use a formula to move from one term to another. Don't let that word throw you off.

> Each term, such that $n > 2$, in a sequence is found by using a recursive formula: doubling the previous term and adding some number x. If the first term in the sequence is 2, and the 6th term is -29, what is x?

Here our best bet is to work backwards. This problem sounds confusing, but just follow the pattern. Remember n just refers to what number term we're on. By saying the formula is used from the 2nd term on, we allow the first term to just be stated or defined.

Let's work our way up to the 6th term from the first one, one term at a time, using x for itself.

$$\text{Each Term} = 2(\text{previous term}) + x$$

$$
\begin{aligned}
\text{First Term} &= 2 \\
\text{Second Term} &= 2(2) + x \\
&= 4 + x \\
\text{Third Term} &= 2(4 + x) + x \\
&= 8 + 2x + x \\
&= 8 + 3x \\
\text{Fourth Term} &= 2(8 + 3x) + x \\
&= 16 + 6x + x \\
&= 16 + 7x \\
\text{Fifth Term} &= 2(16 + 7x) + x \\
&= 32 + 14x + x \\
&= 32 + 15x \\
\text{Sixth Term} &= 2(32 + 15x) + x \\
&= 64 + 30x + x \\
&= 64 + 31x
\end{aligned}
$$

Now we simply set that 6th term equal to -29:

$$
\begin{aligned}
64 + 31x &= -29 \\
31x &= -93 \\
x &= -3
\end{aligned}
$$

Answer: -3.

PACING TIP: should you encounter an atypical sequence on a late problem (after number 50-55), the problem could take some time to complete. You may want to prioritize other problem types that are not as time consuming.

SEQUENCES/SERIES QUESTIONS

1. Stacy and Stephanie are building a 4-level square pyramid out of wooden blocks for a school project. Each of the levels consists of a consecutive, perfect square amount of blocks with the top level having 4 blocks. How many blocks must Stacy and Stephanie use?

   A. 54
   B. 90
   C. 29
   D. 28
   E. 14

2. Christine started a savings account in order to save up money to get laser eye surgery to correct her vision. She deposits $25 into her account the 1st month. Each following month, she deposits $15 more than the amount she deposited the previous month. So, Christine deposits $40 the 2nd month, $55 the 3rd month, and so on and so forth. She makes her final deposit of $235 the 15th month. What is the total amount for Christine's 15 deposits?

   A. $1950
   B. $1715
   C. $2200
   D. $2465
   E. $600

3. The first term of an arithmetic sequence is 10 and its common difference is -3. What is the sum of the first 10 terms of this sequence?

   A. -35
   B. -50
   C. -32
   D. 0
   E. -3

4. How many terms are there between 23 and 72, inclusive, in the given arithmetic sequence?

   $$2, 9, 16, 23, \ldots, 72$$

   A. 9
   B. 6
   C. 8
   D. 7
   E. 10

5. The first 3 terms of an arithmetic sequence are $11\frac{1}{2}$, $9\frac{5}{16}$, $7\frac{1}{8}$ respectively. What is the fifth term of the sequence?

   A. $4\frac{15}{16}$
   B. $\frac{9}{16}$
   C. $2\frac{3}{4}$
   D. 5
   E. $5\frac{15}{16}$

6. The 9th, 10th, and 11th terms of an arithmetic sequence are 47, 55, and 63, respectively. What are the first 3 terms of the sequence?

   A. $-25, -16, -7$
   B. $-24, -17, -9$
   C. $-9, -1, 7$
   D. $-9, -2, 5$
   E. $-17, -9, -1$

7. What is the sum of the first 55 terms of the arithmetic sequence $2, 4, 6 \ldots$?

   A. 3,080
   B. 3,192
   C. 1,540
   D. 110
   E. 38

8. During his first week as a salesman, Barney sold 33 products. He challenged himself to sell 3 more products each successive week than he had sold the week before. If Barney meets, but does not surpass his goal, how many products does he sell during his 15th week?

   A. 145
   B. 78
   C. 72
   D. 75
   E. 63

9. What is the sum of the first 3 terms of an arithmetic sequence in which the 6th term is 23 and the 11th term is 30.5?

   A. 46.5
   B. 55.5
   C. 51
   D. 53.5
   E. 1.5

10. Consecutive terms of a certain arithmetic sequence have a positive common difference. The sum of the first 3 terms of the sequence is 135. Which of the following values CANNOT be the first term of the arithmetic sequence?

    A. 44.7
    B. 42.5
    C. 35
    D. 44
    E. 50

11. The first 4 terms of a geometric sequence are $0.66, -1.98, 5.94$ and -17.82. What is the ninth term?

    A. 4330.26
    B. −1443.42
    C. −12990.78
    D. 18247.68
    E. −9.9

12. If 3 is the first term and 243 is the fifth term of a geometric sequence, which of the following is the third term?

    A. 81
    B. 27
    C. 9
    D. 54
    E. $\frac{3}{5}$

13. The first and second terms of a geometric sequence are x and $\triangledown x$, respectively. What would be the 999th term in the sequence?

    A. $\triangledown^{1000} x$
    B. $\triangledown^{999} x$
    C. $\triangledown^{998} x$
    D. $\left(\triangledown x\right)^{998}$
    E. $\left(\triangledown x\right)^{999}$

14. The first 3 terms of a geometric sequence are 2, 7, and 24.5. What is the next term in the sequence?

    A. 85.75
    B. 49
    C. 42
    D. 29.5
    E. 36.75

15. The formula to find the sum of an infinite geometric series with the first term a and common ratio $|r|<1$ is $\frac{a}{1-r}$. The sum of a given infinite geometric series is $\frac{2}{3}$ and the common ratio is $-\frac{1}{2}$. What is the fourth term of this series?

    A. $-\frac{1}{8}$
    B. 1
    C. $-\frac{1}{2}$
    D. $\frac{1}{4}$
    E. $-\frac{1}{32}$

16. In a certain number sequence, each term after the 1st term is the result of adding 3 to the previous term and then multiplying the sum by 4. The 4th term in the sequence is 700. What is the first term?

    A. 40
    B. 9.95
    C. 20.2
    D. 10
    E. 7

17. The decimal representation of $\frac{5}{14}$ repeats and can be expressed as $0.357142857142857142857\ldots$. What is the 500th digit of this decimal?

    A. 1
    B. 2
    C. 8
    D. 7
    E. 5

CHAPTER 6

SEQUENCES/SERIES QUESTIONS

18. A triangular number, T_n, when $n > 0$, is a triangular array of dots with n points on each side. The figure below shows the first 4 triangular numbers. What is the value of T_{50}?

Term	Value
1	1
2	3
3	6
4	10

    A. 1275
    B. 2750
    C. 2550
    D. 125
    E. 147

19. The sum of the first 20 positive odd integers is 400. Which of the following is the sum of the first 40 positive odd integers?

    A. 1,580
    B. 1,600
    C. 800
    D. 1,523
    E. 4,400

20. Which of the following terms is not in the geometric sequence 4, −10, 25 … ?

    A. −62.5
    B. −390.625
    C. 2441.406
    D. 156.25
    E. 976.5625

21. Which of the following integers must be a factor of the sum of any 4 consecutive integers?

    A. 2
    B. 3
    C. 4
    D. 5
    E. 6

CHAPTER 6

ANSWER KEY

1. A 2. A 3. A 4. C 5. C 6. E 7. A 8. D 9. C 10. E 11. A 12. B 13. C 14. A
15. A 16. E 17. E 18. A 19. B 20. C 21. A

ANSWER EXPLANATIONS

1. **A.** The first level has 4 blocks, and $4=2^2$ So, the next level has $(2+1)^2=3^2 \to 9$. The third level has 4^2 blocks and the final 4th level has 5^2 blocks. Adding the total number of blocks together to make the pyramid, we get $2^2+3^2+4^2+5^2=4+9+16+25 \to 54$ blocks.

2. **A.** Because each deposit increases by a fixed amount each time, the list of deposit amounts is an arithmetic sequence. The first deposit amount is $25 and after n months, Christine deposits $235, so we know her first and last deposit amounts. We want to calculate the sum of all 15 deposits, which is $25+40+55...+235$. Remember the sum of arithmetic sequence is just the average times the number of items. To find the average, we can simply average the first and last values that we already know: $\left(\frac{235+25}{2}\right)=\$130$. We now multiply this by the number of deposits total, 15: $15(130)=1950$. Alternatively, we can use the formula on page 61.

3. **A.** To find the sum of the first 10 terms in an arithmetic sequence, remember sum equals average times number of items in the list. To find the average, we'll need the first and last terms (remember the average of an arithmetic sequence can be found by averaging the first and last terms). We know the first term but don't know the last term (the 10th) so we'll start with that. The nth term, a_n, in an arithmetic sequence can be calculated as $a_n=a_1+(n-1)d$ with first term $10=a_1$, common difference $d=3$, and the number term we want to solve for $n=10$. So, the 10th term in the sequence is $10-3(10-1) \to 10-27 \to -17$. Now find the average of the set by averaging the first and last terms: $(10+(-17))\div 2 = -3.5$ and then multiply by the number of terms $n=10$. $-3.5(10)=-35$, answer A. If you prefer, use the formula: $Sum=\frac{(First\ term+Last\ term)}{2}(n)$ where $n=$ the number of numbers in the sequence. Plugging in we get $Sum=\frac{(10-17)10}{2} \to -35$.

4. **C.** To find the common difference of the arithmetic sequence $2, 9, 16, 23...72$ subtract any two terms in a row, such as the first and second terms: $9-2=7$. Since each term is separated by 7, we add 7 with each additional term: we add it once to get to 9, two times to get to 16, three times to get to 23, etc. Notice the number of times we add 7 is one less than the term, i.e. 9 is the 2nd term and we add one 7. 16 is the 3rd term and we add two 7s. So we can see that the last term, 72, would equal the first term plus 7 times n minus 1, where n is the number of the term 72: $72=2+(n-1)7$. This is the same as the formula for the nth term of an arithmetic sequence, but you can see how we can think it through to arrive at the formula. We simplify: $72-2=(n-1)7 \to 70=7(n-1) \to 10=n-1 \to 11=n$. So 72 is the 11th term. But we're not done!! Don't put 11! We want the number of terms BETWEEN 23 and 72 inclusive. We can count out that 23 is the 4th term. Between the 4th term and 11th we have the 5, 6, 7, 8, 9, 10th terms, or six terms, plus the two "inclusive" terms on the ends, the 4th and 11th terms. That makes a total of 8 terms, B. We could also solve the last part by knowing that the number of items in an inclusive list of consecutive numbers (i.e. from 4 to 11 inclusive) is the difference of the numbers plus one $(11-4+1=8)$. Be very careful on problems like this though, better to quickly count from $5-10$ than make a careless error.

5. **C.** To find the common difference of the arithmetic sequence $11\frac{1}{2}, 9\frac{5}{16}, 7\frac{1}{8}...$ subtract any two terms in a row, such as the first and second terms: $9\frac{5}{16}-11\frac{1}{2}=9\frac{5}{16}-11\frac{8}{16} \to -2\frac{3}{16}$. So the 4th term in the sequence is $7\frac{1}{8}-2\frac{3}{16}=4\frac{15}{16}$ and the 5th term is $4\frac{15}{16}-2\frac{3}{16}=2\frac{12}{16} \to 2\frac{3}{4}$. When the terms are small enough like this, don't bother with formulas.

6. **E.** To find the common difference of the arithmetic sequence, subtract any two terms in a row: $55-47=8$. So, if the 9th term in the sequence is 47, then the first term in the sequence is 8 terms before the 9th term. To move one term down, we subtract 8, so to find the 1st term, we want to move 8 terms down, so we subtract 8 eight times. $47-8(8)=47-64 \to -17$

SEQUENCES/SERIES   ANSWERS

To find the 2nd and 3rd terms, add eight to each. The first 3 terms are -17, $-17+8$, and $-17+8(2)$ or -17, -9 ,and -1.

7. **A.** To find the sum of the first 55 terms, remember sum equals average times number of items in the list. The average is typically most easily found by averaging the first and last terms. We know the first term, so let's work to find the value of the 55th or last term. To do so we can use the formula for the nth term in an arithmetic sequence (or derive it as I do in question 4's explanation): $a_n = a_1 + (n-1)d$, where the first term in the sequence $a_1 = 2$,the common difference is d, and the term we want is $n = 55$. To find the common difference of the arithmetic sequence, subtract any two terms in a row: $4-2=2$. Now we can calculate the 55th term of the sequence by plugging in: $2+2(55-1)=2+2(54) \to 110$. Now we find the average of the 1st and 55th terms: $\left(\frac{110+2}{2}\right) = 56$. Now we multiply 56 times the number of numbers in the list, 55: $(55)(56)=3080$, answer A. If you prefer, use the formula $Sum = \frac{(First\ term + Last\ term)}{2}(n)$ where $n =$ the number of numbers in the sequence. Plugging in we get $Sum = \frac{(2+110)55}{2} \to 3080$.

8. **D.** Because we add a constant number of sales each week, this is an arithmetic sequence. To solve, use the formula for the nth term in an arithmetic sequence (or derive it as I do in question 4's explanation) to find the 15th term: $a_n = a_1 + (n-1)d$ where the first term in the sequence is a_1 ,the common difference is d, and the term we want is $n = 15$. From the problem, we find the first term a_1 is 33 and the common difference, d, is 3, so: $n_{15} = 33 + 3(15-1) \to 33 + 42 \to 75$.

9. **C.** Since the 6th term is 23, the 11th term is 30.5, and the 11th and 6th terms are 5 common differences away, the common difference in the sequence is $\frac{30.5-23}{5} = \frac{7.5}{5} \to 1.5$. So, the first term of the sequence—which is 5 common differences away from the 6th term—can be calculated by $23-1.5(5)=23-7.5 \to 15.5$ The 2nd and 3rd terms are then $15.5+1.5=17$ and $15.5+1.5(2)=18.5$. So, the sum of the first three terms of the sequence are $15.5+17+18.5=51$. We could also have found the 2nd term, the average of the 1st and 3rd terms and median/mean of the first three terms, and simply multiplied it by three to get the sum of the first three terms.

10. **E.** The mean of an arithmetic sequence is also the median. Because there are 3 terms in the set, the 2nd term is the median and mean. Find the mean of the terms by dividing their sum, 135, by 3: $\frac{135}{3} = 45$, so the 2nd term and the median is 45. The first term cannot be greater than the median/second term, 45, so choice E is impossible because $50 > 45$.

11. **A.** The common ratio of the geometric sequence can be calculated by taking any term and dividing it by the term before it, such as the 2nd term divided by the first: $-\frac{1.98}{0.66} = -3$. As the 4th term is -17.82 and 9th term is 5 ratios greater than the 4th term, the 9th term can be calculated as $n_9 = -17.82 \times (-3)^5 \to 4330.26$.

12. **B.** Since 3 is the first term and 243 is the 5th term, we can multiply by the comon ratio four times to get to the 5th term from the 1st, so the common ratio of the geometric sequence can be calculated as d in the equation $3d^4 = 243$ So, $d^4 = 81 \to d = 3$. Multiply the 1st term times the common ratio 3 twice to get the 3rd term. That is, $n_3 = 3 \times 3^2 \to 27$.

13. **C.** Since the first and second terms are separated by a factor of ∇, ∇ is the common ratio of the geometric sequence. This means that the 999th term, which is 998 terms away from the 1st term, is $\nabla^{998} x$.

14. **A.** The common ratio of the geometric sequence can be calculated by dividing the 2nd term by the 1st term: $\frac{7}{2} = 3.5$. So, the next term in the sequence is this common ratio times the given 3rd term, 24.5: $24.5(3.5) = 85.75$.

15. **A.** We first solve for the first term, a, by plugging $S = \frac{2}{3}$ and $r = -\frac{1}{2}$ into the formula $S = \frac{a}{1-r}$. $\frac{2}{3} = \frac{a}{1-\left(-\frac{1}{2}\right)} \to \frac{2}{3} = \frac{a}{\frac{3}{2}} \to \frac{2}{3} = \frac{2}{3}a \to a = 1$ So, the 4th term of the sequence, which is 3 common ratios greater than the first term, is $1\left(-\frac{1}{2}\right)^3 = -\frac{1}{8}$.

16. **E.** Each term in the sequence can be calculated as $n_i = 4(n_{i-1}+3)$. Isolating the n_{i-1} ,we get the formula $n_{i-1} = \frac{n_i}{4} - 3$ We are given that $n_4 = 700$. So, $n_3 = \frac{700}{4} - 3$. We can now run this operation on n_3 to get n_2 and then again on n_2 to

CHAPTER 6

get the first term. (Feel free to solve down each ugly fraction and simplify in your calculator at each step).

$$n_1 = \frac{\frac{\frac{700}{4}-3}{4}-3}{4}-3 \to \frac{\frac{172}{4}-3}{4}-3 \to \frac{40}{4}-3 \to 7$$

17. **E.** This is a complex style remainder problem. Typically in a remainder problem, we have a pattern that repeats. Here, that pattern is offset first by a number, and then begins. If you look at the decimal, the repeating pattern does not start with the first digit of the decimal, 3. Rather, it starts with the 2nd digit of the decimal, and as a result, our problem is a bit trickier. That repeating pattern is then 6 digits 571428 repeating in that order. Write out labels for each term to visualize:

$$5 \quad 7 \quad 1 \quad 4 \quad 2 \quad 8$$
$$2^{nd}, \ 3^{rd}, \ 4^{th}, \ 5^{th}, \ 6^{th}, \ 7^{th},$$
$$8^{th}, \ 9^{th}, \ 10^{th}, \ 11^{th}, \ 12^{th}, \ 13^{th}$$

We can think of it like this: The number 8 always appears on a term that is ONE MORE than a multiple of 6. Because there are 6 repeating digits, we must correlate everything to that idea of multiples of 6. I.e. $7 = 6+1$, $13 = 2(6)+1$, etc. To find the 500th digit, we'll first find the number closest to 500 that we can that is a multiple of 6, then we'll move up one in my pattern and be at the number 8. We can find the number divisible by 6 closest to 500 using divisibility rules: we know the digits must sum to something divisible by 3 and the number must be even. 500 is not a multiple of 6. Then we try 498 (the next even number below 500). $4+9+8 = 21$. Because 21 is divisible by 3, so is 498. Because 498 is also even it is thus divisible by 6. Now we count forward to find the number 8 in the pattern at 499. Then we count forward in the pattern to get to the number 5 in my pattern at 500. Answer: 5 Alternatively, we could know in the pattern the number 2 is always on a multiple of 6 and count up from there two spots.

18. **A.** Each triangular number is simply the sum of the term n plus all the previous terms, i.e.

$T_n = n + (n-1) + (n-2) \ldots + (n-n)$, thus to calculate T_{50}, the 50th term, we need to calculate the value of $1 + 2 + 3 + \cdots + 49 + 50$, which is an arithmetic series. The fastest way to do so is with the formula for the sum of an arithmetic series, $Sum = \frac{(F+L)n}{2}$ where F = the first number in the sequence, L = the last number in the sequence, and n = the number of numbers in the sequence. We can also think of this formula as rooted in the average formula; simply find the average of the series (first term plus last term divided by two) and multiply this average times the number of terms. Remember, Average=Sum/Number of Terms or Sum=(Average)(Number of Terms). The average of 1 and 50 is 25.5, and then we multiply this by the number of terms in the series, 50. $25.5(50) = 1275$. Alternatively, plugging into the formula: $F = 1$, $L = 50$, and $n = 50$ and $Sum = \frac{(1+50)50}{2} = 1275$.

19. **B.** The sequence of odd numbers is $1, 3, 5, 7\ldots$ is an arithmetic sequence. Because we add two with each subsequent term, we know the common difference is 2 and can use the formula for the sum of a sequence as outlined in problem 18's explanation. Here, though, we must calculate the 40th term before using the formula. Let's use the formula for finding the nth term of a sequence: $a_n = a_1 + (n-1)d$ where a_n is the nth term (what we want), a_1 is the 1st term (here 1), n is the number of the term (here 40), and d is the common difference (here 2). Plugging into the formula we get $n_{40} = 1 + (40-1)2$, which simplifies to $1 + 39(2) = 1 + 78 = 79$. Now to find the set's average, we average our first (1st) and last (40th) terms: $\frac{1+79}{2} = 40$. The problem tells us we want the first 40 terms. Now we just multiply our number of terms, 40, times our average, which is also 40, to get the sum: $(40)(40) = 1600$. Note, we can ignore the sum of the first 21 terms they give us. When using the formula, incorporating that information won't speed anything up. If you struggle with formulas, you can approach the problem by writing out and finding patterns or stacking numbers as we outlined earlier in the chapter.

20. **C.** The common ratio of the geometric sequence can be calculated by taking any term and dividing it by the term before it, because (1st term)(common ratio)=2nd term and so on. $-\frac{10}{4} = -2.5$. So, the terms in the geometric sequence can all be represented in the form $4(-2.5)^n$ for some integer n. (True, the formula typically includes $n-1$, but that doesn't really matter here. We just want to know if a value is possible, not solve for any specific n. So long as the exponent is an integer, it is some value in the sequence). Now let's think about what happens when we take -2.5 to an integer power: the very last digit in our calculation, assuming no rounding occurs and $n > 1$, must end in a value that is a multiple of 5, because

SEQUENCES/SERIES   ANSWERS

our divisibility rule for 5 states that all multiples of 5 end in 0 or 5. When we multiply -2.5 to some integer power by 4, then that product, too, must end in a multiple of 5. The only answer choice that does not end in a multiple of five is C. Knowing the properties of numbers and what happens when our last digit is five, Answer C cannot be in the sequence. You could also use logs and the answer choices to backsolve, ignoring the negative sign in the -2.5 as most calculators can't handle logs with negatives. That method is more time consuming, though. For more similar problems see **Chapter 2, Properties of Numbers** in this book.

21. **A.** If we let x represent the first integer, then the 4 consecutive integers can be represented as $x, x+1, x+2, x+3$. The sum of these numbers is then $x+x+1+x+2+x+3 = 4x+6 = 2(2x+3)$. Because we can factor out a two, the sum of 4 consecutive integers is always divisible by the factor 2.

PART TWO: PROBABILITY AND STATISTICS

CHAPTER 7

AVERAGES

SKILLS TO KNOW

- Mean (Average), Median, Mode, and Range
- Stem-and-leaf plot
- Set transformations (adding/subtracting/multiplying/dividing members of a set)
- Average rate word problems
- Standard deviation

NOTE: Average problems also appear in **Chapter 8: Data Analysis** in this book. Average rate problems are also covered in **Chapter 9: Ratios, Rates & Units in Book 1**.

MEAN, MEDIAN, MODE, RANGE

Mean (Average)

MEAN and AVERAGE are the same thing: the average value of a set of numbers, found by adding all of the numbers in a set and then dividing by the number of items in the set (below, left). **It is also useful to know a slightly reworked version of this equation that solves for the sum (below, right).** You can pick up speed if you know both versions.

THE AVERAGE FORMULA

$$AVERAGE = \frac{SUM\ OF\ ALL\ ITEMS}{NUMBER\ OF\ ITEMS} \text{ or } SUM = (average)(\#\ of\ items)$$

The average of $6, 9, 15 = \frac{6+9+15}{3} = 10$

Median

If you line up a set of numbers in **numerical (chronological) order,** and you find the number physically in the middle of this list, that number is called the MEDIAN. If there are an even number of items in a list, then average the two middle values.

The median of $2, 6, 9, 15, 17$ is 9, because 9 is physically in the middle of the list.

The median of $2, 6, 9, 10, 15, 17 = 9.5$, because the average of 9 & 10 is 9.5, and 9 & 10 are physically in the middle of the list.

Mode

The mode occurs most often in a set. I like to think that **MO**de and **MO**st both start with "**MO**" to remember this one. If you have more than one number that occurs the most (i.e. if three numbers occur five times each), **you can have MULTIPLE modes, unlike median and mean.**

Mode of the set $\{5,6,6,6,7,7,7\} = 6$ *AND* 7

AVERAGES — SKILLS

Range
The range of a set of values is the difference between the greatest value and the least value.

> The range of this set: 2, 6, 9, 10, 15, 17 is found by subtracting the least number from the greatest one: $17 - 2 = 15$

Average/Mean Word Problems
Here's a basic example of an "average" word problem. (NOTE: most ACT problems aren't this easy)

> In Los Angeles, the daily high temperatures in degrees Fahrenheit (F) over one week of June were $72°, 72°, 75°, 82°, 88°, 97°,$ and $104°$.
>
> To the nearest degree Fahrenheit, what was the mean daily high temperature for that week?
>
>   A. 72°   B. 74°   C. 82°   D. 84°   E. 85°

All we have to do here is add all the numbers and divide by the number of temperatures—easy:

$$\frac{72+72+75+82+88+97+104}{7} = \frac{590}{7} = 84.285\ldots$$

Answer: **D.**

What's not so easy are problems that actually give you the sum, the number of terms, and maybe a few numbers that contribute to the sum. These kind of "average" problems are much more common. On the ACT, we can solve these more complex "average" problems in a couple ways:

METHOD 1: USE ALGEBRA
One way to solve them is to plug into your original average equation and then roll up your Algebra sleeves!

> Mario has taken 4 of 6 equally weighted tests in his Biology class and has an average score of exactly 87 points. What must he score on the 5th and 6th test, on average, to bring his average up to 90 points?
>
>   A. 96   B. 100   C. 93   D. 99   E. 94

To solve, we'll write down the formula, plug in what we know, and solve for what we don't know. Then we will build out another equation using the formula, and again solve. We repeat this process, building out multiple versions of the average equation, until we get what we need.

He's averaged 87 points (average) on 4 tests (number of items). Let's plug that into our equation:

$$\text{AVERAGE} = \frac{\text{Sum of All Items}}{\text{Number of Items}} \quad \text{so} \quad 87 = \frac{\text{Sum of First Four Tests}}{4}$$

As you can see, I use plain words to represent what goes where—these are easier to understand than variables and the technique helps me more clearly set up my equation. Okay, so I can't solve for the individual test score, but I CAN solve for the sum of the first four test scores. (If I memorized the version of the average equation that solves for the sum, I can skip the previous step):

$$SUM = (average)(\# \text{ of items}) \quad \text{so} \quad 87(4) = \textit{Sum of First Four Tests} = 348$$

Now let's go back to what we need. We need to know the average of the last two tests. We also know he's got 2 tests left (out of 6 tests) to bring his grade up to a 90 (desired average). Let's start by creating a new equation incorporating in the information we know, again using our formula, but this time for the final grade equation. We'll fill in what we don't know with English.

$$90(\textit{Desired Average}) = \frac{348(\textit{Sum of First Four Tests}) + \underline{} + \underline{}(\textit{Sum of Last Two Tests})}{6(\textit{Number of Tests})}$$

Now I am going to think about the last two test scores—those two blanks. To make this easy, let's assume* they are identical and equal to their average, and call that average n. If two values are equal, they are also their own average.

*Alternatively, I could reason that "sum=average times number of items" means **2n** represents the sum of the two values.

$$90 = \frac{348 + 2n}{6}$$
$$90(6) = 348 + 2n$$
$$540 = 348 + 2n$$
$$192 = 2n$$
$$n = 96$$

He needs an average of 96, because two tests of score 96 would get him the points he needs!

Answer: **A.**

METHOD 2: NO ALGEBRA METHOD
Another way to solve many of these problems involves no algebra at all.

The first step I take with this method is to start with some assumptions that could be true. First, I know his average on 4 tests is 87 points. I don't know what he got on each test, but I know that one possibility would be that he actually scored 87 on every single test. That's my first assumption. I write out blanks for each number I "know" and the two that I don't know and fill them in as so:

$$\underline{87} \ \underline{87} \ \underline{87} \ \underline{87} \ \underline{} \ \underline{}$$

My goal is to actually make this list look like six 90's in a row—if he averaged 90 it would be as if he scored 90 on each test. With that in mind, now I add to each score that already exists "3" points—that is what he needs to "catch up" to 90 points for each of those tests, i.e. I'm three points short on each of those four tests, so he needs three more points for each 87. Then for the last two tests, he will need the equivalent of 90 points for each. If I add the "catch up" three points for the first four tests, plus 90 points from the last two, that is the total amount of points he will need to have averaged 90:

87 87 87 87 __ __	This row represents the points he has so far on four tests.
+3 3 3 3 90 90	This row represents the number of points he still needs.
90 90 90 90 90 90	This is what I actually want his grade card to look like.

So now I add up all the numbers in that 2nd horizontal row:

AVERAGES SKILLS

$$3+3+3+3+90+90=192$$

I also know he needs to score 192 points in two test sittings—so the number of points, on average, per test he needs to score is 96 points per test (192 divided by 2):

$$192 \div 2 = 96$$

I can also think of this as redistributing the "catch up" points from the four 3's—I need 12 more points and I can distribute 6 of these on the first additional 90 point test and 6 on the 2nd to get 96 points per test remaining. Make sense?

Answer: **A.**

Median Problems

What is the median of the following set?
42, 33, 85, 60, 15, 29

A. 33   B. 37.5   C. 44   D. 60   E. 72.5

MISTAKE ALERT: The first rule of medians: put everything in order! DO NOT FORGET TO PUT YOUR NUMBERS IN ORDER!

42, 33, 85, 60, 15, 29 becomes 15, 29, 33, 42, 60, 85

Now cross off equal numbers of numbers on the left and right to find the center values:

~~15~~, ~~29~~, 33, 42, ~~60~~, ~~85~~

Because we have an even number of items in our set, we can now find the average of 33 and 42:

$$\frac{33+42}{2} = \frac{75}{2} = 37.5$$

Answer: **B.**

SPEED TIP: To save time, you can count "up" or "down" to find the median in a single direction. For example, in the problem above, you can see there are six numbers. Instead of hashing off numbers from both sides, instead count up to the halfway point, here the 3rd and 4th numbers, the average of which will be the median. This is particularly helpful if you have an exceedingly long list of numbers in which to find a median and know the number of items in a list (hint: that number is often cited in the text of the word problem). See the Stem & Leaf plot problem on the next page for an example.

Mode Problems

What is the mode of the following set?
0, 1, 3, 4, 8, 9, 9, 1, 2, 6, 3, 1, 5

A. 1   B. 3   C. 3.5   D. 4   E. 9

To identify the mode, we also want to rearrange numbers in order so we can be sure which occur most often—if the mode is very apparent though we can just count in place.

0, 1, 1, 1, 2, 3, 3, 4, 5, 6, 8, 9, 9

Here 1 occurs three times. It is the mode.

Answer: **A.**

STEM-AND-LEAF PLOT

Tammy surveyed the ages of 30 adults at her most recent family reunion using the stem-and-leaf plot below. What is the median age of these thirty adults surveyed?

Stem	Leaf
2	1, 2, 8, 9, 9
3	1, 2, 2, 2, 3
4	1, 1, 1, 2, 6, 7
5	1, 2, 5, 7, 7, 7, 8, 8, 8
6	4, 5, 7, 9, 9

Key:

Stem	Leaf
2	1

Represents "21"

Wondering what a stem and leaf plot is!?! Look at the KEY! The ACT® is really nice about giving you clues when it offers obscure ways of displaying information. The **1, 2, 8, 9, 9** in the first row stands for **21, 22, 28, 29, 29**—the stem is the tens place, and the leaf is the ones place.

In the example above, I know there are **30** adults, so I know the median is the average of the 15th and 16th numbers. I count upwards 14, and circle the next two terms (15 & 16) then find their average.

Stem	Leaf	
2	~~1, 2, 8, 9, 9~~	
3	~~1, 2, 2, 2, 3~~	
4	~~1, 1, 1, 2,~~ 6, 7	—— 15th and 16th numbers
5	1, 2, 5, 7, 7, 7, 8, 8, 8	
6	4, 5, 7, 9, 9	

46 and **47** are the 15th and 16th items in the list. Their average is **46.5**.

Answer: **46.5**.

SET TRANSFORMATIONS

Each element in a data set is divided by 3, and each resulting quotient is then increased by 8. If the median of the final data set is 21, what is the median of the original data set?

A. 13 **B.** $9\frac{2}{3}$ **C.** 39 **D.** 44 **E.** 55

AVERAGES — SKILLS

With set transformations, think about what happens to a set of numbers—remember a median is often an actual number in a set when there are an odd number of items—and if we divide, multiply, add or subtract to the whole set, what we do to the median is the same as what we do to each number. Thus we can do to the median what we do to the set and imagine that it will hold its position*. What we essentially have in a number set is a giant inequality:

$$a < b < c < d < n < f < g < h < k$$

As long as we're dealing with the median, the relationship will hold because even if we had to flip the sign, it's still in the middle. We can **multiply, divide, add or subtract** to modify all the elements and n will remain in the center position. For problems involving the mean, these relationships also hold, likewise with problems involving medians with even numbers of members*. I could prove this all algebraically, but don't want to spend too much time on these as they're not all that common.

For this problem, we can simply do unto the median as to the rest of the set:

Divide it by 3, add 8, get the new data set version, 21:

$$\frac{n}{3} + 8 = 21$$
$$\frac{n}{3} = 13$$
$$n = 39$$

Answer: **C**.

NOTE: This rule generally works for most set transformation problems on the ACT, but know that if more complex operations were performed to a set, say applying exponents, radicals, or absolute value, these principles might not hold.

AVERAGE RATE WORD PROBLEMS

For more on rates, see **Chapter 9: Ratios, Rates & Units** in Book 1.

I commonly see students struggle with the concept of average rates. An average rate is always the TOTAL of the first amount divided by the TOTAL of the second amount.

AVERAGE RATE FORMULA

$$AVERAGE\ (ELEMENT\ A)\ per\ (ELEMENT\ B) = \frac{TOTAL\ OF\ ALL\ ELEMENTS\ A}{TOTAL\ OF\ ALL\ ELEMENTS\ B}$$

For example, if you're asked for Martha's average speed on her way to school in miles per hour, and you're given she goes 15 mph for 15 minutes and she goes 30 mph for 30 minutes, remember that PER means divide, and we want TOTAL miles over TOTAL hours, or TOTAL distance over TOTAL time.

MISTAKE ALERT: We do NOT want to simply average the two rates that she drove!

AVERAGE SPEED FORMULA

$$Average\ Speed = \frac{TOTAL\ DISTANCE}{TOTAL\ TIME}$$

If you see the word "average" but you're really looking at a rate (anything with the word "per" or "for each/every"), remember you need the total of each independent element to find that average. For example, if we want to find the average retail price of a (or PER) jelly donut sold by a particular donut chain, we add the TOTAL retail sales of all the donuts and divide by the TOTAL number of donuts sold. This is similar to the SUM over number of ITEMS– the concept covered earlier, but adjusted to be the sum total of the first amount over the sum total of the second.

For each of 3 years, the table below gives the number of auctions held at an auction house, the number of items sold, and the auction house's income.

Year	Auctions	Items sold	Income
2007	45	405	$1,206,903
2008	38	266	$682,024
2009	42	504	$1,568,448

1. To the nearest dollar, what is the average income the auction house made on an item sold in 2007?

2. If, on average, an item between 2007-2009 cost the auction house $1,400 to acquire, process, and sell at auction, disregarding any other expenses, what was the average amount of profit per auction over the three-year period?

1. We want the average income **per item**, NOT the average overall income. DO NOT add up all the income column and divide by 3! If you have the word "per" translate that into a division bar! Remember we want the TOTAL income over the TOTAL number of items.

Divide the 2007 income by the number of items sold to get the average per item.

$$\frac{2007\ income}{\#\ of\ items\ sold\ in\ 2007} = \text{average selling price \textbf{per item}, so } \frac{\$1,206,903}{405} = \$2,980$$

Answer: $2,980.

2. Here we have to do more work. Let's figure out what we NEED:

"the average amount of profit per auction:"

Using the pattern I describe above, that will be TOTAL profits divided by TOTAL number of auctions:

$$NEED = \frac{TOTAL\ PROFITS}{TOTAL\ NUMBER\ OF\ AUCTIONS}$$

Total number of auctions is easy— that's three numbers from the auctions column added together. Total profits will be more difficult. We know income but profit is different. Profit subtracts costs, so we need the costs in addition to income to figure out total profits. Costs is just a way of saying the "cost of doing business" or "expenses."

$$NEED = \frac{TOTAL\ INCOME - TOTAL\ COSTS}{TOTAL\ NUMBER\ OF\ AUCTIONS}$$

What makes this even more difficult is that we are told an average unit cost, not the total costs. So we must convert that average unit cost to calculate the total costs of doing business. If we know a unit rate, though, we can multiply by the number of units to get to the sum total.

Remember, $\frac{SUM}{\#\ of\ items} = average$ and by extension: $SUM = (average)(\#\ of\ items)$

We need the **sum**, or total costs (i.e. what it costs to do business), and can find that by using the **average** cost per item and multiplying by the **number of items**:

$$\text{Total Costs} = (\text{Average Cost Per Item})(\text{Total \# of Items at Auction})$$

The average expense cost per item is given $\$1,400$, and we can find the total number of items at auction by adding each number in the items sold column of the chart:

$$\begin{aligned}\text{Total Costs} &= (\$1400)(405+266+504) \\ &= (\$1400)(1175) \\ &= \$1,645,000\end{aligned}$$

Let's head back to our "NEED" equation and fill in what we know. We can find the *total income* by adding all the elements in the income column, we just solved for *total costs*, and again the *total number of auctions* we can find by adding all numbers in the number of auctions column:

$$NEED = AVG\ PROFIT\ PER\ AUCTION = \frac{TOTAL\ INCOME - TOTAL\ COSTS}{TOTAL\ NUMBER\ OF\ AUCTIONS}$$

$$= \frac{(\$1,206,903 + \$682,024 + \$1,568,448) - \$1,645,000}{(45 + 38 + 42)}$$

$$= \frac{\$1,812,375}{135}$$

$$Average\ Profit\ Per\ Auction = \$14,499$$

Answer: $\$14,499$.

STANDARD DEVIATION

Since 2014, the ACT has occasionally asked about **standard deviation**. Most likely, you won't actually have to know how to solve for the standard deviation, but you do need to know the concept. Essentially standard deviation measures how tightly packed a set of data is about the mean. If the data values are all very close to the mean, (Say, {49, 49, 49, 50, 50, 50}) the set has a small standard deviation. If the data values are very far away from the mean (Say {0, 0, 0, 99, 99, 99}), the set has a large standard deviation.

You can often ballpark estimate the value of a single standard deviation in a set of numbers that are **"normally distributed"** as the amount you add or subtract from your mean such that **about 2/3 (more accurately 68%) of your data set falls in that range.** For example, if my data set is: {3, 4, 5, 6, 7, 8} my mean is 5.5 and 2/3 of my data is approximately between 4-7, so my standard deviation will be about 1.5 (it's actually 1.87). Again, this is NOT a precise calculation, it won't work for irregular data, and outliers will also affect standard deviation, but it is a way to eyeball a standard deviation amount if you have to. **Normally distributed** data generally takes the shape of a bell curve, with the most data centering around the mean. I could also observe that the data set I just used as an example is NOT bell shaped because it is a list of consecutive integers. Bell shaped data is plentiful around the mean and less common when values are further from the mean.

For data sets A: **103, 107, 165, 189, 240** and B: **160, 161, 161, 161, 161**, which of the following statements must be true?

I. The standard deviation of set A is less than that of set B

II. The standard deviation of set A is equal to that of set B

III. The standard deviation of set A is greater than that of set B

IV. The mean of set A is less than that of set B

A. III only   **B.** II only   **C.** I only   **D.** II and IV only   **E.** IV only

Here we first can analyze the two data sets. Because IV asks for an observation of the mean, we can calculate that as the sum of all numbers in each set divided by five:

For the first set, we calculate:

$$\frac{(103+107+165+189+240)}{5} = \frac{804}{5} = 160.8$$

For the second set we calculate:

$$\frac{(160+161+161+161+161)}{5} = \frac{804}{5} = 160.8$$

Thus the means are equal and IV is not true. We thus know the answer must be I, II, or III.

The standard deviation measures how tightly packed the data is about the mean. I can clearly see that the second data set is all clustered tightly about the mean, 160.8. However, the first set is much more spread out, ranging from 103 to 240. Thus the standard deviation of the first set A will be larger than that of set B, as I will have to travel farther from the mean to include about 68% of the data.

Thus III only, Choice A is correct.

AVERAGES QUESTIONS

1. In a science class, Bob scored 95 on one test, 89 on another test, and had an average test score of 83 in the class before taking those two tests. If Bob's average test score for the entire class, including all tests, is 85, and each test he takes in the course is weighted equally, how many tests has Bob taken in the class?

    A. 3
    B. 9
    C. 7
    D. 14
    E. Cannot be determined from the given information

2. Mary's test average after 8 tests is 80. Her score on the 9th test was 71. If all 9 tests are weighted equally, what is Mary's test average for all 9 tests?

    A. 71
    B. 75.5
    C. 77
    D. 79
    E. 80

3. Theo has taken 6 of 7 equally weighted tests in his Chemistry class and has an average score of exactly 81 points. What must he score on the 7th test to bring his average up to 83 points?

    A. 95
    B. 93
    C. 85
    D. 83
    E. 82

4. The mean of 6 numbers is 27. The smallest of the 6 numbers is 12. What is the mean of the other 5 numbers?

    A. 27
    B. 30
    C. $30\frac{3}{4}$
    D. $41\frac{2}{3}$
    E. 42

5. The 7 positive integers a, a, a, a, a, b, c have an average of a. Which of the following equations must be true?

    A. $b = c$
    B. $b + c = a$
    C. $b + c = 2a$
    D. $c = -b$
    E. $b + c = 0$

6. Each element in a data set is increased by 3 then divided by 7. If μ is the mean of the final data set, what is the mean of the original set?

    A. $\frac{\mu}{7} + 3$
    B. $\frac{\mu + 3}{7}$
    C. $\frac{\mu - 3}{7}$
    D. $7\mu - 3$
    E. $7\mu + 3$

7. The average of a set of 8 numbers is 13. When a 9th number is added to the set, the average increases to 16. What is the 9th number?

    A. 16
    B. 20
    C. 29
    D. 32
    E. 40

8. A music concert was rated on a 5 point scale by the audience. 10% gave a 1, 18% gave a 2, 33% gave a 3, 20% gave a 4, and 19% gave a 5. To the nearest tenth, what is the average rating given by the audience?

    A. 2.5
    B. 2.8
    C. 3.2
    D. 3.4
    E. 4.6

9. In a town of 600 people, 250 males have an average age of 43, and 350 females have an average age of 38. To the nearest whole number, what is the average age of the entire town?

   A. 38
   B. 39
   C. 40
   D. 41
   E. 42

10. Each of 16 students took a test and received a whole number of points. The median of the scores was 78, and 25% of the students scored 74 or below. No student received a score of 78. How many students scored 75, 76, or 77?

    A. 2
    B. 3
    C. 4
    D. 5
    E. 8

11. Nick has 30 collectible coins. He paid $36.50 for each coin 2 years ago. The average value of each coin is currently $38.15. To the nearest cent, how much *more* must the average value per coin rise for the combined value of these 30 coins to be $350.00 more than Nick paid for them?

    A. $8
    B. $8.50
    C. $10.02
    D. $11.67
    E. $300.50

12. A data set has 20 elements. A second data set of 20 elements is obtained by adding 7 to each element of the first set. A third data set of 20 elements is obtained by multiplying each element of the second data set by 3. The median of the third data set is 63. What is the median of the first data set?

    A. 7
    B. 14
    C. 21
    D. 196
    E. 210

13. What is the median of the following test scores?

    $\{56, 93, 64, 78, 92, 83, 88, 40, 61\}$

    A. 64
    B. 73
    C. 78
    D. 83
    E. 92

14. Each number on a list of 8 numbers is multiplied by 13 to produce a 2nd list of 8 numbers. Each of the 8 numbers on the second list is increased by 4 to produce a 3rd list of 8 numbers. The median of the 3rd list is x. What is the median of the first list?

    A. $\dfrac{x-4}{13}$
    B. $\dfrac{x}{13}$
    C. $x - 4$
    D. $\dfrac{x}{13} - 4$
    E. Cannot be determined from the information given

15. List B contains all the elements of List A and also including integers x, y, and z, where $x \geq 43, y = z, y \leq 18$. What is the median of List B?

    List A: $\{12, 17, 18, 30, 30, 36, 42, 43, 48, 48, 51\}$

    A. 27
    B. 30
    C. 33
    D. 36
    E. 39

16. Which of the following data sets has the largest standard deviation?

    A. 13, 14, 15, 16, 17, 18
    B. 12, 17, 19, 24, 27
    C. 9, 15, 27, 85, 98, 99
    D. 10, 10, 10, 103, 103, 103
    E. 27, 27, 27, 27, 27, 27

17. The following table shows the total number of cabinets built in a workshop over 50 consecutive days. What is the average number of cabinets built per day, rounded to the nearest tenth?

Number of cabinets built in a day (Output)	Number of days with this output
0	5
1	11
2	10
3	18
4	6

A. 2
B. 2.1
C. 2.2
D. 3
E. 3.3

Use the following information to answer questions 18-20.
Mr. Rivera gives his 20 students a progress report once a month that gives a student their scores on tests, quizzes, and homework, and the average score of all 20 students. The following is a progress report for Isaac Shemtov.

Student: Isaac Shemtov			
Task	Possible Points	Student score	Class average
Homework #1	100	89	93
Homework #2	100	94	90
Quiz #1	100	75	80
Quiz #2	100	87	85
Quiz #3	100	100	95
Quiz #4	100	89	78
Test #1	100	85	83
Test #2	100	93	85
Test #3	100	95	90

18. Isaac is about to take Test #4. What score does he need to get an average test score of 93?

A. 91
B. 93
C. 95
D. 97
E. 99

19. In Mr. Rivera's class, a homework assignment that has not been turned in receives a score of 0. Of the 20 students in his class, what is the maximum number of students who could have not turned in homework #2?

A. 0
B. 1
C. 2
D. 3
E. Cannot be determined from the given information

20. What is Isaac's average quiz score to the nearest point?

A. 86
B. 87
C. 88
D. 89
E. 90

21. The 6 positive integers $a, a, b, b, c,$ and d have an average of c. What is the value of $a+b$?

A. $\dfrac{5c-d}{2}$

B. $5c+d$

C. $\dfrac{5c+d}{2}$

D. $3c+d$

E. Cannot determine from the given information.

22. In a biology course, a student scored 94 on one test, 100 on another test, and 90 on each of the other tests. The student's final test average was 91.75. How many tests did the student take?

A. 6
B. 7
C. 8
D. 9
E. 10

23. What is the median of the following set?
$$-5, 8, 3, 17, 7, 12, 100, 7, 0, 20$$

A. 7
B. 7.5
C. 8
D. 9.5
E. 16.9

24. The median of a set of data containing 12 items, all of different values, was found. Three data items, *a*, *b*, and *c*, were added to the set. *a* was greater than the original median, but less than all of the original data greater than the median. *b* was greater than all of the original data. *c* was equal to the original median. What is the new median of the set?

    A. *a*
    B. *b*
    C. *c*
    D. The average of *a* and *c*
    E. Cannot be determined from the given information.

25. What is the mode of the following data?

    $-3, -1, 0, 3, 5, 5, 8, 9, 10$

    A. 0
    B. 4
    C. 5
    D. 5.7
    E. 6

26. Mr. Harker organized his students' test grades into a stem and leaf plot in order to see the frequency of different letter grades. What is the range of the set of test grades?

Stem	Leaf
6	7 9
7	2 3 6
8	0 0 2
9	1 4 5 9

    A. 2
    B. 9
    C. 32
    D. 40
    E. 166

27. To decrease the mean of 5 numbers by 2, by how much would the sum of the 5 numbers have to decrease?

    A. 2
    B. 5
    C. 7
    D. 10
    E. 17

28. A data set has 4 members. The mode of the data set is both 6 and 8. What are the mean and median of the data set, respectively?

    A. 2, 7
    B. 6, 8
    C. 7, 7
    D. 24, 6
    E. 24, 7

29. What is the mode of the data given below?

    $12, 26, 13, 7, 20, 35, 38, 13$

    A. 12.5
    B. 13
    C. 16.5
    D. 20.5
    E. 31

30. The stem and leaf plot below shows the number of walk-ins to The Doctor's Office during a 30-day observation period. What is the median number of daily walk-ins?

Stem	Leaf
2	2 3 3 4 7 9
3	1 1 4 4 5 5 9
4	0 0 2 3 7 7 7 8
5	0 3 4 6 6
6	1 3 7 8

    Key: 2 | 2 = 22

    A. 41
    B. 42
    C. 43
    D. 44
    E. 47

31. Mr. Eames was hired for a new job and was requested to keep a log of all his working hours. He recorded the number of hours he worked each day in a table, as shown below. What was the mean number of hours he spent per day?

Day	Monday	Tuesday	Wednesday	Thursday	Friday
Hours Worked	7	4	6	4	9

    A. 4
    B. 6
    C. 7
    D. 9
    E. 13

32. Which of the following statements is correct concerning the data set below?

    $36, 47, 47, 47, 59, 72, 89, 89, 98$

    A. Its mean is 67
    B. Its median is approximately 63.8
    C. Its range is 98
    D. Its mode is 47
    E. Its median is not a member of the data set

AVERAGES QUESTIONS

33. The signs of a and d are positive, but the signs of b and c are negative. If it can be determined, what are the signs of the mean and median of the four numbers, respectively?

    A. Both positive
    B. Both negative
    C. Both neither (zero)
    D. The mean is neither, the median is negative
    E. Cannot be determined from the given information

34. The 8 consecutive integers below add up to 332.

    $$x-3$$
    $$x-2$$
    $$x-1$$
    $$x$$
    $$x+1$$
    $$x+2$$
    $$x+3$$
    $$x+4$$

    What is the value of $x+2$?

    A. 40
    B. 41
    C. 42
    D. 43
    E. 44

35. A new show was in town for the past 7 days. The average attendance of the slowest and busiest night was 224 people. The average attendance for the other 5 days was 311. How many people attended the show over the course of 7 days?

    A. 1704
    B. 1405
    C. 2093
    D. 2003
    E. 1967

36. A printer can print 15 pages per minute when on slow mode and can print 1200 pages per hour when on fast mode. If the printer is run for half an hour on slow mode and 50 minutes on fast mode, what is its average printing speed for that interval of time, in pages per minute? (Round to the nearest whole number)

    A. 17
    B. 18
    C. 20
    D. 21
    E. 756

37. The mean of 8 integers is 53, and the median of the integers is 50. If 6 of the integers are 45, 49, 51, 60, 65, 70, and the remaining other integers are sequentially next to each other in the list of all the values, then which of the following could be one of the other integers?

    A. 47
    B. 43
    C. 38
    D. 34
    E. None of the above

38. In a data set containing a certain number of distinct values, the minimum value is replaced with a much smaller value to create a new data set. Which of the following statements accurately describes the change in mean and median from the old data set to the new data set?

    A. Both the median and mean will decrease
    B. The median will decrease, but the mean will remain the same
    C. Both the median and mean will remain the same
    D. The mean will decrease, but the median will remain the same
    E. The mean will decrease, but the median will increase

39. For data sets A: 13, 13, 13, 88, 88, 88 and B: 40, 45, 50, 50, 65, 70 which of the following statements must be true?

    I. The standard deviation of set A is greater than that of set B
    II. The standard deviation of set A is equal to that of set B
    III. The mean of set A is greater than that of set B
    IV. The mean of set A is less than that of set B

    A. I only
    B. IV only
    C. I and IV only
    D. II and III only
    E. None of the above

40. A researcher is collecting data on tulips. He records data from 20 tulips with heights between 5 and 7 inches. The researcher then mistakenly adds an incorrect datum for another tulip, recording it as a 96 inch tall tulip. Which of the following statistics calculated from the recorded flower heights changed the least after the researcher added the erroneous 96 inch tall tulip measurement?

    A. Mean
    B. Standard Deviation
    C. Range
    D. Median
    E. Cannot be determined without more information

ANSWERS — AVERAGES

ANSWER KEY

1. B	2. D	3. A	4. B	5. C	6. D	7. E	8. C	9. C	10. C	11. C	12. B	13. C	14. A
15. C	16. D	17. C	18. E	19. C	20. C	21. A	22. C	23. B	24. A	25. C	26. C	27. D	28. C
29. B	30. A	31. B	32. D	33. E	34. D	35. D	36. B	37. B	38. D	39. C	40. D		

ANSWER EXPLANATIONS

1. **B.** If the average of Bob's test scores for the entire class is 85, the sum of his test scores can be represented as $85x$, where x = the number of tests he took in total. We know two of his test scores are 95 and 89, and the average of the test scores before those two tests was 83. So, the sum of his test scores can also be represented by $83(x-2)+95+89$. Setting the two expressions equal to each other, we get $85x = 83(x-2)+95+89$. Distributing the 83 on the right hand side, we get $85x = 83x - 166 + 95 + 89$. Combining like terms, we get $2x = 18$. So, $x = 9$. Answer B is correct. If you got 7, you likely made the variable the number of tests before the next two scores were added in. Always double check that you've solved for what you need.

2. **D.** Mary's average after 8 tests was 80, so the sum of those 8 test scores is $80 \times 8 = 640$. Now, adding the 9th test score, we have $640 + 71 = 711$. To average her 9 test scores, we take the sum of the 9 test scores and divide by 9. $\frac{711}{9} = 79$.

3. **A.** Theo's average after 6 tests was 81, so the sum of those 6 test scores is $81 \times 6 = 486$. If he wants the average of 7 tests to be 83, then the sum of the 7 tests should be $83 \times 7 = 581$. Since we already know the sum of his first 6 test scores, the 7th test score can be calculated as the difference between these totals: $581 - 486 = 95$.

4. **B.** If the mean of 6 numbers is 27, then the sum of the 6 numbers is $27 \times 6 = 162$. We know that the smallest number is 12, so if we take away that number, we know that the sum of the remaining 5 numbers is $162 - 12 = 150$. This means that the average of those 5 numbers is $\frac{150}{5} = 30$.

5. **C.** We can use the formula for finding the average in order to solve this problem (sum over number of terms equals average). The average, a, is equal to $\frac{a+a+a+a+a+b+c}{7}$. Through algebraic calculation, we can find the $7a = 5a+b+c$. Therefore, $2a = b+c$.

6. **D.** If every element in a data set is increased by 3 and then divided by 7, then the average would also be increased by 3 and divided by 7. So, if μ is the mean of the final data set, then $\mu = \frac{x+3}{7}$ is true for the original mean $= x$. Solving for x now, we get $7\mu = x+3$ so $x = 7\mu - 3$.

7. **E.** If the average of 8 numbers is 13, then the sum of the 8 numbers is $8 \times 13 = 104$. When a 9th number is included, the average changes to 16, so the sum of those 9 numbers is $9(16) = 144$. The 9th number is the difference of the sums $144 - 104 = 40$.

8. **C.** In this case—to make the math easier—we can view each percent of the audience as one count (one person). If 10% of the audience gave the concert a 1, then the concert has $10*1 = 10$ points from the 10%. Multiplying out the other percentages and points, we have $18 \times 2 = 36$, $33 \times 3 = 99$, $20 \times 4 = 80$, and $19 \times 5 = 95$. Adding these points together, we get the total number of points accumulated by the ratings. We get $10 + 36 + 99 + 80 + 95 = 320$. The average can then be calculated by dividing this sum by the total of 100 counts. $\frac{320}{100} = 3.2$.

9. **C.** Since the average of 250 males is 43, the sum of their ages is $250 \times 43 = 10750$. Likewise, the average of 350 females is 38, so the sum of their ages is $350 \times 38 = 13300$. Adding these two sums together we get $10750 + 13300 = 24050$. To find the average of the total 600 people, we divide this sum by 600 to get $\frac{24050}{600} = 40.08$. Rounding this to the nearest whole number, we get the average age $= 40$.

10. **C.** If 25% of the students scored 74 or below, then 4 students scored 74 or below. If the median score was 78, but no one scored a 78, then the 8th and 9th numbers average out to be 78. This means that the 5th, 6th, 7th, and 8th numbers are either 75, 76, or 77. This means 4 students scored 75, 76, or 77.

CHAPTER 7

AVERAGES ANSWERS

11. **C.** $350.00 more than what Nick paid for the coins is $350 + 36.5(30) = 1445$. Thus each coin must be worth $\frac{1445}{30} = 48.17$. Now subtract the current value, $38.15. $48.17 - 38.15 = 10.02$ more than the coins' current value.

12. **B.** Basic operations on a set of elements don't change the order of the set, so the median for a previous set of numbers can simply be found by applying those operations to the newest median in reverse order. Here, we divide 63 by 3 and then subtract 7. $63 \div 3 = 21$. $21 - 7 = 14$. The median of the first set is 14. We can also solve this question algebraically. $x_1, x_2, x_3 \ldots x_{20}$ denotes the 20 elements in the first data set, the second data set is $x_1 + 7, x_2 + 7 \ldots x_{20} + 7$ and the third data set is $3(x_1 + 7), 3(x_2 + 7) \ldots 3(x_{20} + 7)$. The median of the third data set is 63, so the average of the two middle terms is 63. $\frac{3(x_{10} + 7) + 3(x_{11} + 7)}{2} = 63$. We manipulate this equation to find the median of the first data set, which is $\frac{x_{10} + x_{11}}{2}$. Distribute the 3's and simplify, $\frac{3x_{10} + 3x_{11} + 42}{2} = 63$. Multiply by 2 on both sides, $3x_{10} + 3x_{11} + 42 = 126$. Subtract 42 on both sides, $3x_{10} + 3x_{11} = 84$. Divide by 6 on both sides, $\frac{x_{10} + x_{11}}{2} = 14$.

13. **C.** Putting the scores in numerical order, we get $\{40, 56, 61, 64, 78, 83, 88, 92, 93\}$. Now, cancelling out the largest and smallest numbers one by one, we arrive at the median. $\{\cancel{40}, \cancel{56}, \cancel{61}, \cancel{64}, 78, \cancel{83}, \cancel{88}, \cancel{92}, \cancel{93}\}$. So, the median is 78.

14. **A.** As explained in question 12, because simple operations don't change the order of the elements in the set, the old median can be found by applying the operations to the new median in reverse order. For this problem, take x and subtract 4 to get the median of the second list: $x - 4$. Then, divide it all by 13 to get the median of the first list: $\frac{x - 4}{13}$.

15. **C.** Integer x comes somewhere after 43 in list B, and both y and z come before 18. Because these reference points are at least a few numbers away from the curret median, we do not need to know their specific locations to find the median. Let us arbitrarily place the integers into the list as such: $\{12, y, z, 17, 18, 30, 30, 36, 42, 43, 48, 48, x, 51\}$. It does not matter where we place $x, y,$ and z as long as they satisfy their inequalities, again, because they are far enough from the center to not impact the median in an unpredictable way. Cancelling out the largest and smallest numbers one by one, we are left with $\{30, 36\}$. Since we have two numbers, we take the average between them and get the median: 33. If this problem used reference points that were not so clearly on the left or right of the median, only then would we be unable to find a solution.

16. **D.** The standard deviation of a data set can be defined as how far apart data in the set are from the mean. Since the data sets in choices A, B and E all have values that are relatively close together, we can go ahead and eliminate them; their standard deviations are small. This leaves C and D. Larger standard deviations result from spread out data – i.e. a set like 1, 1, 5, 5 will have a larger standard deviation than a set like 1, 2, 5, 6. Because the first set, Choice D, has two distinct clusters of data, its data are farther from the mean and its standard deviation thus larger than that of choice C.

17. **C.** The average number of cabinets built per day is equal to the TOTAL number of cabinets built over the time period divided by the TOTAL number of days. We multiply each of the possible number of cabinets built in a day by the number of days when that output was reached and add together each of our results to get the total number of cabinets built: $5 \times 0 + 11 \times 1 + 10 \times 2 + 18 \times 3 + 6 \times 4 = 109$. We divide by the number of days: $\frac{109}{50} = 2.18$, which rounds to 2.2.

18. **E.** Isaac's average test score after he takes another test will be $\frac{85 + 93 + 95 + n}{4}$ where n is his score on the newest test. We can find out what n must be to get an average of 93 by setting the equation equal to 93: $\frac{85 + 93 + 95 + n}{4} = 93$. This becomes $85 + 93 + 95 + n = 372$. Isolating n, $n = 99$.

19. **C.** We can find the maximum number of students who could have not turned in homework #2 by assuming all the other students (except maybe one) scored 100% on the assignment. The more students who get a perfect, the more they make up for slacker kids who turn nothing in. To maximize slackers, we need as many perfects as possible to offset their non-contributions, that is, until we get the points we need to reach the necessary class average, at which point everybody left won't need to turn anything in. If the average of the class is 90, and there are 20 kids in the class, we need to amass

$20 \times 90 = 1800$ points between all the kids. Now divide this by 100: $1800 \div 100 = 18$. So if 18 kids get a perfect 100, we'll have what we need for the average and the remaining 2 kids of the total 20 in the class can slack off and turn nothing in. Thus the answer is 2.

20. **C.** Note: The chart includes quizzes and tests and homework; we only want quizzes. Isaac's average quiz score is $\frac{75+87+100+89}{4} = \frac{351}{4} = 87.75$, which rounds to 88.

21. **A.** The average of the 6 integers is $\frac{a+a+b+b+c+d}{6} = c$. This becomes $a+a+b+b+c+d = 6c$. We simplify to $2(a+b)+c+d = 6c$. Isolate $(a+b)$: $2(a+b) = 5c-d$. $a+b = \frac{5c-d}{2}$.

22. **C.** Let n be the number of tests the student scored 90 on. We can thus express the sum of these tests as $90n$. The average score of all the tests will be $\frac{91+100+n(90)}{n+2}$. Setting this equal to 91.75 and restructuring the equation gives us $94+100+n(90) = 91.75(n+2)$. Simplify: $90n+194 = 91.75n+183.5$. Simplifying further, $1.75n = 10.5$. Then, $n = 6$. Since the student took 2 tests besides the ones she scored 90 on, the total number of tests is $n+2 = 8$.

23. **B.** Reorganizing the set in ascending order gives us $-5, 0, 3, 7, 7, 8, 12, 17, 20, 100$. Crossing off the lowest and highest number one by one leaves us with $7, 8$. Since we have two numbers left, we average them to get 7.5.

24. **A.** Let's sketch the set: o o o o o o   o o o o o o. There is a space between the two sets of 6 shapes that make up the lower and higher halves of the set. Since there is an even number of items in the set, the median will be the average of the two shapes in the center (the shapes adjacent to the gap). We will place c there since it is equal to this original median: o o o o o o c o o o o o o. Next, place a. It is greater than c (the original median), but less than the items greater than it: o o o o o o $c a$ o o o o o o. Finally, add b, which is greater than all of the original data: ⌀ ⌀ ⌀ ⌀ ⌀ ⌀ a ⌀ ⌀ ⌀ ⌀ ⌀ ⌀. Crossing off the highest and lowest items one by one gives us a. The median is a.

25. **C.** The <u>mo</u>de appears <u>most</u> often. In this case, 5 occurs twice, and all other values occur once, making 5 the mode.

26. **C.** The stem in a stem and leaf plot places the tens place in the stem and the ones place in the leaf of each data point. Thus, a number which has a 3 in the stem column and a 2 in the leaf column represents a 32. The range of a set of data is the difference between the greatest and smallest number. Therefore, the range for this stem plot would be $99-67 = 32$.

27. **D.** If we decrease the mean by some amount, doing so is equivalent to decreasing each number in the set by that same amount. Remember, a set of numbers with a particular mean could be multiple instances of the average (i.e. if three numbers have an average of 7, all three could be $7\ 7\ 7$). In such a case, it's easy to see that to decrease the mean we'd take away the same amount we want off of the mean from each item in the set. We can thus take the decrease amount, 2, and multiply it by 5, the number of numbers, to get 10. If that's a bit of stretch for your brain cells, pick simple numbers and plug them into the problem. Let's use $1, 2, 3, 4,$ and 5. The mean of this set would be $\frac{15}{3}$, or 3. If we decrease this mean by 2, we get a new mean of 1. This means that the new sum, or x, divided by the number of numbers, 5, is now equal to 1. Through algebraic calculation, if $\frac{x}{5} = 1$, we know that the new sum, x, would have to be 5. From the original sum of 15 to the new sum of 5, there would have to be a decrease of 10.

28. **C.** In a data set of four values, 2 of the values must be 6 and the other 2 of the values must be 8 in order to have a mode of both 6 and 8. If there were only 1 instance each of 6 and 8, and the 2 remaining values were different numbers, then each value would occur exactly once and <u>every value in the set would be a mode</u>. Thus, the data set is $6, 6, 8, 8$. The mean of the set is $\frac{28}{4}$, or 7. The median of the set is the average of the two terms in the physical center of the above list, 6 and 8, which is equal to 7.

29. **B.** The <u>mo</u>de of a set is the number that occurs the <u>most</u>. In this set the number 13 occurs twice, and all other numbers occur only once. The mode is 13.

AVERAGES — ANSWERS

30. **A.** The median is the number in the physical middle of a sequentially ordered set, or in the case of a set with an even number of values, the average of the two middle terms. Thanks to the stem and leaf plot, the data set is already in order. All that is left to do is to begin crossing out numbers from the beginning, 22,23,23, until we get to the midpoint, the 15th and 16th terms: 40 and 42, which we average to get the median: 41. (See page 79 for more)

31. **B.** The mean is the statistical term for the average. To find the mean in a data set, we add all the numbers, then divide by the total number of members in the set. In this case, $\frac{4+4+6+7+9}{5}=6$.

32. **D.** One value appears more often than any other values in the data set: 47. This makes it the mode.

33. **E.** A common mistake would be to assume that two negative and two positive numbers always cancel each other out by summing to zero, when that is not always the case. We know that when summing positive and negative numbers, the answer could be positive *or* negative. Diving by the total number of numbers to get the mean will not change the sign of that sum. Thus, the mean's sign cannot be determined. The median, because we are dealing with an even count of numbers (4 numbers), is the mean of the two center numbers in the set. This mean is again the average of a combination of positive and negative numbers, so its sign cannot be determined either. Another approach would be to make up several cases of numbers that have differeing means, medians, etc. For example -1000, -999, 1, and 2 would create a negative mean and median. But -1, -2, 2000, and 3000 would create a positive mean and median.

34. **D.** To sum the 8 consecutive integers, add up the x values and constant values separately, and then add those two sums together. Adding up the x values, we find 8 $x's$, so their sum is $8x$. Then we add the constants: $-3-2-1+0+1+2+3+4$. $-3-2-1$ cancels out with $1+2+3$, so the sum of the constants is equal to 4. Now, we can add the x values and constant values together to get the total sum. This, we are given, is equal to 332. Solving this equation, we get $8x+4=332 \rightarrow 8x=328 \rightarrow x=41$. We are asked to find the value of $x+2$ which is $41+2=43$.

35. **D.** Remember average times number of items equals the sum. The average of the slowest and busiest nights was 224 which means that the total attendance in those two days was $2\times 224 = 448$. The average of the other 5 days was 311 which means that the total attendance for those 5 days was $5\times 311 = 1555$. Then, the total attendance for the week is $448+1555=2003$. Remember, $sum = average(\# \, of \, items)$.

36. **B.** Remember, average rates are always total amount (here pages) over total time. If the printer prints on slow mode for 30 minutes, then it prints a total of $30 \, minutes \times \frac{15 \, pages}{1 \, minute} = 450$ pages during that time. If the printer prints on fast mode for 50 minutes, or $\frac{5}{6}$ of an hour, then it prints a total of $\frac{5}{6} \, hours \times \frac{1200 \, pages}{1 \, hour} = 1000$ pages during that time. Therefore, the printer prints a total of $1000+450=1450$ pages over a total of $30+50=80$ minutes; its average speed over that time interval is $\frac{1450}{80} \approx 18$ pages per minute.

37. **B.** Let y and x represent the two numbers we are not given in the set. If the mean of the set is 53, then the sum of the set divided by 8 must be equal to 53. We can set this up as an equation to solve for the sum of the unknowns: $\frac{45+49+51+60+65+70+x+y}{8}=53 \rightarrow \frac{340+x+y}{8}=53 \rightarrow 340+x+y=424 \rightarrow x+y=84$. If the median of the set is 50, then the two middle values of the set must average out to 50. Because there are eight values, and four of these (51, 60, 65, 70) are greater than 50, the two middle values must be 49 and 51, as the right middle value must be 51 (the lowest of the 4 values greater than 50) and only 49 will pair to make an average of 50: $\frac{49+51}{2} \rightarrow \frac{90}{2}=50$. We already have a value of 49 in the list. At this point, we know both x and y must be 49 or less. We also know their sum is 84, so their average is 42, and they must be sequentially next to each other. Let's experiment placing x and y around where 42 would be in the list. Say we try $x,y,45,49,51,60,65,70$ as the order. Here, we could come up with pairs that

sum to 84: (42, 42), (41, 43), (40, 44) or (39, 45). From these sets we can see that 43 is an option. But why won't other choices work? We can't make both values greater than 45, as $45(2) = 90$, and we know they must sum to 84. If a single value were greater than 45, it would need to pair with a number lower than 45, and then the values would not be sequentially next to each other in a list of all values as 45 would come between them. Thus we know the maximum possible value of either x or y is 45, and plugging this into our equation $x + y = 84$, we can find the minimum value of either x or y it could pair with to make 84: $45 + (x \text{ or } y) = 84 \rightarrow (x \text{ or } y) = 39$. Therefore, one of the other integers must be in the range from 39 to 45, and the only option given in this range is 43.

38. **D.** If the new data set has a much smaller value in place of the minimum of the old data set, then it must have a lower mean than the old data set, because the sum of the data values would decrease while the number of data values remains the same. However, it would have the same median, because regardless of how small the minimum data value gets, it still remains the smallest. For instance, if we reduce the minimum of the set 2,3,4 from 2 to 1, the median of the set would still be 3. Thus, the mean will decrease and the median will remain the same.

39. **C.** The mean of set A is $\frac{13+13+13+88+88+88}{6} = 50.5$ and the mean of set B is $\frac{40+45+50+50+65+70}{6} \approx 53.33$. Therefore, the mean of set A is less than that of set B, and IV is correct while III is incorrect. The standard deviation of a data set can be defined as how much said data set's values vary from the mean. Therefore, set A has a relatively large standard deviation as all of its data values are very far from 53.55, whereas set B has a relatively small standard deviation, as the majority of its data values are clustered around 50.5. Thus, the standard deviation of set A is greater than that of set B, and I is correct while II is incorrect. Therefore, both I and IV are correct.

40. **D.** If the researcher adds a 96 inch tulip measurement to his data on 20 tulips, then the range of his data would increase greatly because the new maximum height of his tulips would be much higher than a value between 5 and 7 inches. The mean and standard deviation would also increase, as the addition of 96 would greatly increase the sum of the data while also only adding 1 to the total number of values and it would also create more variance in the data values by adding a value far outside of the range from 5 to 7 inches. However, the median of the data would not change greatly, as the median value would still be located within the range from 5 to 7 inches even after the addition of a 96 inch tulip.

CHAPTER 8

DATA ANALYSIS

> ## SKILLS TO KNOW
> The three major pitfalls in data analysis:
> - TMI (Too much information!)
> - Failing to understand what data you have
> - Failing to understand what you need to solve for

Data analysis may sound simple enough: look at graphs, charts, or a cluster of information and do the problem as you would any word problem. In fact, many of the skills you'll need to solve data analysis questions aren't in this chapter. Instead, they're part of the chapters on averages; percents; word problems; ratios, rates, and units; and any other chapters that contain word problems. Even so, on the ACT®, many students struggle with these questions for reasons specific to how the data is presented, and thus I've given these problems their own chapter.

The most common data analysis questions are "multi-part questions," i.e. sets of two, three, or four questions that use figures, graphs and charts in addition to preliminary information as a basis for further word problems.

Three issues often emerge when solving data based problems:

1. **TMI (Too much information!):** Students often become overwhelmed by the flood of information and don't know which numbers to crunch. Conversely, they may overlook important details because there is too much to parse through, so finding what matters is difficult. To make it worse, the ACT® often includes sentences rife with useless information, which only slows down and confuses students. For this reason, **I recommend focusing on questions first and data second.** Don't overwhelm yourself with irrelevant facts! Figure out what you need then go back and look for the pieces necessary to solve.

2. **Failing to understand what data you actually have:** Too often, students don't actually READ the labels on graphs and charts, or don't understand what information they are actually given. Be careful. For example, if you need the number of students on a committee, you may actually be given the percent in a chart, not the number. Many students will lift a value off the chart (i.e. the percent of students, not the number of students) and think they got the question right. Instead they needed to read for the total and calculate the number of students by multiplying the whole by the percent. Other times students will need to process chart information with some other secondary step. In any case:

ALWAYS READ CHART LABELS CAREFULLY!

If the chart has labels on the left and right or on the top and bottom, figure out what everything is before you lift any values from it. Likewise, if there is a chart or graph title or key of any sort, read it! This issue emerges big time on the science section; it occurs on the math section as well.

3. **Failing to understand what the question actually wants.** Occasionally, data problems use confusing language when asking for a probability, ratio, or other calculation Many students will then calculate the incorrect value because they thought the question was asking something different. You might, for example, have a chart of professional musicians in different genres and what their favorite music to listen to is. A classical musician who likes rock is not the same as a rock musician who listens to classical. The ACT® can try to create somewhat dizzying information so that you don't know quite what you're looking for or looking at.

At other times, data questions are a reworking of very traditional math, but in a format you don't recognize. For example, you may have a question that simply requires you to turn two variables of data into a linear equation. But you won't see x or y—instead you'll see two other variables. What you need to realize in these situations is that the work itself is much easier than it looks.

Other questions will require precise understanding of what "math" words mean. For example, "per" means divide. If you don't know this, you may get confused when the question asks for something "per" something else. A question may offer information about the speed and time a submarine travels and ask for the distance it travels, and you may need to know distance is usually rate times time, but your chart may have vertical depth of a submarine over time, which isn't the same as "distance." Again, because the information is presented graphically or in a chart, these problems look different and may throw you off. **Systematic carefulness when reading questions is important.**

OVERCOMING TMI

The NASDAQ composite is an index of stocks listed on the NASDAQ stock market. The chart below gives the NASDAQ closing values from July 5th to July 22nd of 2016.

Date	Closing Value	Date	Closing Value
7/5	4,822.90	7/14	5,034.06
7/6	4,859.16	7/15	5,029.59
7/7	4,876.81	7/18	5,055.78
7/8	4,956.76	7/19	5,036.37
7/11	4,988.64	7/20	5,089.93
7/12	5,022.82	7/21	5,073.90
7/13	5,005.73	7/22	5,100.16

What is the percent increase in closing values over the business-week of July 11th to July 15th to the nearest tenth of a percent?

The biggest mistake students make on this question is that they become so overwhelmed by all the numbers in the chart, they don't actually crunch the right numbers. Typically, this comes down to a lack of attention to some part of the chart or the question itself. Here, we need just the percent increase from July 11th to July 15th. Many students, though, will assume the chart mentions exactly what they need, and only what they need, and will simply take the smallest value and the highest value on the chart as a whole and find the percent change between those.

We have about 2-3 weeks worth of stats, but only need one week of percent increase.

Closing value on 7/11: 4,988.64

Closing value on 7/15: 5,029.59

Thus we must pluck $4988.64 as our starting number and $5029.59 as our ending one, and then apply the formula to solve for percent change:

$$\frac{new - old}{old} = \frac{percent}{100}$$

$$\frac{4988.64 - 5029.59}{4988.64} = \frac{x}{100}$$

Percent change: $\frac{40.95}{4,988.64} \approx .0082 \approx 0.8\%$

Answer: **0.8%**.

> The table below gives the conversion rate of 1 bitcoin (BTC) to US Dollars on January 1st for 4 consecutive years.
>
> | Year | 1 BTC in USD |
> |------|--------------|
> | 2013 | $13.01 |
> | 2014 | $745.45 |
> | 2015 | $315.64 |
> | 2016 | $433.09 |
>
> What is the mean of the exchange rates on January 1st for the 4 years listed in the table, in US Dollars per BTC to the nearest $0.01?

What most students struggle with on a question like this is figuring out what you need to find. We have language in the question that clues us into what we need:

The mean, or average "US Dollars per BTC."

Remember "per" means divide and is code that you're looking for a rate.

Now this is where it's a bit confusing. Usually, when we have "Average Rate" problems, we create something of a weighted average. For example, in a distance problem we would do total distance over total time—i.e. if we wanted to find the average rate of travel for the entirety of a road trip and we just added all the speeds and divided by 4 the answer would be wrong! We always consider time and distance so that the average is weighted according to which rate is practiced for the longest amount of time (it would thus be "weighted" toward the rate that is in play for the longest amount of time).

But here, we can't calculate the "total dollars" exchanged or the "total bitcoin" redeemed. We can't create an average exchange rate that is weighted because we don't know the trading volumes, i.e., we don't know how many Bitcoins were exchanged or dollars were redeemed, so we cannot actually weight this average.

When we don't have information, we can assume the method that requires that information isn't what the question wants. Still, it's a good idea to scan all information before making a conclusion about which method the question requires.

I can also carefully read the question: what is the mean "of the exchange rates"—we aren't looking for an average rate, we're looking for the mean of the rates, which sounds more like the problem wants us to add all the rates and divide by four than "the average rate of exchange" or another wording would. Thus, I can assume that we need just the average of the four unit rates.

Because "per" means divide, I need the rates in dollars per bitcoin. Since the chart gives me this unit rate (the number of US dollars in a single bitcoin), the numbers given on the chart are already in the format I need. Still, I need to think this through. On another chart, you might have a unit rate of bitcoins to $1, and then you'd have to convert to dollars per bitcoin before making my calculation.

In any case, the question is easy once I know what I need to do: add the four conversion rates and divide by four:

$$\frac{\$13.01 + \$745.45 + \$315.64 + \$433.09}{4} = \frac{\$1507.19}{4} = \$376.80$$

GETTING YOUR FACTS STRAIGHT

Getting your facts straight is of utmost importance on three-part-questions. You'll also still need to remember the two ideas above—all three of these "trouble" spots are at play throughout the questions below. Frankly, there's not only TMI, but graphs and charts that can be confusing to read, units that switch up all over the place, and language that calls for "closest" answers as opposed to exact ones. I'll walk you through this three-part question below. It is inspired by real ACT® questions, but I've cranked the difficulty on the first two questions to a slightly higher level (typically only 1-2 questions of the three are so tough) to show you the "worst" that can happen.

DATA ANALYSIS   SKILLS

A submarine (submarine X) travels underwater at an average rate of $7.75\frac{m}{s}$ from a military base to a dock several kilometers away. A single one-way trip is completed by submarine X in 32 minutes. Submarine X emits a sonar wave at a frequency of 1200 Hz. A graph showing this submarine's nautical depth in meters as a function of time in seconds over the course of this journey is given below. Submarine X reaches a maximum depth of 160 meters. A second submarine (submarine Y) travels from the dock to the military base at an average rate of $11.95\frac{m}{s}$. It emits a sonar wave at a frequency of 1300 Hz.

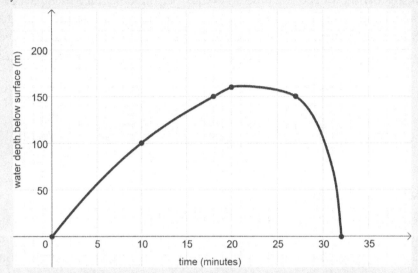

Which of the following is closest to the total distance, <u>in kilometers</u> to the nearest tenth, traveled by submarine X in a single one-way trip between the military base and the dock?

    A. 0.4   **B.** 0.7   **C.** 14.9   **D.** 23.0   **E.** 24.8

This first question requires unpacking a few things. In other words, keeping my facts straight!

1. Find what you need. When I see I need distance, my instinct is to look at the chart. It relays nautical depth in meters (a similar unit of measure). I recognize, however, that DEPTH is not the same as DISTANCE. Distance also involves lateral/horizontal movement. Knowing to avoid the measurement in the chart will keep you from a long road of confusion. On multiple part problems, know much of what you're given will be irrelevant on any given question. In fact, some of what you encounter will be completely useless filler to distract you (sonar anyone?).

2. Know that we're dealing with submarine X, NOT submarine Y. One of the most common errors students make is using the wrong numbers because the question has TMI. Keep your facts straight and the question straight: we need the distance traveled by submarine X one-way. So let's list out facts. I have meters per second, but time in minutes:

$$7.75\frac{\text{meters}}{\text{sec}}=\text{rate}$$
$$32\text{ min}=\text{time}$$
$$\text{need}=\text{distance in km}$$

We need these to be the same "time" measures. Let's do that first, then worry about kilometers later.

98   CHAPTER 8

3. Focus on distance. Distance traveled equals rate times time. ($d = rt$) I will convert the minutes to seconds so both are in the same "language." (I set up my problem using dimensional analysis. I cover this in **Ratios, Rates and Units** in Book 1 if you aren't familiar with it).

$$32 \text{ minutes} \times \frac{60 \text{ seconds}}{1 \text{ minute}} = 1920 \text{ seconds}$$

4. Convert if needed. In "kilometers" is underlined. Whenever units are underlined, they are typically different from the units given in the problem.

So my task is to find the distance traveled knowing it equals rate times time. I will find this first in meters, then worry about converting to kilometers.

$$\text{Now I can do } distance = rate \times time : 7.75 \overbrace{\frac{\text{meters}}{\text{second}}}^{rate} \times \overbrace{1920 \text{ seconds}}^{time} :$$

$$\frac{7.75 \text{ meters}}{1 \text{ sec}} \times 1920 \text{ seconds} = 14880 \text{ meters}$$

Now I'll convert this to kilometers.

There are **1000** meters in a kilometer, so I move the decimal point three to the left:

14.880 km, which rounds to **14.9**.

Answer: **14.9**.

> Submarine X reaches a nautical depth at or below 150 meters for approximately what fraction of time during a complete one-way trip from the military base to the dock?
>
> F. $\frac{5}{32}$ G. $\frac{1}{5}$ H. $\frac{7}{25}$ J. $\frac{18}{25}$ K. $\frac{28}{32}$

This question looks innocent but is quite treacherous. That's because of the word "below." The issue is, in real life, a depth is <u>BELOW</u> another depth if it is <u>GREATER</u> numerically. I.e. a depth of **160** is BELOW (or lower in the water) than a depth of **150**. But the number **160** is greater than the number **150**. Call it ironic. Call it confusing. But it's pretty wicked for most students. To make things worse, the area BELOW our reference point in water depth is represented as ABOVE **150** on the actual graph. Yes. Very confusing. You must keep the facts straight here, and not be thrown by the counter-intuitive presentation of data.

On the graph from about **18** minutes to about **27** minutes we have the curve above **150** (i.e. numbers ranging from **150** to **160**), which represents a depth at or below **150**. That's about **9** minutes. We take **9** minutes and divide it by **32**, and get about **0.281** but that's not an answer, nor is $\frac{9}{32}$. So our best bet is to back-solve what looks closest to **0.281** by dividing the available fractions in our calculator to convert to decimals: $\frac{1}{5}$ and $\frac{7}{25}$. 1 divided by 5 equals **0.2** while 7 divided by 25 equals **0.28**. Our answer is thus H.

If you put J, you likely mixed up the concept of "below" when dealing with water depth.

If you put F or K, you got too hung up on the idea that there are 32 minutes in the trip.

Answer: **H**.

> Which of the following values is closest to the average slope, in meters per minute, of the graph on the interval between 10 minutes and 20 minutes
>
> A. $\frac{3}{4}$ B. 3   C. 6   D. 8   E. 10

Here, the biggest mistake students make is counting boxes for rise over run without paying any attention to units! You can't count boxes if it's a data problem! You must always double check the labels on your coordinate plane first: these are the "facts" you must keep straight.

Here we see the time is in increments of 5, while the depth in increments of 10. We go from 100 to 160 over the course of a span of 10 minutes. Thus we travel 60 meters in 10 minutes, or 6 meters per minute, our slope.

Many students, however, erroneously get $\frac{3}{4}$ instead. They count 1.5 boxes up (rise) and two boxes over (run), but because the boxes represent different amounts, this method won't work. Other students accidentally divide 60 by 20 (the end point of the "run") because they forget to subtract 10 from 20 so as to calculate the amount of time in the range they are considering.

Still other students don't even know where to start with this question. They see slope, but in the absence of the letter "x" and the letter "y" don't know what to do. Just remember slope is always rise over run, but that you must calculate each numerically.

Always read your chart labels.

Answer: **C**.

Use the following information to answer questions 1–3.

A local coffee shop is selling various packs of coffee beans from different countries in various weights. The shop is taking inventory of all of its packs of coffee beans. The shop has the same total pounds of Vietnamese and Colombian coffee. The table below gives the numbers for the different packs of coffee. All the packs except the 2 pound (lb) packs of Vietnamese coffee have been counted.

Coffee Countries	Number of 2 lb packs	Number of 5 lb packs	Number of 8 lb packs
Indonesia	50	20	0
Colombia	0	20	25
Brazil	30	30	10
Vietnam	?	0	20

1. How many more 2 pound packs than 8 pound packs of Vietnamese coffee does the shop have?

    A. 50
    B. 40
    C. 35
    D. 25
    E. 10

2. The shop decides to repackage the 2 pound packs of Vietnamese coffee into 5 pound bundles at a discounted price. If each bundle will sell for $30, how much money can the shop make if it sells all of its 5 pound bundles?

    A. $330
    B. $460
    C. $590
    D. $840
    E. $900

3. Britney buys $\frac{1}{4}$ of the 5 pound packs of Indonesian coffee for $60.

    How much did she pay per pound of Indonesian coffee?

    A. $12
    B. $2.40
    C. $0.60
    D. $0.96
    E. $1.60

Use the following information to answer questions 4 and 5.

The table below gives the prices for Magic Carpet carpet cleaning.

Size of house	Steam clean	Steam clean + protective finish
One-story	$40	$65
Two-story	$60	$85

4. Huang owns 12 houses that need to have their carpets steam cleaned with a protective finish before he can rent them out. Huang pays a total of $840. How many two-story houses did he clean?

    A. 0
    B. 2
    C. 3
    D. 8
    E. 9

5. Huang has also had the carpet in his personal residence (two stories) steam cleaned 5 times without the protective finish. He received a special offer for being a loyal customer, so the 5th cleaning was at a 20% discount. How much did Huang pay, on average, for 5 steam cleanings?

    A. $81.60
    B. $72.00
    C. $62.40
    D. $61.60
    E. $57.60

DATA ANALYSIS — QUESTIONS

Use the following information to answer questions 6–8.

Pete's Printer Shop sells different types of printer toner. The table below shows the estimated number of pages per toner cartridge and the price for three different toner cartridges.

Type of toner	Estimated number of pages that can be printed	Price per toner cartridge
A	500	$14.00
B	400	$12.50
C	325	$9.00

6. Which of the following is closest to the average price of toner per sheet printed for Toner B?

   A. $0.03
   B. $12.50
   C. $14.00
   D. $32
   E. $35.71

7. Pete's Printer Shop has 40 type A toners, 50 type B toners, and 25 type C toners. If Pete picks two toners at random, what is the probability, to the nearest hundredth, that both are type C?

   A. 0.04
   B. 0.05
   C. 0.06
   D. 0.22
   E. 0.43

8. 3 months ago, customers bought 4 times fewer type A toners than type C toners and 3 times fewer type A toners than type B toners. If Pete's Printer Shop sold 160 toner cartridges, how much did the shop make?

   A. $3200.00
   B. $2050.00
   C. $1893.33
   D. $1820.00
   E. $1750

Use the following information to answer questions 9 and 10.

Every day, Robert and Aubrey run from one end of their neighborhood to the other and back along a 2 kilometer-long street that runs the length of the neighborhood.

9. One day, Robert is feeling ambitious and attempts to complete his daily run in 20 minutes. Aubrey is a faster runner than Robert, and runs each kilometer 30 seconds faster than Robert that day. How long will Aubrey's round-trip run take that day?

   A. 9 minutes
   B. 18 minutes
   C. 19 minutes
   D. 19.5 minutes
   E. 38 minutes

10. After a year of running, Aubrey has improved his time. He can complete his daily run in 16 minutes. What is Aubrey's speed, to the nearest hundreth, in meters per second?

    A. 2.08
    B. 4.17
    C. 8
    D. 0.24
    E. 125

Use the following information to answer questions 11 and 12.

Windy City Wireless has new cell phone plans that vary in price according to how much cellular data is included. The following graphic advertises several of the plans:

Windy City Wireless Cell Phone Plans
500 megabytes for $59.99 per month
1000 megabytes for $74.99 per month
3000 megabytes for $109.99 per month
All plans include unlimited calling and texting services
For each additional megabyte, there is a charge of $1.10

11. If Windy City Wireless has a basic plan with unlimited calling and texting but no cellular data that costs $39.99, how much does each included megabyte cost on the $59.99 plan?

   A. $0.12
   B. $0.08
   C. $0.04
   D. $18.18
   E. $0.002

12. Brandon was unsure of how many megabytes of data he would use per month so he bought the $74.99 plan. Unfortunately, he underestimated his data usage and has to pay extra for each additional megabyte over his plan's allowance. In the month of July, if Brandon uses m megabytes where $m > 1000$, which of the following expressions gives the amount on his bill?

   A. $.1.1(74.99m - 1000)$
   B. $1.1(74.99 - m)$
   C. $74.99 + 1.10(m - 74.99)$
   D. $.74.99 + 1.10m$
   E. $74.99 + 1.10(m - 1000)$

Use the following information to answer questions 13 and 14.

The table below shows the old and new rates for a taxi-alternative service, DriveMeAroundTown. The total cost of a trip is equal to the flat rate plus the cost of the distance traveled.

	Old Rate	New Rate
Flat rate per trip	$8.50	$10.00
Cost per mile		
First 5 miles	$0.35 per mile	$0.50 per mile
Every additional mile	$0.50 per mile	$0.60 per mile

13. Ricardo is going to use the DriveMeAroundTown service to go to the museum. He knows it is more than 5 miles from his house, but he does not know exactly how far it is. Using the new rates, which of the following expressions shows how much Ricardo will have to pay for his trip? (Use d as the variable for distance to the museum in miles).

   A. $8.50 + 0.35(5) + 0.5d$
   B. $10.00 + 0.5(5) + 0.6d$
   C. $10.00 + 0.5(5) + 0.6(d-5)$
   D. $8.50 + 0.35(5) + 0.5(d-5)$
   E. $0.5(5) + 0.6(d-5)$

14. How much more does a 13-mile trip cost with the new rates than with the old rates?

   A. $3.05
   B. $1.50
   C. $1.55
   D. $14.55
   E. $17.30

Use the following information to answer questions 15–17.

The Junior Soccer League team of Forrest Hills is raising money for an end-of-the-year pizza party. To cover the party costs, they need $600, and they have two options for fundraising.

Option 1: The kids can sell cookies in the community. There is a $20 startup fee and a cost of $0.60 in baking supplies per box. The cookies sell for $3 per box, and there are 15 cookies per box.

Option 2: The kids can wash cars in the community. After a startup fee of $40, the kids can wash cars for $7 with a cost of $2 per car for washing supplies.

15. How many cars must the kids wash to meet half of their goal?

    A. 64
    B. 68
    C. 56
    D. 52
    E. 62

16. If the kids choose the cookie option, at minimum, how many total cookies must they sell to meet their goal?

    A. 3625
    B. 3750
    C. 3885
    D. 11625
    E. 10875

17. There are 16 kids on the team. If 75% of the kids donate $10 towards the pizza party, how many boxes of cookies would they have to sell to meet their goal?

    A. 209
    B. 192
    C. 309
    D. 167
    E. 42

18. Jen asks 150 students questions about traveling.

Questions	Yes	No
1. Have you ever traveled on a train or airplane?	115	35
2. If you answered Yes to Question 1, did you travel on an airplane?	79	36
3. If you answered Yes to Question 1, did you travel on a train?	48	67

How many students have been on both a train and an airplane?

   A. 31
   B. 103
   C. 127
   D. 12
   E. 13

19. A teacher asked all of the students in the freshman about the number of siblings and/or pets they had. The results are given on the table below. How many students answered that they had 1 or more pets?

		1 or more pets?	
		Yes	No
1 or more siblings?	Yes	53	87
	No	73	45

   A. 140
   B. 136
   C. 126
   D. 73
   E. 53

20. To gather information about changing public opinions on foreign policy, a research company called 500 households randomly selected from the census and gathered answers to a pre-prepared survey. 6 months later, another 500 households randomly selected from the census were surveyed and their answers were compared to the previous answers. Which of the following phrases best describes the company's methodology?

   A. Randomized census
   B. Randomized experiment
   C. Non-randomized experiment
   D. Randomized sample survey
   E. Non-randomized sample survey

ANSWERS DATA ANALYSIS

ANSWER KEY

1. A 2. D 3. B 4. C 5. E 6. A 7. B 8. E 9. B 10. B 11. C 12. E 13. C 14. A
15. B 16. C 17. A 18. D 19. C 20. D

ANSWER EXPLANATIONS

1. **A.** The shop has the same amount of Vietnamese and Colombian coffee, and the total number of pounds of Colombian coffee is $20(5)+25(8)=300$ lb. This means there are also 300 pounds of Vietnamese coffee. So, we can write the total number of pounds of Vietnamese coffee as $300=20(8)+2x$ where x is equal to the number of 2 lb packs. Simplifying this equation, we get $300=160+2x \rightarrow 140=2x \rightarrow x=70$ 2 lb packs of Vietnamese coffee. There are 20 8 lb packs of Vietnamese coffee, so there are $70-20=50$ more 2 lb packs than 8 lb packs of Vietnamese coffee.

2. **D.** From our work in question 1, we know there are 70 2 lb packs of Vietnamese coffee, so the shop can make $\frac{140}{5}=28$ 5 lb bundles of Vietnamese coffee. If each bundle sells for 30, then the shop can make $28(30)=840$ dollars.

3. **B.** There are 20 packs containing 5 lb Indonesia coffee, so if Britney bought $\frac{1}{4}$ of the 20 packs, she bought $20\left(\frac{1}{4}\right)=5$ packs for 60. Each of those five packs contains 5 lb Indonesia coffee, so she bought a total of $5(5)=25$ pounds and spent $\frac{60}{25}=2.4$ dollars for each pound. Remember we take the total cost and dive by total pounds to get the cost per pound.

4. **C.** We look at the second column because Huang needs steam cleaning and protective finishing for all 12 houses. We can represent the amount of money he spent on the one-story houses as $65x$ and the amount of money he spent on the two-story houses as $85y$ where $x =$ the number of one-story houses he has and $y =$ the number of two-story houses he has. He pays a total of 840, which means $840=65x+85y$. We also know that he owns a total of 12 houses, so $x+y=12$. We can now subtract y on both sides and write $x=12-y$. Substituting this in for x in the equation $840=65x+85y$, we get $840=65(12-y)+85y$. This simplifies to $840=780-65y+85y \rightarrow 840=780+20y \rightarrow 60=20y \rightarrow y=3$. So, Huang has 3 two-story houses.

5. **E.** Huang is only getting steam cleaning for his two-story house, so we are looking at the bottom-left cell in the table. For his first 4 cleanings, he pays the regular fee of 60, and on the 5th cleaning, he gets 20% off the 60 dollars, which means he pays $0.8(60)=48$ dollars. So, in total he paid $60(4)+48=288$ for the 5 cleanings. This means he paid an average of $\frac{288}{5}=57.6$ dollars per cleaning.

6. **A.** We look at the second row for the data regarding Toner B. It tells us that an estimated 400 pages can be printed for a cartridge that costs 12.50. The average cost **per** page is cost divided by pages: $\frac{12.50}{400}=\$0.03$.

7. **B.** Pete's Printer Shop has a total of $40+50+25=115$ toners. 25 of these are type C toners. So, there is a $\frac{25}{115}$ chance that one of the toners he picks is type C. Then, there is a $\frac{24}{114}$ chance that he picks a second toner that is also type C since he already picked a type C toner out without replacement. So, the probability of both events happening (he picks two toners that are both type C) is $\frac{25}{115}\left(\frac{24}{114}\right)=0.0458 \approx 0.05$. See **Chapter 10** for more on Probability.

8. **E.** Pete's Printer Shop sold a total of 160 toners, so $A+B+C=160$ where $A =$ number of type A toners sold, $B =$ number of type B toners sold, and $C =$ number of type C toners sold. He sold 4 times fewer type A toners than type C toners, so $C=4A$, and he sold 3 times fewer type A toners than type B toners, so $B=3A$. Substituting in $C=4A$ and $B=3A$ into the equation $A+B+C=160$, we get $A+3A+4A=160 \rightarrow 8A=160 \rightarrow A=20$. This means $B=3A=3(20)=60$ and $C=4A=4(20)=80$. So, looking at the table for the prices of the respective toners, we find the total amount of money the shop made was $20(14)+60(12.5)+80(9)=280+750+720=1750$.

CHAPTER 8 105

DATA ANALYSIS ANSWERS

9. **B.** Since Aubrey runs a kilometer 30 seconds faster than Robert, and the length of the daily run is $2(2km) = 4km$, Aubrey can run the trip $4(30) = 120$ seconds, or 2 minutes (as 120 seconds is 60 seconds times 2) faster than Robert can. Robert's time is 20 minutes, so Aubrey's time is 20 minutes $-$ 2 minutes $=$ 18 minutes.

10. **B.** The round-trip run is twice the length of the road one-way, or 4 km long, and Aubrey can run it in 16 minutes. We want a rate of distance over time, so set up a dimensional analysis row (see **Chapter 9** in **Book 1** for more help), placing the distance on the top of a fraction and the initial time on the bottom and multiply across, inserting the necessary conversion factors (1000 meters in 1 km and 1 min for every 60 seconds): $\frac{4km}{1} \times \frac{1000m}{1km} \times \frac{1}{16min} \times \frac{1min}{60sec} = 4.17 \frac{m}{sec}$.

11. **C.** The difference between the $59.99 plan with 500 megabytes of data and the $39.99 plan without cellular data is $59.99 - 39.99 = 20.00$. This means that each megabyte in the $59.99 plan costs $\frac{\$20.00}{500} = \$0.04$.

12. **E.** Brandon's cellular data usage under 1000 megabytes is covered in his $74.99 plan, so we only need to add the amount that he needs to pay for his usage over 1000 megabytes. We are given that he uses m megabytes, so he uses $m - 1000$ megabytes of data that his plan does not cover. For each additional megabyte, there is a charge of $1.10, so he must pay $\$1.1(m - 1000)$ in addition to his $74.99 plan. That means he has to pay a total of $74.99 + 1.10(m - 1000)$ dollars.

13. **C.** Since $d > 5$, we can represent the additional miles after the first 5 miles as $d - 5$. So, Ricardo will have to pay the flat rate plus 0.50 per mile for the first 5 miles plus 0.60 per mile for every additional mile. Looking at the column on the right for the new taxi rates, we calculate that he has to pay $10.00 + 0.50(5) + 0.60(d - 5)$ for his trip.

14. **A.** For a 13 mile trip with the old rates, one would have to pay $8.50 + 0.35(5) + 0.50(13 - 5) = 8.50 + 1.75 + 4 = 14.25$. For a 13 mile trip with the new rates, one would have to pay $10.00 + 0.50(5) + 0.60(13 - 5) = 10.00 + 2.50 + 4.8 = 17.3$. So, the difference between the old and new rates for a 13 mile trip is $17.3 - 14.25 = \$3.05$.

15. **B.** Half of their goal is $\frac{600}{2} = 300$ dollars. So, if they wash cars, they will need to make $300 + 40 = \$340$ to reach half of their goal and cover the expense of the $40 startup fee. The kids can wash cars for $7 with a cost of $2 per car for washing supplies, which means they make $\$(7 - 2) = \$5$ per car wash. So, in order to make $340, they must wash $\frac{340}{5} = 68$ cars to meet half their goal.

16. **C.** The kids' goal is $600. To sell cookies, they must cover a $20 startup fee, so they need to make $600 + \$20 = \$620$ in order to reach their goal and cover the startup fee. Each box sells for $3 but given the $0.60 cost of ingredients per box, the kids only keep $3.00 - 0.60 = \$2.40$ for each box they sell. Dividing their goal of $620 by $2.40 yields 258.33. However, since they sell the boxes in whole numbers, they must sell 259 boxes, or they won't make enough to cover the party. There are 15 cookies per box, so the total number of cookies is $259(15) = 3885$ cookies.

17. **A.** 75% of 16 kids is $0.75(16) = 12$ kids. So, 12 kids donate $10, which means they have a total of $12(10) = 120$ in donations. If they want to sell cookies, they must make a total of $600 + 20 = 620$ to reach their goal and cover the startup fee of $20. Since they already have $120 in donations, the total amount they have to make decreases to $610 - \$120 = \$500$. Each box sells for $3 but given the $0.60 cost of ingredients per box, the kids only keep $3.00 - 0.60 = \$2.40$ for each box they sell. This means they would have to sell at least $\frac{500}{2.4} = 208.33$ boxes. They can't sell a partial box, so they must sell 209 boxes to reach their goal.

18. D. 79 students responded yes to airplanes and 48 students responded yes to trains, and $79+48=127$. However, we were given that only 115 students responded yes to either airplanes or trains. So, we know that $127-115=12$ must be the number of students who have been counted twice in questions 2 and 3 since they have been on both airplanes and trains. Draw a Venn Diagram (described in **Word Problems** in **Book 1**) to help you visualize the situation. You can also make up variables for each segment in a Venn Diagram ($a, b,$ and c) and then create algebraic equations. For example, let $a =$ trains yes only, $b =$ both trains and airplanes yes, and $c =$ airplanes yes only. $a+b=79$ and $b+c=48$. $a+b+c=115$. If we take the first two equations and add the left sides together and the right sides together, we get $a+b++b+c=79+48$. This simplifies to $(a+b+c)+b=127$. Now I can substitute into the sum of $a+b+c$ with 115 to get $115+b=127$. Thus the overlap, b, must be 12.

19. C. We look at the "yes" column under "1 or more pets" and see that there are two values, 53 and 73. Since we only want to know how many students answered yes to having one or more pets, the number of siblings they have is irrelevant. So, we add up both values under the "yes" column. $52+73=126$ students have one or more pets.

20. D. Answer choice (A) is wrong because a census targets the entire population. The company only calls a sample, or portion, of the population. Answer choices (B) and (C) are wrong because in an experiment, the tester changes variables between two groups and compares the differences. The research company does not change any of the variables between the two groups. A survey is a process in which a group of people answers questions; it is not an experiment. This leaves us with answer choices (D) and (E). We know that the answer is (D) because the houses are selected "randomly," making this a "randomized" sample survey.

CHAPTER 9

COUNTING AND ARRANGEMENTS

SKILLS TO KNOW

- The Fundamental Counting Principle
- Independent & Dependent Events
- Factorial
- Unique Elements/Order Matters
- Permutations (order matters)
- Order Doesn't Matter
- Combinations (order doesn't matter)
- Finding the number of possible arrangements (hybrid problems)
- Spatial arrangements

NOTE: This chapter is among the most in-depth in the series. **If you're short on time or not in gear for epic prep, focus on the Fundamental Counting Principle, a few examples in Permutations, and a few examples in Combinations.** Then try the problem set (or the odds) and reassess if you want to review more. This problem type is important as it occurs on almost every post-2014 ACT I've seen. Beware, older ACT tests may not include these items.

Oftentimes on the ACT®, you'll need to calculate the number of possibilities for a certain outcome. For example, maybe you need to count up all the possible sandwiches you can build at a sandwich shop. Maybe you need to know how many ways you can arrange the letters in CHILLAXIN. Or maybe you're choosing four peers for an Ultimate Frisbee team, and you want to count the number of possible teams.

In all of these situations, you are counting. That is what this chapter is all about. Regardless of what kind of counting problem you are working on, it's important to understand a general principle of counting:

THE FUNDAMENTAL COUNTING PRINCIPLE

> If there are *m* ways to do one thing, and *n* ways to do another thing, and assuming that each "thing" is unique (or that the order of these things matters), there are *m* times *n* ways to do both.

In other words,
  STEP 1: figure out the number of possibilities of each unique condition, choice, or event
  STEP 2: multiply those possibilities together

When counting, you constantly ask yourself, HOW MANY OPTIONS DO I HAVE? Then multiply the results of your thinking to get the answer.

Students are selling gluten-free, probiotic-enriched, artisanal frozen yogurt ("fro-yo") sundaes at a football game. Each sundae consists of one flavor of frozen yogurt, one sauce, and one topping. If there are 4 flavors of fro-yo, 3 sauces, and 2 different toppings, how many different sundaes could be created?

In middle school, you likely created a tree diagram to solve questions like these.

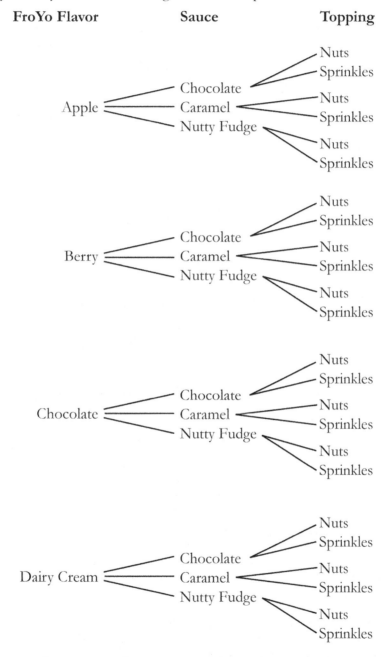

That works, but there's a faster way. Because FroYo, Sauce and Topping choices are all unique choices we can use the Fundamental Counting Principle.

I create a blank for each unique "event" or choice, and ask myself, "how many options do I have?"

$$\underline{} \quad \underline{} \quad \underline{}$$
FroYo   Sauce   Topping

Based on our counting principle, we just put the number of options for each in the blanks, and multiply together.

$$\underline{\ 4\ } \times \underline{\ 3\ } \times \underline{\ 2\ } = 24 \text{ sundaes possible}$$
FroYo   Sauce   Topping

As you can see, this is something of a condensed "tree" diagram. The tree lists out the number of items, and at each branching we essentially multiply by the number of options. Using the counting principle and a few blanks, however, is more efficient.

Answer: **24 sundaes.**

INDEPENDENT AND DEPENDENT EVENTS

When you ask yourself "how many options do I have?" the number of options you have depends on whether events are **independent** or **dependent**.

Independent Events

Independent events don't affect the outcome of each other. For example, if I select a bingo letter at random, and then put it back in the bucket before I select another, each selection of a letter is an independent event. If I choose a particular letter the first time, nothing is stopping me from choosing it again the second time. The probability of picking that particular letter is the same on both draws. Common independent events include coin flips (the coin "resets" each time you flip) and dice rolls (each number is available on each roll).

TIP: Whenever you see the words **"with replacement"** or **"can repeat"** in a word problem, you are dealing with independent events.

How many numbers exist between 300 – 600 that only include odd digits?

$$\underline{} \times \underline{} \times \underline{}$$

My choice of the digits is independent. What I choose for the first digit doesn't affect what digits are available for the 2nd or 3rd digits. Let's use our Fundamental Counting Principle and some blanks to solve this problem. For the first blank, I can choose 3 or 5 (4 is even so it won't create an odd digit number, 6 only creates 600, an even number with all even digits), so I have two choices:

$$\underline{\ 2\ } \times \underline{} \times \underline{}$$

For the 2nd blank, **I don't "remove" 3 or 5 from my available options**, as each digit can repeat. I can choose any digit of 1, 3, 5, 7, or 9. That's 5 choices.

$$\underline{\ 2\ } \times \underline{\ 5\ } \times \underline{}$$

For the 3rd blank, I can again choose any digit of 1, 3, 5, 7, or 9. That's again 5 choices. Because

CHAPTER 9

these events are independent, I am free to put the exact same digit I chose in the 2nd blank if I want to. I don't need to subtract out any digits I've already chosen. There are no additional limits.

$$\underline{2} \times \underline{5} \times \underline{5}$$

I multiply $2(5)(5)$ to get 50 possible numbers in all.

Answer: 50.

TIP: In general, with independent events, don't automatically "subtract" one possible choice as you go to the next blank.

Dependent Events

Dependent events affect the outcome of subsequent events. If I offer you a dessert from a tray with two vegan oatmeal cookies, one chocolate chip cookie, and a crispy rice treat, and you take the chocolate chip, the next person to choose a cookie can't have that cookie. Your choosing of a cookie and the girl-sitting-next-to-you's choice are dependent events: what you choose impacts what she can choose. If you see words in the problem describing each outcome as **"different"** or **"distinct,"** or are told that letters or numbers **"cannot repeat"** you likely are dealing with dependent events.

> Three students each select a different book to read from a list of 10 summer reading books. How many different ways can the students select their books?

Because the students are selecting "different" books, we know these events are dependent. Let's make up some names to wrap our heads around the situation. If June selects <u>Othello</u>, then Max cannot select <u>Othello</u>. We can set up blanks for each student, and think about the number of choices we have:

$$\underset{\text{June}}{\underline{10}} \times \underset{\text{Max}}{\underline{9}} \times \underset{\text{Silas}}{\underline{8}} = 720 \text{ choices}$$

June has 10 choices, then whatever she picks, Max can't pick, so Max has 9 choices. Silas can't pick either of the two books June and Max picked, so he has 8 choices. Each choice reduces the number of books available for the next choice.

Answer: 720.

TIP: In general, with dependent events, subtract one option as you go to the next blank.

FACTORIAL

Denoted by an **exclamation point**, a **factorial** is the product of a positive integer and each positive integer that is less than that integer.

DEFINITION OF A FACTORIAL

If n is a positive integer, then $n! = n(n-1)(n-2)(n-3)...$

Example: $5 \times 4 \times 3 \times 2 \times 1$ is written $5!$ and is read 5 factorial.
By definition, $0! = 1$.

CHAPTER 9

COUNTING/ARRANGEMENTS SKILLS

 CALCULATOR TIP: Most graphing calculators have the factorial function built in. On a TI-84, hit MATH then select the PROB menu and key down to the exclamation point!

> If x and n are positive integers greater than 3, and x is two more than n, then
>
> $$\frac{x!}{n!} = ?$$
>
> **A.** $(x-n)!$ **B.** $n+2$ **C.** $(n)(n-1)$ **D.** $(x)(x+1)$ **E.** $(n+2)(n+1)$

To solve this problem, we must understand what factorial means. Knowing that, we can make up some numbers that adhere to the situation to help make the problem easier to solve.

 TIP: Whenever you see variables in answer choices, you can make up numbers to help solve.

Let's say x is 7 and n is 5.

$$\frac{7!}{5!} = \frac{7 \times 6 \times 5 \times 4 \times 3 \times 2 \times 1}{5 \times 4 \times 3 \times 2 \times 1} = \frac{7 \times 6 \times \cancel{5} \times \cancel{4} \times \cancel{3} \times \cancel{2} \times \cancel{1}}{\cancel{5} \times \cancel{4} \times \cancel{3} \times \cancel{2} \times \cancel{1}}$$

We can see that everything cancels out except the 7 times 6.

$$7 \times 6 = 42$$

We get 42. But I look up and now have to figure out which answer yields 42 when I plug in $x=7$ and $n=5$.

A. $(x-n)!$	$(7-5)! = 2! = 2$		No.
B. $(n+2)$	$(5+2) = 7$		No.
C. $(n)(n-1)$	$5(5-1) = 20$		No.
D. $(x)(x+1)$	$7(7+1) = 56$		No.
E. $(n+2)(n+1)$	$(5+2)(5+1) = 42$		Yes.

Answer: **E.**

We can also think of this algebraically: I know $x = n+2$, so I can substitute $n+2$ in for x in my original expression. (This adheres to the equation→expression pattern you may remember from our Algebra chapters.)

$$\frac{x!}{n!} = \frac{(n+2)!}{n!} = \frac{(n+2)(n+1)(n)(n-1)(n-2)\ldots(2)(1)}{(n)(n-1)(n-2)\ldots(2)(1)}$$

Here we can see what will happen: from the term n downward, all the terms on the top match all the terms on the bottom and will cancel out:

$$\frac{(n+2)(n+1)\cancel{(n)}\cancel{(n-1)}\cancel{(n-2)}\ldots\cancel{(2)}\cancel{(1)}}{\cancel{(n)}\cancel{(n-1)}\cancel{(n-2)}\ldots\cancel{(2)}\cancel{(1)}} = (n+2)(n+1)$$

Using the algebraic method can be a bit faster, but may be more challenging; do what works for you.

UNIQUE ELEMENTS/ORDER MATTERS

Before we get too deep into counting and arrangement problems, you must be able to identify what kind of problem you are trying to solve. One of the first things you need to analyze when confronting one of these problems is whether the items you are counting or arranging ARE UNIQUE and thus ORDER MATTERS or whether they are NON-UNIQUE or ORDER DOESN'T MATTER.

So far, all of the problems in this chapter have dealt with situations in which **either the elements counted are unique OR in which the order of the elements matters.** When one or both of these conditions are true, we can use the **Fundamental Counting Principle** (AKA set up blanks and multiply).

What do I mean by "unique" and "order matters"?

Anytime we have events that occur and they are different in nature, we consider them **unique.** Examples: choosing a hat and choosing a scarf. Choosing a meat, a bread, and a cheese for a sandwich.

We call problems in which order matters "permutations." Any time we create a particular order of items or select a few items from a larger set to arrange in order or in unique positions, we have a permutation. Examples: Choosing an order for the Battle of the Bands, Choosing three dogs to win 1st, 2nd, and 3rd place in a dog show, Choosing four officers (President, Treasurer, Secretary, Vice President) from a group of eligible students.

We can treat **unique events** and those in which **order matters** the same: **create blanks, fill them with the applicable number of choices, and multiply.**

PERMUTATIONS

Again, **permutations** involve finding the number of ways you can arrange **in order** a certain set or number of items. For easy problems (until say question 30), the ideas above are all you need to solve permutations. For those aiming for a 27+ score, you may need to know a bit more, though, about permutations and how you can solve them.

General Permutations

What I'll call "general permutations" require you to count the number of ways to arrange a particular number of items *in order* taken from a set of unique items.

GENERAL PERMUTATIONS FORMULA

The formula for general permutations (with distinct items that are dependent events) is:

$$_nP_r = \frac{n!}{(n-r)!}$$ where n is the number of items taken r at a time.

Sue has 4 picture books and must choose three, in order, for a toddler story time event at the library. How many ways can she do so?

If using this formula, our variable n would equal the number of items we are choosing from, 4, and r would equal the number we are selecting and placing in order, 3. You could solve by plugging these numbers into the formula.

$$_4P_3 = \frac{4!}{(4-3)!}$$

And then solving:

$$_4P_3 = \frac{4 \times 3 \times 2 \times 1}{(1)} = 24$$

But formulas have limits. This particular formula will only work if you have dependent events, order matters, and every item you select is unique. Thinking through all that makes my head hurt. True, more formulas exist for other conditions, but that's just more to memorize.

An easier way to approach this question is to use blanks and solve the problem manually, as explained earlier. Just ask yourself at each blank: how many choices do I have?

$$\underline{4} \times \underline{3} \times \underline{2}$$
First Book   Second Book   Third Book

We then multiply these amounts together $(4 \times 3 \times 2)$ to get 24.

Any time order matters, setting up blanks will work. When I ask myself "how many choices do I have," I automatically consider whether the events are dependent (i.e. you lose one possibility with each successive blank) or independent (i.e. you have just as many choices each time). As long as I correctly answer that question, I can intuitively fill in blanks rather than strain my brain with formulas, worried about which formula works when.

You can also solve "general permutations" like this one with your calculator.

We are taking 4 books 3 at a time, and order matters, so we use the function $_nP_r = {_4P_3}$.

On a TI-83 or TI-84, First enter the value of n (in this case "4") on your display. Then, press MATH, go to the PROB menu, and find $_nP_r$. Select it and hit enter. Then enter the number for r, here that is 3. Click enter and your calculator will compute the answer: 24.

For other calculators, do a quick internet search to determine if this function is available or how to access it.

A car can seat 6 people. 6 friends want to take a road trip. How many unique ways can they be seated?

Let's solve it with blanks first. Line up the choices, placing each choice in order on a blank, and multiply:

$$\underline{6} \times \underline{5} \times \underline{4} \times \underline{3} \times \underline{2} \times \underline{1} = 720$$

We have six "slots" or seats and for the first seat, six people to choose from. After one person is assigned a seat, we have five to choose from for the next seat, and so on.

We can also solve using the formula method:

In this problem, $n=6$ because we have 6 choices, and $r=6$, since we're choosing 6 seats.

$$_nP_r = \frac{6!}{(6-6)!}$$
$$= \frac{6 \times 5 \times 4 \times 3 \times 2 \times 1}{(0)!}$$

But what is $0!$? Well, by definition it is 1 as we mentioned earlier. But that is yet one more thing to memorize and to confuse you...can you tell I don't love formulas? We can essentially ignore the bottom of the fraction as it equals one and solve out the numerator of the fraction to get 720.

Your calculator can likely do this problem, as well. On a TI-84, enter 6 as your value for n, then go to MATH then hit PRB and select $_nP_r$ from the menu. Click enter and finally enter your value 6 for r.

Answer: 720.

> What expression gives the number of permutations of 9 objects taken 3 at a time?
>
> **A.** $\dfrac{9!}{(9-3)!}$ **B.** $\dfrac{3!}{(9-3)!}$ **C.** $\dfrac{9!}{(9-3)!3!}$ **D.** $\dfrac{(9-7)!}{(3)!}$ **E.** 9^3

Ok, ACT®, you got me. This is the one kind of question for which the general permutation formula **does** come in handy.

Using the same formula as above, the answer is $\dfrac{9!}{(9-3)!}$, because out of the 9 choices we have, we are choosing 3 of them. Still, in a pinch, we could use our blanks and back solve (calculating each individual answer): we need $(9)(8)(7)=504$. Choice A simplifies to 504.

Answer: **A**.

WHEN TO AVOID FORMULAS: Permutations with Independent Events & Repeating Elements

You can't use the previous formula when items you are arranging can repeat or when some of the items you are arranging are identical (while others may not be). Though formulas do exist for some of these cases, I don't recommend using them if you don't already know them.

> How many three-digit combinations can you make on a combination lock that includes three digits zero through nine?

Again we make three blanks and ask ourselves for each element of the combination, how many choices do we have? Here, numbers can repeat so for each digit we can choose from ten options (zero through nine is ten different digits).

$$\underline{10} \times \underline{10} \times \underline{10} = 1000 \text{ ways}$$

For this problem, we have the same number of choices in each blank; we don't "take away" any options as we go.

CHAPTER 9

> How many three digit numbers exist?

For this problem I can't use the formulas, because we don't have the same number of options for each blank. Here, the first digit could only be one of nine digits, because it can't be zero. I only have 9 choices, 1–9, as zero is not an option.

$$\underline{9} \times \underline{} \times \underline{}$$

But the 2nd and 3rd blanks have 10 possibilities, because 0 is now an option, in addition to 1–9.

$$\underline{9} \times \underline{10} \times \underline{10} = 900$$

Answer: **900**.

Permutations & Repetition

You may have noticed that earlier I discussed the idea of "unique" choices. Sometimes our choices are not totally unique, even if order matters, such as when some elements repeat. For example, finding the number of ways to arrange the letters in the word MOOD would be a task in which order matters; however, I see that the two O's are not unique. Because these elements are essentially identical, the problem is more complex and we must account for that repetition.

> How many ways can you arrange the letters in the word **CHILLAXIN**?

We have 9 letters to arrange in order, but two of them are L's and two are I's. That makes things a bit complicated.

Basically, if I start off approaching this as a typical permutation, and I pretend each letter is unique (all 9 of them, one C, one H, two I's, two L's, an A, an X and an N), I would have:

$$\underline{9} \times \underline{8} \times \underline{7} \times \underline{6} \times \underline{5} \times \underline{4} \times \underline{3} \times \underline{2} \times \underline{1}$$

But there's a problem. We're treating the problem as if the two I's are unique. Let's imagine one I is upper case and one is lower case so we can visualize how this impacts our calculation. We're treating CHiLLAXIN and CHILLAXiN as two separate arrangements when in reality they only represent a single way of arranging the letters: our I's are NOT unique. We've counted two different options as possibilities that should actually only be counted once. To account for this difference, we must "divide out the repeats," i.e. divide our original calculation by 2.

The same is true of the letter L. Below I've used bold to distinguish L number 1 and L number 2:

CHI**L**LAXIN and CHIL**L**AXIN would be the same arrangement, but with our permutation would be counted twice. Again the two L's double our count unnecessarily. So I need to divide by 2 <u>a second time</u> to account for the two L's.

Thus my answer is:

$$\frac{9\times8\times7\times6\times5\times4\times3\times2\times1}{2\times2}=90720$$

Answer: **90720**.

But what if letters repeat more than two times?

How many ways can one arrange the letters in **SCISSORS**?

I have four S's one C, one I, one O and one R. The four S's are not unique.

Again, let's start with the traditional permutation for the 8 letters as if the S's WERE unique:

$$8 \times 7 \times 6 \times 5 \times 4 \times 3 \times 2 \times 1$$

Here, once more, I'm overcounting my options by pretending order matters. This time, I'll visualize one outcome as $S_1S_2S_3S_4$CIOR. That's the same as $S_1S_4S_2S_3$CIOR or any other version that rearranges those first four letters in any combination. To eliminate these repeats, I now want to know how many ways can I arrange 4 unique items in those first four slots, i.e. to arrange 4 items ($S_1S_3S_2S_4$) taken 4 at a time. I can calculate $_nP_r$ (4!) or set up four slots and ask how many options I have for each slot:

$$4 \times 3 \times 2 \times 1 = 24$$

Regardless of where the 4 S's land, every word in my original count will always be one of 24 "identical" arrangements, because the four slots that contain S's can be rearranged in 24 different ways. I'm "overcounting" by a multiple of 24.

Thus to find our answer, we divide our original calculation by **4!** or **24** to eliminate the repeats:

$$\frac{8 \times 7 \times 6 \times 5 \times 4 \times 3 \times 2 \times 1}{4 \times 3 \times 2 \times 1} = \frac{8 \times 7 \times 6 \times 5 \times \cancel{4} \times \cancel{3} \times \cancel{2} \times \cancel{1}}{\cancel{4} \times \cancel{3} \times \cancel{2} \times \cancel{1}}$$

This reveals an important principle:

Whenever you want to divide out repeats, and p is the number of times an element in an ordered arrangement repeats, divide your original permutation by p!

We can see that everything cancels except:

$$8 \times 7 \times 6 \times 5 = 1680$$

Answer: **1680**.

There is also a **formula for permutations that involve repetition:**

The number of permutations of n items of which p are alike and q are alike is:

$$\frac{n!}{p!q!}$$

Permutations with limitations

If you're not aiming for a 34+ on the math, feel free to skip this one (it's rare on the ACT).

Whenever you have limitations on what can go where, **fill the most restricted slots first**.

CHAPTER 9

COUNTING/ARRANGEMENTS SKILLS

> Mary is arranging five books on her shelf: Geometry, Algebra, US History, English and Biology. If she doesn't want her two math books on either end of the shelf, how many ways can she arrange them?

When permutations have restrictions, you cannot use a formula. Set up your blanks, and then fill the most restricted slots first. That means the first and last slot.

$$\underline{3}\ \underline{}\ \underline{}\ \underline{}\ \underline{2}$$

For the first restricted slot, we can choose from three books (US History, English, Biology). For the last restricted slot, we'll choose from the two remaining that we didn't put in the first slot. Now we have one of these three left over. Now add back in the two math books to this one remaining book. Now I have three books to choose from as I start to fill the three center slots, then two books, then one book:

$$\underline{3} \times \underline{3} \times \underline{2} \times \underline{1} \times \underline{2} = 36$$

Answer: 36.

What do I mean by "order doesn't matter"?

Anytime we are looking to count events in which **order doesn't matter**, counting possibilities generally becomes more complicated.

Finding the number of ways I can have two dice rolled and sum to 5 is a situation in which **order doesn't matter**. Each die can be in either order (i.e. I can roll a 2 then a 3 or a 3 then a 2).

When we have elements that are **"non-unique"** or **"order doesn't matter,"** we generally start by calculating the situation as if order *does* matter, and then dividing out the repeats.

COMBINATIONS

We call arrangements in which order doesn't matter "combinations." A combination typically involves choosing a certain number of items from a set. Typical combinations involve pulling from a single group to create a separate (typically smaller) group. Examples: Choosing three pizza toppings from 10, selecting four winners who all win the same prize from a raffle, picking four people to attend a quiz bowl meet from a team of 8.

Combinations without repetition are most common. These occur when every item you choose from is unique. For example, if four girls are vying for two spots on the prom committee, Jenny can't be both members of the committee. Thus there is no repetition, as all items in our final selection are unique.

SKILLS COUNTING/ARRANGEMENTS

COMBINATIONS FORMULA USING UNIQUE ELEMENTS

The formula for combinations when choosing from unique elements is:

$$_nC_r = \frac{n!}{r!(n-r)!}$$

Where n is the number of choices we have and r is how many elements we select. We can also say we take n items r at a time, or n choose r.

But alas, formulas are hard to remember! So let's learn how to not use the formula! (Or at least how to derive it ☺).

TWO STEPS to solving combinations (with unique elements that don't repeat):
1. Solve as if it were a permutation: make blanks, fill in number of choices, multiply.
2. Divide out the "repeats."

> Panda Salad Emporium is offering a salad trio dish, which invites customers to choose one portion of three different salads from their 6 summer salad options. Customers can choose between Chinese Chicken, Sesame Ginger Spinach, Fruit Salad, Tuna Salad, Edamame Breeze Salad, and Pasta? Pasta! Salad.
>
> How many different salad trios are possible?

Step 1: Run the permutation (pretend order matters…)

$$\underline{6} \times \underline{5} \times \underline{4} = 120$$

Step 2: Divide out the "repeats"
Now let's think. I now have a list of all the salad trios as if order matters. But there's a problem. C S F seems like the same trio as S F C and C F S, etc. How many repeats would this make?

We figure out how many ways can you arrange three salads in a row, if you know which salads they are.

$$\underline{3} \times \underline{2} \times \underline{1} = 6 \text{ or } 3!$$

Thus we've counted each arrangement we want six times, when we actually only want to count each arrangement once. In other words, we overcounted by a multiple of 6 times! If we divide by 6, we can "divide out" these repeats:

$$\frac{120}{6} = 20$$

Answer: **20**.

 TIP: the number of "repeats" in a combination are always the number of blanks factorial.

In other words, to find a combination, first find the permutation, then divide by the number of blanks factorial.

CHAPTER 9

COUNTING/ARRANGEMENTS SKILLS

CALCULATOR TIP: You can also solve this or other simple combination problems using your calculator. On a TI-83 or TI-84, First enter the value of n, the items you are choosing from, (in this case "6") on your display. Then, press MATH, go to the PROB menu, and find $_nC_r$. Select it and hit enter. Then enter the number for r, how many items you are selecting at a time; here that is 3. Press enter and your calculator will compute the answer: 20.

For other calculators, do a quick internet search to determine if this function is available.

Complex Combination/Permutation problems

The ACT® generally avoids really tricky arrangement problems, but with the influx of so many arrangement problems as of June 2014, students aiming for top scores should be ready for complex situations in this category. If you're aiming for a 34+, give this next problem a go! (Be warned, I will be using shortcuts and assume you're adept with all the above ☺).

> A deck of Tujeon cards, traditional playing cards from Korea, contains eighty cards in eight suits. Each suit contains nine numeral cards and one General (jang) card. In how many ways can someone select six cards of a single suit?

This question is trickier. We need to solve it either piecemeal as a combination OR as a permutation and then manually divide out the repeats. I will solve both ways.

Method 1: Add together all the possible cases.

I'm going to first approach this problem one suit at a time. First, I'm going to ask: how many ways can I select six cards from a single group of 10 cards (one full suit)? Order doesn't matter, so that would be 10 choose 6, $_{10}C_6$. I could also solve by doing, $10\times 9\times 8\times 7\times 6\times 5$ and dividing by $6!$ (the repeats, or number of blanks factorial).

On my calculator, I input $_{10}C_6$ (key "10," then MATH > PROB menu > $_nC_r$, then "6") to get 210. Now, I'm going to have that many options for each of the eight suits, so I multiply 210 by 8 to get 1680.

Method 2: Approach as a permutation, then take out the repeats.

Here, I'll imagine I'm choosing 6 of one suit from all 80 cards to start.

In my first blank (pretending order matters), I can chose ANY of the 80 cards. If my goal is to pick 6 cards of a single suit, I can pick any card of any suit to start. I'm not restricted in my choice until the 2nd blank, when that card must match the suit of the first card I chose.

Now for the 2nd blank, I have 9 choices. Regardless of what my first choice was, I now have to "follow suit" and select a card from the same suit. As all suits have 10 cards, and I've already burned one of the cards in that first blank, I have 9 left to choose from. Then 8 to choose from in the third blank, then 7 and so on…

$$80\times 9\times 8\times 7\times 6\times 5 = 1209600$$

Now I must divide out the repeats. The repeats are equal to the number of blanks factorial, because I can rearrange this permutation in 6! ways and still have all the same elements. I can use the built in factorial function in my calculator to make this fast and easy: 6! = 720

I divide my first permutation (1209600) by my repeats (720):

$$\frac{1209600}{720} = 1680$$

Answer: **1680**.

Combinations & Repetition

Combinations with repetition rely on multiple duplicate items in our pool we choose from. For example, if we shop for five spiral notebooks at Sav-R-Store, the store may have many colors of the same notebook on sale. We can buy black notebooks, red notebooks, green notebooks or blue ones, but we could also choose all black, 2 black and 3 red, etc. Which exact black notebook I pick doesn't matter, they are mass-produced and all essentially the same. Most of the time, these kind of situations are NOT dependent events (they might be partially dependent if stock is low on something, say you only have 2 red notebooks left). If Sav-R-Store has 200 of each color in stock, I can choose as many of any as I want, or choose three different colors, so that would be independent. **When the items you choose from are NOT unique** (like pens, dice added to form a sum, combinations of postage from multiple rate stamps, coins that are gathered not flipped, notebooks at the store, etc.) **be careful and avoid formulas unless you are 100% sure you're using the right one. Often these problems must be computed manually** or approached creatively because they are potentially complicated.

> Jessica has 3 coins in her pocket. She knows they are some combination of quarters, dimes, nickels, and pennies, but doesn't know which type(s) she has. How many different combinations of coins could she have in her pocket?

For this problem, we're best off finding the solution manually. By manually, I mean we're at least using blanks, or possibly counting out individual cases.

Because I can have repeats, dividing out the repeats and using blanks may not work. For instance, if order mattered, I know I would have 4×4×4 options, or 64 permutations. But I can't simply divide by 3! Not only would that give me a non-integer answer, but it doesn't encompass the number of repeats because QQQ would only be counted once in my "order matters" calculation, but PQQ, QPQ, and QQP would be a single combination counted three times.

Thus my best solution is to write out all the possibilities, but do so in a systematic way.

Three of a kind: QQQ, PPP, NNN, DDD—**4 options**

Two of a kind: QQ(P/N/D), PP(Q/N/D), NN(Q/P/D), DD(Q/P/N)—**12 options**

If I have two quarters, I can then have one penny/one nickel/one dime as my third coin. That makes 3 options.

The same is true when I have two pennies, two nickels or two dimes. I have three options in each of these cases, too. I could also think of this as a permutation. I'm choosing "two" coins but order matters as the first coin is doubled and the second isn't. $_4P_2 = 12$ if I use my calculator or $\underline{4} \times \underline{3}$ if I write it out.

One of a kind: QPN QPD QND PND—**4 options**

When I say "do manually," I sometimes will speed up using permutation or combination principles once I get into the problem. For this last case, one of a kind, I realize that this task is actually a good old regular combination! I am taking 4 unique items 3 at a time, and nothing repeats! I could do this part quickly by entering $_4C_3$ in my calculator, which equals 4. Using this combination function will keep me from careless errors that occur more often with manual listing.

Fun Fact: As you might notice, from Three of a Kind and One of a Kind above, choosing 3 items from a group of 4 has the same number of options as choosing 1 item from 4. Think about it: choosing 3 to select is the same as choosing 1 to leave out. In other words, $_nC_r = {_nC_{(n-r)}}$. You don't have to know memorize this, but it's a fun fact that may be exploited in a tough question on the test.

Now I just add together my options from each case: $4 + 12 + 4 = 20$.

Answer: 20.

MISCELLANEOUS PATTERNS & SPATIAL ARRANGEMENT PROBLEMS

Sometimes problems involve patterns but the traditional mold of combinations and permutations doesn't work so well. For these problems, draw it out, use logic, make a chart, or try to figure it out manually. (For similar linear arrangement questions, see the problem set of **Chapter 11: Lines & Angles** #12 in this book or questions in **Chapter 6: Distance and Midpoint**, Book 1).

Four points W, X, Y, and Z, lie on a circle with a circumference of 10 units. W is 2 units clockwise from X. Z is 8 units counter-clockwise from W. Y is 6 units clockwise from W. What is the order of the points starting with X and going clockwise around the circle?

Here I draw out a circle, and mark off ten segments separated by hash marks, with each segment representing a unit. Be sure to make 10 segments NOT ten hash marks!

Now I start drawing out each option one at a time, systematically. I know clockwise means around to the right, and counterclockwise around to the left. I count each "hop" manually for every unit I move left or right. See my work on the following page:

SKILLS COUNTING/ARRANGEMENTS

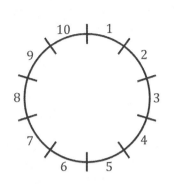

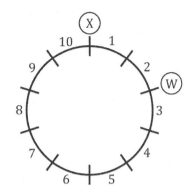

W is 2 units clockwise from X

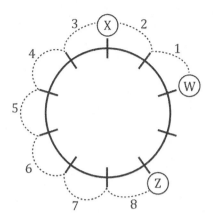

Z is 8 units counterclockwise from W

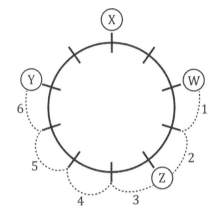

Y is 6 units clockwise from W

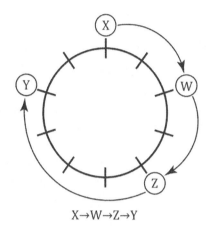

X→W→Z→Y

What is the order starting with X, going clockwise?

Answer: X,W,Z,Y.

CHAPTER 9

COUNTING/ARRANGEMENTS — QUESTIONS

1. Given 8 points, no 3 of which lie on the same straight line, what is the maximum number of straight lines that can be drawn through pairs of the 8 points?

   A. 16
   B. 28
   C. 32
   D. 56
   E. 64

2. Six friends are standing in line to buy movie tickets. In how many different ways can they stand in line?

   A. 1
   B. 21
   C. 120
   D. 720
   E. 360

3. Dwayne and 3 of his friends go on a carnival ride that can seat 4 people from front to back. They only go on the ride once, but Dwayne is curious in how many different ways he and his friends could sit on the ride. How many different ways can they sit on the ride?

   A. 10
   B. 6
   C. 40
   D. 1
   E. 24

4. If 5 letters are taken from the word COMPUTER without repeating a letter, how many different orderings of these letters are possible?

   A. 120
   B. 6875
   C. 6720
   D. 3125
   E. 25

5. Four points, X,Y,Z,W, lie on a circle that has a circumference of 23 units. X is 4 units clockwise from W and Z is 15 units counterclockwise from W. Y is 9 units counterclockwise from X. Starting counterclockwise from X, what is the correct order of points?

   A. X,Z,Y,W
   B. X,Z,W,Y
   C. X,Y,Z,W
   D. X,W,Y,Z
   E. X,W,Z,Y

6. California license plates consist of 1 number (1 to 9) followed by 3 letters (from the 26-letter alphabet) followed by 3 more numbers (0 to 9). How many possible license plates are there?

   A. $9 \times 10^3 \times 26^3$
   B. $10^4 \times 26^3$
   C. $10 \times 9 \times 8 \times 7 \times 26 \times 25 \times 24 \times 23$
   D. $9 \times 9 \times 8 \times 7 \times 26 \times 25 \times 24 \times 23$
   E. $9^4 \cdot 26^3$

7. The School Store has in-stock 15 packages of blue pens, 12 packages of black pens, and 20 of red pens. Ramsey will select 3 packages of pens to purchase. How many different selections of 3 pen packages are possible?

   A. 3
   B. 6
   C. 9
   D. 10
   E. 27

8. In a polygon, diagonals can be drawn between every vertex of the polygon and every other vertex of the polygon, excluding the two adjacent vertices. A regular polygon has equal side lengths and equal angle measures. How many distinct diagonals can be drawn in a 6-sided regular polygon?

   A. 6
   B. 9
   C. 12
   D. 18
   E. 36

9. Sabrina has been instructed to make up three linear equations and then graph them on the coordinate plane. The equations do not necessarily need to form distinct lines. If she graphs any such three linear equations, what are all the possible numbers of distinct regions that these lines could divide the plane into?

   A. 2,3,4,6,7
   B. 2,4,5,6,7
   C. 3,4,5,6,7
   D. 1,2,4,6,7
   E. 3,4,5,6,7

10. A 1st grade teacher has 30 students. She needs to choose 3 students for 3 different leadership positions: class president, class secretary, and class treasurer. If she picks three students at random to fill the positions, which of the following gives the number of different possible results for choosing these 3 class officers?

   A. $30 \times 29 \times 28$
   B. 30^3
   C. $29 \times 28 \times 27$
   D. $3 \times 2 \times 1$
   E. 3^3

11. Asa has 4 pairs of shoes, 5 pairs of pants, and 4 shirts which can be worn in any combination. He needs to choose an outfit to wear to his friend's birthday party. How many different combinations consisting of one item in each category are possible?

   A. 9
   B. 13
   C. 20
   D. 40
   E. 80

12. Buildings A, B, C, and D all lie on the same city street though not necessarily in that order. The street runs straight from east to west. If Building B is 5 blocks east of building D, building A is 8 blocks west of building C, and building B is 5 blocks west of building C, what is the relationship between building A and D?

   A. Building A is 3 blocks east of Building D
   B. Building D is 3 blocks east of Building A
   C. Building A is 2 blocks east of Building D
   D. Building D is 2 blocks east of Building C
   E. Cannot be determined

13. In a poetry contest, a group of 6 poets stand in a circle. A designated poet is chosen to speak first. He or she then must call out the name of another poet to speak, but that poet cannot be to her immediate right or left, and cannot be the last poet who spoke. Given this arrangement, what is the earliest speaking turn that the first poet could possibly speak again?

   A. 3rd
   B. 4th
   C. 5th
   D. 6th
   E. 7th

14. At a school, one junior and one senior are named student of the month. If there are 40 juniors and 50 seniors, how many different 2-person teams of 1 junior and 1 senior are possible?

   A. 10
   B. 90
   C. 200
   D. 1000
   E. 2000

15. A programmable lock allows users to make an alpha numeric pin consisting of one upper case letter in the first position (A through Z) followed by two digits (each digit can be any digit zero through 9). How many different such pins can be made?

   A. $9 \times 2 \times 26^3$
   B. $10^2 \times 26$
   C. $26 \times 25 \times 24$
   D. $9 \times 8 \times 26$
   E. 26×10

16. On a camping trip, the Lee family rents a paddle boat that seats two people. The Lee family has two adults and four children, one of whom is an infant who cannot go on the paddle boat. At least one adult must remain on shore to care for the infant. Given these restrictions, how many different pairs of two family members could ride together on the paddle boat?

   A. 6
   B. 9
   C. 10
   D. 12
   E. 14

17. At a rice bowl shop, Hanna can choose her own rice, protein and vegetable. For rice she can choose from 3 styles of rice, for protein, she can choose tofu, chicken, or beef, and she can choose one of 4 different vegetables. How many possibilities are there for Hanna's three choices for her rice bowl?

   A. 10
   B. 12
   C. 36
   D. $3^2!4!$
   E. $3!3!4!$

COUNTING/ARRANGEMENTS   QUESTIONS

18. Serial codes on a line of toys consist of 3 digits from the 10 digits 0 through 9, 4 letters taken from the 26 letters, A through Z, followed by another 2 digits from the 5 digits 0 through 4. Which of the following expressions gives the number of distinct serial codes that are possible given that repetition of both letters and digits is allowed?

   A. $10^5 26^4$
   B. $10^3 26^4 5^2$
   C. $(10 \times 3)(26 \times 4)(5 \times 2)$
   D. $3^{10} 4^{26} 2^5$
   E. $10!^3 \, 26!^4 \, 5!^2$

19. Which of the following expressions gives the number of permutations of 27 objects taken 6 at a time?

   A. $(27)(6)$
   B. $(27-6)!$
   C. $\dfrac{27!}{6!}$
   D. $\dfrac{27!}{(27-6)!}$
   E. $\dfrac{27!}{(6!)(27-6)!}$

20. Students at a university are assigned identification codes consisting of numbers followed by letters. The codes consist of 3 digits out of 10 possible digits followed by 3 out of 26 possible letters. No code will repeat digits or letters. How many codes are possible?

   A. $10^3 \times 26^3$
   B. $9 \times 8 \times 7 \times 26 \times 25 \times 24$
   C. $10 \times 9 \times 8 \times 26 \times 25 \times 24$
   D. $10 \times 9 \times 8 \times 25 \times 24 \times 23$
   E. $3 \times 10 \times 3 \times 26$

21. Amy is making her schedule for next semester's classes. She needs to take six specific classes, but in no particular order. How many different possible schedules can Amy make if she takes those six classes, assuming she has six academic periods each day?

   A. 36
   B. 120
   C. 360
   D. 720
   E. 46,656

22. Which of the following expressions gives the number of distinct permutations of the letters in QUADRATIC?

   A. $9!$
   B. $8!(2)$
   C. $\dfrac{8!}{2}$
   D. $\dfrac{9!}{2}$
   E. $(9)(8)(7)(6)(5)(3)(2)(1)$

23. The factorial of a number, notated $n!$, is the product of all positive integers less than or equal to n. For example, $4! = 4 \times 3 \times 2 \times 1 = 24$. What is $\dfrac{7!}{4!2!}$?

   A. 15
   B. 70
   C. 105
   D. 210
   E. 420

24. If n is an integer greater than 3, what expression is equivalent to $\dfrac{(n+4)!(n-2)!}{((n+3)!)^2}$?

   A. $\dfrac{n+3}{(n+4)(n+2)(n+1)(n)(n-1)}$
   B. $\dfrac{n+4}{(n+3)(n+1)(n)(n-1)}$
   C. $\dfrac{n+4}{(n+3)(n+2)(n+1)(n-1)}$
   D. $\dfrac{n+3}{(n+3)(n+2)(n+1)(n)}$
   E. $\dfrac{n+4}{(n+3)(n+2)(n+1)(n)(n-1)}$

25. If $\dfrac{(x+1)!}{(x-1)!} = 42$ for positive x, then $x! = ?$

   A. 6
   B. 7
   C. 5040
   D. 720
   E. 26

ANSWERS COUNTING/ARRANGEMENTS

ANSWER KEY

1. B   2. D   3. E   4. C   5. D   6. A   7. D   8. B   9. A   10. A   11. E   12. C   13. B   14. E
15. B   16. B   17. C   18. B   19. D   20. C   21. D   22. D   23. C   24. E   25. D

ANSWER EXPLANATIONS

1. **B.** This is a combination: order doesn't matter and we are taking two points at a time from 8 or "8 choose 2." Formally, we write this as $_8C_2$. You can solve this in your calculator (on the TI-84 hit "8" then MATH then PROB then nCr then "2" and enter to get the answer). We'll unpack this manually, too, though, as it's important you understand how these work. If we labeled our points A-H and made a list of the combinations it would look something like:

    AB, AC, AD, AE, AF, AG, AH...

    This is not a permutation, but we'll start off by figuring out the number of arrangements if order DID matter, and then divide out the repeated terms. Each of the 8 points can pair with any of the 7 others in the list. So pretening order matters we calculate the number of pairs by multiplying the number of choices per point times the number of points: 7 x 8= 56. But right now, we're counting BA and AB as different lines—that's not ok. Because there are two terms, there are two ways to arrange any given pairing (i.e. if I know point A and point B are involved, there are 2 ways to arrange them). As such we need to divide by 2 to divide out all the repeated terms: $56 \div 2 = 28$.

2. **D.** Use the fundamental counting principle. As order matters, this is a permutation: We draw out six blanks: __ __ __ __ __ __ —then fill them in with numbers of how many students we have to choose from to fill each blank, and then we multiply together. With each step we have one less person to choose from, as a single person can't be in two positions:

    $$6 \times 5 \times 4 \times 3 \times 2 \times 1 = 720$$

    We can alternatively use the $_nP_r$ function on a calculator ($n=6, r=6$).

3. **E.** This question is a bit of a trick question—Dwayne and THREE OF HIS FRIENDS is actually FOUR people—Dwayne plus the other three. So we're taking 4 people 4 at a time in a permutation. Method is the same as in question 2—so $4 \times 3 \times 2 \times 1 = 24$.

4. **C.** Here, order matters (how many different ORDERINGS) because where letters appear in a word depend on order. Thus this is a permutation—there are 8 distinct letters in COMPUTER and we're choosing five at a time, or $_8P_5$. Imagine five slots for each letter we are choosing, and we would have $8 \times 7 \times 6 \times 5 \times 4 = 6720$. Remember the letters are unique and none repeat—so we subtract one option as we go.

5. **D.** We can sketch a circle and place W arbitrarily to begin. Place X slightly clockwise and label the distance between them 4. Then place Z a good distance counterclockwise from W. We know that it does not go past X because the distance clockwise from W to get Z, 15, plus the distance counterclockwise from W to get X, 4, is less than the full circumference of the circle, 23. Thus, they do not overlap. Then, starting from X, we place Y 9 units counterclockwise, which goes past W, since that is only 4 units. Thus, our order starting counterclockwise from X is X, W, Y, Z.

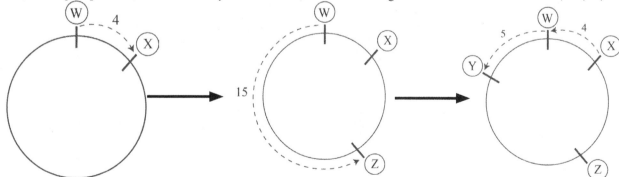

6. **A.** Here order matters, but because of the special conditions (letters and numbers) and the fact that numbers CAN repeat, we cannot use the permutations button on a calculator or the general permutations formula to solve. Instead, set up blanks:

    __ __ __ __ __ __ __
    \# Letter Letter Letter \# \# \#

CHAPTER 9   127

COUNTING/ARRANGEMENTS   ANSWERS

Now for the first blank we have 1 to 9—that's NINE numbers (1, 2, 3, 4, 5, 6, 7, 8, 9). So we write a "9" in the first blank. In the next three blanks, we have 26 three times. In the last three blanks, we actually have TEN numbers—as this part calls for numbers ZERO through nine, not 1-9—(1-9 is nine plus zero makes ten options). Altogether:

$$9 \times 26 \times 26 \times 26 \times 10 \times 10 \times 10$$

Now that the answer choices are in exponent form, we group our like terms together:

$$9 \times 10 \times 10 \times 10 \times 26 \times 26 \times 26$$

That's one nine times three 10's or 10^3 times three 26's or $26^3 = 9 \times 10^3 \times 26^3$. Be sure to READ CAREFULLY!

7. **D.** In this case, order doesn't matter. Picking two packs of red pens first and a black package last is the same as picking black first, and two reds second—one either buys the packs or doesn't. Given the fact that we CAN repeat terms (i.e. we could buy 3 packs of all black pens—once we pick black it's not "out"), we cannot use traditional permutation or combination equations to solve this problem. The store, furthermore, has plenty of pens in stock—i.e. if we're only buying 3 packs, there are more than enough of EVERY type of pen—so we don't need to worry about that as a limiting factor. Because the numbers are small, this is easy enough to just write out—but try to be systematic (organized). Here, X is used for black, R for red and B for blue:

THREE OF A KIND: All Black, All Red, All Blue (3 options)
TWO OF A KIND: XXR, XXB, RRX, RRB, BBX, BBR (6 options)
ONE OF EACH: XRB (1 option)

Altogether, that's 10 options.

8. **B.** In this problem, order doesn't matter—the diagonal from vertex A to vertex C is the same as a diagonal from vertex C to vertex A. So it's not a permutation, but more of a combination. However, we don't count the lines between a point and the adjacent points. Because of this restriction, we can't just use the combination function in our calculator (6 vertices choose 2), but we do have to divide out the "repeats" at the end of the problem (which we do WHENEVER order doesn't matter). First, figure out how many diagonals can be drawn from any given vertex (picture below). From the drawing it's clear that we can draw THREE diagonals per vertex.

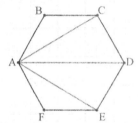

Every vertex is connected to three diagonals, so $6 \times 3 = 18$. But we're not done yet! 18 includes REPEATS—i.e. we've counted the diagonal from vertex A to vertex C AND a diagonal from vertex C to vertex A, etc. Now, divide 18 by 2 to get the answer: 9. One can also memorize the formula for number of diagonals in a polygon with n sides: $\dfrac{n(n-3)}{2}$.

9. **A.** The best way to solve something like this is to draw it out. To divide the plane into 2 regions, draw the 3 linear equations as overlapping to form a single line. To divide the plane into 3 regions, draw 2 lines that overlap as identical equations and another line parallel to them. To divide the plane into 4 regions, draw 3 different parallel lines. To divide the plane into 6 regions, we can draw 2 different parallel lines and 1 line perpendicular to those lines (so they will intersect). To divide the plane into 7 regions, we can form a triangle with the 3 lines, so each line has 2 intersections.

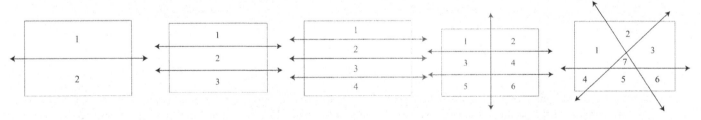

10. **A.** To solve arrangements problems we first must determine whether order matters. In this problem, order matters: assigning Amy to be president instead of secretary is a different outcome. Thus, we can solve this question using the Fundemental Counting Principle described earlier in the chapter. Make a slot for each choice we have to make:

president   secretary   treasurer

Now we, how many people do we have to choose from for each position? For the first position we have 30 people. Now fill in "30" for that first slot (president). For the second blank, ask again, how many people can we choose from? The person we just chose for president can't be secretary, so subtract one option to get 29 people—write that in slot 2. For the third blank, two people are spoken for already, so we have 28 people to choose from. Now we multiply the three numbers together. We're left with: $30 \times 29 \times 28$. Another way to solve this problem is to recognize this is a permutation (an arrangement in which order matters with no repeated terms and with no special conditions) and use a calculator. The permutation function— $_nP_r$ —on the TI-84 plus is found in the MATH menu under PROB. Enter $_{30}P_3$ (30 then key $_nP_r$ then 3) to get the answer.

11. **E.** This is similar to a permutation—order "matters" as each category we are choosing is distinct. Make a slot for each choice we have to make as so:

_____ × _____ × _____
shoes   pants   shirts

Then fill in the blanks with the number of choices we have: 4, 5, 4, and multiply together: $4 \times 5 \times 4 = 80$. Because order "matters" or each slot is distinct, there's no need to worry about repeats. We're done! If this confusing, we might start listing out what is possible and making the problem more real for ourselves. Shoes A would have five pant options (1, 2, 3, 4, 5)—each of those pant options 4 shirt (Red, Green, Blue, White) options, etc. So we could do A-1-Red, A-2-Red, etc. Each shirt has 5 pant options; for shoes A alone we'll have 5×4 or 20 options—for shoes B, C, and D we'll also have 20 options each…together that's 80 options.

12. **C.** For this problem, we draw a picture, taking one command at a time. We draw the 1st and 3rd commands last—as the middle command has two new letters, so we might place them in the wrong order if we do this command first. We solve this using a number line, as shown below. Once we see the distance from D to C is 10 units, we can subtract the 8 units to get that A is 2 units EAST of building D.

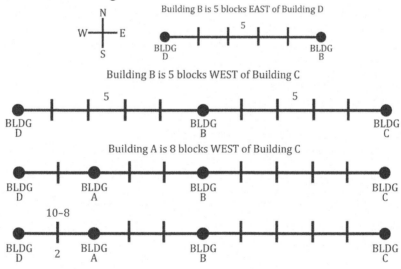

13. **B.** Note: problems like this can be time consuming. Do them last! The best way to solve them is to draw out possibilities, be systematic by always being aware of where choices occur, and know what to aim for. Let's name our poets Amy, Betsy, Candy, Dina, Emily, Farrah – and imagine they are in that order around the circle (so Farrah is next to Amy). Draw these names out in a circle in this order. Now we'll experiment:

FIRST TRIAL: Let's try to do this on the 3rd turn.
1st Turn—Amy speaks—then can choose CANDY, DINA, or EMILY—let's stick with CANDY for now. If she doesn't work we'll potentially have to consider the others.
2nd Turn—Candy speaks
3rd Turn—Amy can't speak because Candy can only pass to someone who is not the LAST person who spoke—which would be Amy. So 3rd turn doesn't work—and this would be true no matter which lady spoke first.

COUNTING/ARRANGEMENTS ANSWERS

SECOND TRIAL: Let's try for the 4th turn.

1st Turn—Amy speaks—then can choose CANDY, DINA, or EMILY—let's stick with CANDY for now. If this doesn't work we'll have to check the other ladies, too.

2nd Turn—Candy speaks—now Candy can't pass to neighbors (Betsy, Dina), or Amy (last speaker)—she can pass to EMILY or FARRAH.

3rd Turn—two options—EMILY or FARRAH—if FARRAH speaks, she's next to Amy, so Amy can't be 4th. Let's try EMILY—if Emily speaks, she can't pass to Candy (last speaker), or Farrah or Dina (neighbors)—AHA! If Emily speaks she CAN pass to Amy on the 4th turn—so Amy can be on the 4th turn.

Obviously, there are other combinations—we could have tried DINA 2nd, and then BETSY or FARRAH would be 3rd—making it impossible for AMY to be 4th—that's why we need to be aware of each choice we have made between multiple options so we can track back and look at other possibilities.

14. **E.** For this problem, we ask ourselves how many juniors we have to choose from (40) and then how many seniors (50)—simply multiply these numbers together. If we were to listed them out we would do so as Amy Jr.—Allan Sr., Amy Jr.—Brian Sr., Amy Jr.—Carl Sr.…Amy would have 50 seniors she could be paired with, and so on for each of the 40 juniors. This is like a permutation in that there is NO overlap between juniors and seniors—i.e. we have 40 different people who are juniors to choose from and none of them can also be seniors—as such order matters. If you put (B) you likely added the numbers together—multiply, don't add. If you put (C) you likely tried to multiply in your head and messed up the decimal place—use your calculator if you need it. If you put (D) you likely thought you had to divide out repeats. No need to divide out any repeats—Amy the junior and John the senior can't be created in reverse order (because John is not a junior).

15. **B.** In a combo lock, order matters: this is a permutation. We first ask "how many options do we have to choose from" for each element, then multiply the number of options together. Remember that there are 26 letters, and 10 digits for each of two digit positions (0-9 is TEN digits—1-9 is nine digits plus zero makes 10). $26 \times 10 \times 10 = 10^2 \times 26$.

16. **B.** Occasionally an arrangements problem has so many restrictions, it's best to work out the logic and list out all the options rather than fully rely on formulas. Here, we've got a family that is going to divide out into groups of two—and only one parent at a time can be on the boat. Now we could also have kids alone on the boat. As far as we can tell, order DOES NOT matter—you're either on the boat together or not. We don't care who sits in what seat. Let's start with the condition that mom is on the boat (and dad is on shore with the baby), then dad is on the boat, then kids only. Let's label the kids A, B, C and mom and dad M, D.

> If mom is on the boat: MA, MB, MC—3 options
> If dad is on the boat: DA, DB, DC—3 options
> If no parents are on the boat: (here I can use a combination to find "3 kids choose 2" for the boat: either do $_3C_2$ on my calculator, 3×2 then divide by 2 to eliminate repeats, or I can list out) AB, BC, CA—also 3 options
> As we can see—we have 9 total options

If you chose (A) you may have added all the members together or missed some options, if you chose (C) you may have added the pair MD (Mom and Dad), which would be impossible (abandoned baby!)—or done $_5C_2$—but this is not a simple combination; it has restrictions (D) would double count the kid pairs, and (E) would be a permutation (if order mattered) of 5 people taken two at a time with no restrictions; again we have restrictions so you can't just multiply 5×4, and order doesn't matter so you must eliminate repeats.

17. **C.** To find the number of possible combinations of rice, proteins, and vegetables, we multiply the number of items in each category together. There are 3 kinds of rice, 3 kinds of protein, and 4 kinds of vegetable, so our answer is $(3)(4)(4) = 36$.

18. **B.** The number of different possibilities is equal to the number of possible numbers or letters that can occupy each position in the code multiplied together. In a 3 digit code using the digits 0 through 9, we have 3 positions with 10 different possible digits in each position, so the number of combinations is $(10)(10)(10) = 10^3$. We have 4 positions with 26 letters as options for each: $(26)(26)(26)(26) = (26)^4$. And we have 2 positions with 5 digits to choose from for each: $(5)(5)$ or $(5)^2$. Thus, the total number of combinations is $10^3 26^4 5^2$.

19. **D.** The number of different permutations will equal to the number of members available to fill the first spot, times the number of members available to fill the second spot, and so on. Since objects can't repeat (e.g. the object in the first posi-

tion can't also be in the third position) every spot has one fewer possible objects to fill it than the one before it. Thus, the number of different permutations equals $(27)(26)(25)(24)(23)(22)$. A shorthand way to express this kind of answer is as the factorial of the members of a set divided by the factorial of the members that won't be used at the end. We don't use any of the 27 objects except for the 6 that we do use, so the number of objects we don't use is $(27-6)$. Our answer is $\frac{27!}{(27-6)!}$. Alternatively, plug into the formula for permutations given n=27 items take r=6 at a time: $_nP_r = \frac{n!}{(n-r)!}$.

20. **C.** The first element in the identification code can be chosen out of any of the 10 digits, so there are 10 possible outcomes for the first digit. The second digit cannot repeat the first digit, so there are 9 possible outcomes for the second digit. The third element cannot repeat the first or second digits, so there are 8 possible outcomes for the third digit. Likewise, there are 26 letters to choose from for the first letter, 25 for the second, and 24 for the third. The total number of possible permutations is the product of these possible outcomes. So, our answer is $10 \times 9 \times 8 \times 26 \times 25 \times 24$.

21. **D.** When creating her schedule, Amy has six time slots to fill six classes. For the first time slot, she can place any class. For the second time slot, she has 5 remaining classes to choose from, for the third time slot she has 4 remaining classes to choose from, and so on until all time slots are filled. The total number of possible permutations of classes Amy can schedule is calculated by $6 \times 5 \times 4 \times 3 \times 2 \times 1$ or $6! = 720$.

22. **D.** There are 8 individual letters used, with one, the 'A', occurring twice to make 9 total letters in the set. If we imagine that each of the 'A's are different, then the number of possible permutations is $9!$. However, since in each of the permutations the 'A's can switch positions without changing the permutation, there are actually only half the number of permutations. Thus, we have $\frac{9!}{2}$ permutations.

23. **C.** Using the definition given in the problem, $7!=7\times6\times5\times4\times3\times2\times1=5054$ and $4!=4\times3\times2\times1=24$ and $2!=2\times1=2$. So $\frac{7!}{4!2!} = \frac{5054}{24\times2} = 105$. Another way to approach the problem is to write out $\frac{7!}{4!2!} = \frac{7\times6\times5\times4\times3\times2\times1}{(4\times3\times2\times1)(2\times1)}$ Then, canceling out equivalent numbers in the numerator and denominator, we get $\frac{7\times6\times5\times \cancel{4}\times\cancel{3}\times\cancel{2}\times\cancel{1}}{(\cancel{4}\times\cancel{3}\times\cancel{2}\times\cancel{1})(2\times1)} = \frac{7\times6\times5}{2\times1}$ which is a lot easier to calculate. $\frac{7\times6\times5}{2\times1} = 7\times3\times5 = 105$. To simplify factorials, you can also use the factorial function in your calculator. On a TI-84 hit MATH then PROB and look for the exclamation point "!."

24. **E.** $\frac{(n+4)!(n-2)!}{(n+3)^2!} = \frac{(n+4)!(n-2)!}{(n+3)!(n+3)!}$. Since we know:
$(n+4)! = (n+4)(n+3)!$ and $(n+3)! = (n+3)(n+2)(n+1)(n)(n-1)(n-2)!$, we can rewrite the expression as:
$$\frac{(n+4)(n+3)!(n-2)!}{(n+3)!(n+3)(n+2)(n+1)(n)(n-1)(n-2)!}$$
Now we can cancel out equal terms from the numerator and denominator to get:
$$\frac{(n+4)(n+3)!(n-2)!}{(n+3)!(n+3)(n+2)(n+1)(n)(n-1)(n-2)!} = \frac{(n+4)}{(n+3)(n+2)(n+1)(n)(n-1)}$$

25. **D.** We can write $(x+1)!$ as $(x+1)(x)(x-1)!$. So, we can rewrite the equation as $\frac{(x+1)(x)(x-1)!}{(x-1)!} = 42$. Canceling out like terms from the numerator and denominator, we get $(x+1)(x) = 42 \rightarrow x^2 + x = 42$. Subtracting 42 on both sides, we get $x^2 + x - 42 = 0$. Now, we factor this to find the zeros to get $(x+7)(x-6) = 0$. So, $x = -7$ or 6. Since x is positive, $x = 6$. So, $x! = 6(5)(4)(3)(2)(1) = 720$.

CHAPTER 10
PROBABILITY

> **SKILLS TO KNOW**
> - Basic probability
> - Finding the probability something will not happen
> - Independent events & dependent events
> - And situations/Or situations
> - Probability & permutations
> - Finding expected values
> - Probability notation, union & intersection

 NOTE: This chapter builds heavily on **Chapter 9** in this book, **Counting & Arrangements**. Be sure you have a handle on Chapter 9 before attempting this chapter.

BASIC PROBABILITY

The **probability** of an event occurring is the likelihood that something will happen.

Probability is expressed as a decimal or fraction between 0 and 1, inclusive. If the probability of an event is 1, it will happen with 100% certainty. The closer the probability of an event is to 1, the more likely it will occur. The closer the probability of an event is to zero, the less likely it is to occur.

When the selection of an outcome is **at random**, we can calculate probability by creating a fraction:

$$\frac{\textit{Number of possible desired outcomes}}{\textit{Number of total possible outcomes}}$$

For example, if we want to know the chances of choosing a blue marble from a box when there are 3 blue marbles and 10 marbles total, that would be $\frac{3}{10}$.

This can also be expressed as:

$$\frac{\textit{Number of "successful" outcomes possible}}{\textit{Number of "successful" outcomes possible} + \textit{Number of "failed" outcomes possible}}$$

For example, if there are 3 blue marbles and 7 other colored marbles, there are 3 ways to succeed and 7 ways to fail if I pick one marble out of the box and want a blue one. The probability is thus:

$$\frac{3}{3+7}$$

 There are 52 cards in a deck of cards. There are 4 suits (spades, clubs, diamonds, and hearts), each with 13 cards, 3 of which are face cards. 2 suits are red, and 2 are black. If a card is drawn at random, what is the probability that it is a red face card?

To solve, we use the fraction formula for finding probability:

$$\frac{\textit{Number of possible desired outcomes}}{\textit{Number of total possible outcomes}}$$

We need to find two things:

1. **The numerator:** the number of red face cards (what we want).

   We have 3 face cards per suit. Two of the suits are red. That means we have $2(3)$ or 6 total red face cards.

2. **The denominator:** the total number of cards in the deck (total possible outcomes).

   Per the question, we have 52 cards in the deck. We now divide the value for #1 above by that of #2:

$$\frac{6}{52} = \frac{3}{26}$$

Answer: $\frac{3}{26}$.

PROBABILITY THAT SOMETHING *WON'T* HAPPEN

Finding the probability something *will* happen can also be solved by calculating the probability that something *won't* happen. In the world of probability, success and failure are mutually exclusive concepts: either something happens or it doesn't; two options exist. As a result, the probability that something will happen and that something won't happen always sum to 1. For example, if there is a $\frac{3}{5}$ chance you'll pick a red marble from a bucket, there is a $\frac{2}{5}$ chance you won't.

 If you need to know the probability that something happens, but it's easier to find the probability of that something not happening, solve for the latter probability and subtract from one. Likewise, if you're asked to find the probability of something *not happening* and it's easier to solve for probability of something *happening*, solve for that and subtract from one.

> There are 100 slips of paper numbered 1 through 100 inclusive in a hat. If one slip is drawn at random, what is the probability the number drawn is not a perfect square?

I know $1, 4, 9, 16, 25, 36, ..., 81, 100$ are the list of perfect squares. I omitted the middle, because I know how each is formed: squaring a number 1 to 10 inclusive, as 1 is 1 squared, and 100 is 10 squared. In between, I'll have $2^2, 3^2, 4^2$, etc. Thus I know there will be 10 of these numbers in the list. As a result, the probability of getting a perfect square as my number on the slip is:

$$\frac{10}{100} = \frac{1}{10}$$

Now I subtract this value from 1 to find the answer: $1 - \frac{1}{10} = \frac{9}{10}$.

Answer: $\frac{9}{10}$.

PROBABILITY SKILLS

INDEPENDENT VS. DEPENDENT EVENTS

In the previous chapter, we defined **independent** and **dependent** events. To review:

Independent events don't affect the outcome of other events. For example, if I flip a penny and get heads, I'm not more or less likely to get heads again on a second flip. Coin flips, dice rolls, and situations with wording such as **"with replacement"** or **"items/digits can repeat"** are typically **independent events.**

Dependent events affect the probability of subsequent events. Drawing three letters from a bag of lettered tiles, choosing people for a team, or selecting songs to sing at a recital are all dependent events. You wouldn't sing the same song twice at your voice recital, so what you pick for the first song affects what you choose for the 2nd. If you had 4 songs to choose from, after you choose one song you'll only have 3 songs to choose from. Often, you'll see words like **"distinct," "unique,"** or **"without replacement"** when encountering problems that involve **dependent events**.

"AND" SITUATIONS

If the probability of two independent events are A and B, then the probability of both A "AND" B occurring (assuming each event is unique, or order matters) **is A times B.** In short, when you have an **"AND"** situation (Event A is true AND Event B is true) **you multiply.**

> Ned is throwing a coin to see if it's heads or tails. What is the probability that he will throw heads three times in a row?

This is an "AND" situation involving independent events. The chance of getting heads once is $\frac{1}{2}$. Since Ned will be doing this three times and order matters (to get three heads in a row we need each subsequent toss to be heads, the first, the second, and the third) we multiply $\frac{1}{2} \times \frac{1}{2} \times \frac{1}{2}$ to get $\frac{1}{8}$.

Answer: $\frac{1}{8}$.

A similar trend emerges with **dependent events**. We still **multiply probabilities together** when both **event A AND event B** are true to find the probability both are true (assuming **order matters**). However, we must take event A into account when we calculate the probability of B. In other words, we multiply **(Probability of Event A) (Probability of event B given event A happening first)**.

> Marlin is choosing three toys at random to take to the beach from her basket of 10 beach toys. If $\frac{1}{2}$ of the beach toys are plastic, what is the probability she chooses all plastic toys?

We start by remembering the fraction that defines probability:

$$\frac{\textit{Number of possible desired outcomes}}{\textit{Number of total possible outcomes}}$$

These are dependent events, so as I work I must take into account the previous choices. I want to find the probability that three particular events occur in a row: she chooses a plastic toy first, a plastic toy second, and a plastic toy third. For all of these events to be true, we have **an "AND" situation**. Each of these events <u>**must occur**</u> in order to get my desired outcome. Thus we can find the probability of each case and then must **multiply these individual probabilities together**.

Her probability of choosing a plastic toy the first time is $\frac{5}{10}$, but the second is $\frac{4}{9}$. When she chooses the 2nd toy, the 1st toy is no longer an option; there are only 4 plastic toys to choose from and 9 toys left in total to choose from. When she chooses the third toy, there is a $\frac{3}{8}$ chance she chooses a plastic one. At this point, she will only have 3 plastic toys left to choose from and 8 toys in total.

As you can see, we reduce the numerator and denominator accordingly as we go: $\frac{5}{10} \times \frac{4}{9} \times \frac{3}{8} = \frac{1}{12}$.

Why can we solve this problem using a "permutation" when it sounds like a combination?

Though order doesn't matter here in one sense (she is selecting a few items from a group), it DOES matter in terms of her selecting a toy that is plastic on each turn, i.e. each moment when she chooses a toy a certain event must occur: she picks a plastic toy each time. Thus we can still treat this as a permutation and not a combination. We could actually solve this problem using combinations as well. For that method, we'd rely again on the fraction that defines probability, and calculate our numerator as the number of ways to choose 3 plastic toys from 5 ($_5C_3$ or "5 choose 3") and then divide that by the number of ways to choose 3 toys from 10 ($_{10}C_3$ or "10 choose 3"). Using our calculator's built in combinations function we get: $\frac{_5C_3}{_{10}C_3}$ which equals $\frac{10}{120}$ or $\frac{1}{12}$.

"OR" SITUATIONS

Sometimes we can find a probability by adding together the probability of all the unique ways we could get what we want. These cases are essentially **"OR" scenarios**. Situation A is true or B is true or C is true, for example. When we have **"OR"** situations in probability, and our elements are unique (i.e. "mutually exclusive") we **ADD the probabilities together.** Add the probability of all the unique cases that produce your desired outcome to find the overall probability of that outcome.

Let's say we want to know the probability of flipping one head and one tail when two coins are flipped. For example, if the chance I get heads then tails on two coin flips is $\frac{1}{4}$ and the chance I get tails then heads on two coin flips is $\frac{1}{4}$, then the chance in two flips of a coin that result in one heads and one tails is $\frac{1}{4} + \frac{1}{4}$ or $\frac{1}{2}$.

WARNING: If you add probabilities together, you must be certain the outcomes included within each probability you've calculated **DO NOT OVERLAP**. Events must be distinct or **"mutually exclusive"** if you're going to add their probabilities. I can't be in 4th grade and in 5th grade, those are mutually exclusive events that do not overlap. But I could be in 4th grade and female. Those are NOT mutually exclusive events. So if I know 1/2 the students at a school are female and 1/5 are in 4th grade, I CANNOT simply add 1/2 plus 1/5 to find the probability of selecting a student at random who is female and/or in fourth grade. If I did, I would double count all the fourth grade girls.

> A multiple-choice quiz has four answer options for each of five questions. What is the probability of choosing answer choices at random and missing exactly one question?

Because getting a question right or wrong is independent of how I did on the last question (assuming I'm guessing at random), the events are independent. To solve this problem, I'll pretend order matters and break it into multiple cases in which order matters that produce the combination we want.

To get four right and one wrong could look like this:

$$\frac{1}{4} \times \frac{1}{4} \times \frac{1}{4} \times \frac{1}{4} \times \frac{3}{4} = \frac{3}{1024}$$

This is the probability that I miss ONE question, and that the question I miss is the LAST question. But there are 5 different orders this could happen in, i.e. the "wrong question" (our $\frac{3}{4}$ in the string above) could also be 1st, 2nd, 3rd, or 4th:

Case 1: I get question 1 wrong:

$$\frac{3}{4} \times \frac{1}{4} \times \frac{1}{4} \times \frac{1}{4} \times \frac{1}{4} = \frac{3}{1024}$$

Case 2: I miss question 2:

$$\frac{1}{4} \times \frac{3}{4} \times \frac{1}{4} \times \frac{1}{4} \times \frac{1}{4} = \frac{3}{1024}$$

Case 3: I miss question 3:

$$\frac{1}{4} \times \frac{1}{4} \times \frac{3}{4} \times \frac{1}{4} \times \frac{1}{4} = \frac{3}{1024}$$

Case 4: I miss question 4:

$$\frac{1}{4} \times \frac{1}{4} \times \frac{1}{4} \times \frac{3}{4} \times \frac{1}{4} = \frac{3}{1024}$$

For each of these cases the fractions are the same as the case I initially wrote out, but just in a different order. Each of the five cases has a probability of $\frac{3}{1024}$.

This is an "or" situation because any of the above options will work and none will overlap. I thus need the sum of all these possibilities. I could add $\frac{3}{1024}$ to itself five times, or simply multiply $\frac{3}{1024}$ by five to get the answer:

$$5\left(\frac{3}{1024}\right) = \frac{15}{1024}$$

Answer: $\frac{15}{1024}$.

USING PERMUTATIONS IN PROBABILITY PROBLEMS

More complex probability problems synthesize your knowledge of counting problems and of probability. To solve these problems, remember probability is always found by finding:

$$\frac{\textit{Total number of outcomes that fulfill the desired parameters}}{\textit{Total number of possible outcomes}}$$

Oftentimes, we can use permutations (or the fundamental counting principle or even combinations) to solve for each of these two values, and then in turn solve for the probability. The ACT® rarely requires you to know how to do these with combinations, so I'll focus on permutations. Still, the same principle would work if the problem you confront involves combinations.

> In the bleachers of a football stadium, 2 boys and 3 girls are seated together in a random order. What is the probability that the 2 boys are seated next to each other?

First, we know we need to solve for two values:
1. Total number of ways to arrange 2 boys and 3 girls such that the 2 boys are always next to each other (numerator of our probability)
2. Total number of ways to arrange 5 kids (I could say 2 boys and 3 girls, but each person is actually unique, so this is really just how to arrange 5 kids; thinking this way makes the math easier).

Both of these elements can be solved using permutations and a bit of creativity.

Let's start with #1:

When I need to keep 2 items next to each other in a permutation, one way I can think of this is Case 1/Case 2. Let's name our boys Brian and Max, and our girls Leah, Wei Wei, and Ann.

Case 1: Brian is seated directly to the left of Max
Case 2: Brian is seated directly to the right of Max

Now I can solve for the number of permutations I have, pretending that Brian and Max are "glued" together and essentially are one person. I'll just calculate the number of ways to arrange for each case and add all the possibilities together.

Case 1: I can choose from the following four taken four at a time:
$$\text{Brian/Max, Leah, Wei Wei, Ann}$$
$$\underline{4} \times \underline{3} \times \underline{2} \times \underline{1} = 24$$

Case 2: I can choose from the following four taken four at a time:
$$\text{Max/Brian, Leah, Wei Wei, Ann}$$
$$\underline{4} \times \underline{3} \times \underline{2} \times \underline{1} = 24$$

I add these together and get 48 (I also could have seen that Case 1 & 2 will have the same number of options, and thus could have simply multiplied 24 by 2). In any case, I know my numerator: 48.

Now for Step #2:
How many ways can I arrange 5 kids, in order? 5 kids taken 5 at a time when order matters is simply:
$$\underline{5} \times \underline{4} \times \underline{3} \times \underline{2} \times \underline{1} = 120$$

120 is my denominator.

Now I simplify the fraction:

$$\frac{\textit{Total number of outcomes that fulfill the desired parameters}}{\textit{Total number of possible outcomes}}$$

$$= \frac{48}{120} = \frac{2}{5} \text{ or } .4$$

EXPECTED VALUES

Finding the expected value is like finding a weighted average. Here's an example.

The probability distribution of the discrete random variable Y is shown in the table below. What is the expected value of Y?

y	Probability $P(Y=y)$
0	0.15
1	0.26
2	0.29
3	0.11
4	0.19

First of all, don't be thrown by the language **"discrete random variable."** That just means that Y is not continuous as a possible value. For example, the number of people in an elevator is a discrete number because you can't have half a person; every number is a whole number. If I listed out probabilities of the number of people in the office elevator on any given trip, my values would all be discrete. Discrete variables don't have to be integers, but the point is that you don't have to worry about a bunch of values in between what is on your chart.

MISTAKE ALERT! Whenever I have problems with a probability chart like this, I always double check that the given probabilities add to one. If not, the chart is not a complete depiction of what is going on and I must account for that. Occasionally these problems will only give you the "first few" or "select" values of this variable and not all of them.

Here, I see that all my probabilities sum to 1. Thus I know to find my expected value, I simply multiply the value of y times its probability of occurring. Then I add all these little values up:

$$0(0.15)+1(0.26)+2(0.29)+3(0.11)+4(0.19)$$
$$=0+0.26+0.58+0.33+0.76$$
$$=1.93$$

The answer should essentially be the weighted average of the values in your chart. If your answer doesn't seem about "average," go back and check your work. Here this makes sense. 1 and 2 occur most often.

Answer: 1.93.

PROBABILITY NOTATION, UNION & INTERSECTION

(**FYI, this is <u>NOT frequently tested</u>.** This section is for overachievers with lots of time only!).

Occasionally, the ACT® may use certain notation to denote probability. <u>Typically, however, the ACT® will define this notation for you if you are expected to use it.</u> That means you shouldn't worry too much about remembering everything below; just be familiar with it.

We say that the probability of Event A occurring is $P(A)$. What that means is that if I write "$P(A)$" that represents the fraction or decimal probability that something happens. Similarly the probability of Event B occurring would be $P(B)$, of Event C, $P(C)$ and so on.

> Let $P(A)$ represent the probability of event A occurring. If event R occurs when three coins are flipped and all are heads, calculate $P(\text{not } R)$.

For this problem, we can first calculate the odds of getting all heads by multiplying the independent probability of each event. Since the probability of flipping heads is $\frac{1}{2}$, we multiply:

$$\frac{1}{2} \times \frac{1}{2} \times \frac{1}{2} = \frac{1}{8}$$

Because we need "not R" we now subtract the probability of getting R from 1:

$$1 - \frac{1}{8} = \frac{8}{8} - \frac{1}{8} = \frac{7}{8}$$

Answer: $\frac{7}{8}$.

"UNION"

You may occasionally see a U-like symbol in probability problems. When it is facing upwards, we call this "union." For example, the probability of A union B is denoted by:

$$P(A \cup B)$$

Union is the probability that **A happens, B happens, or both happen**. It is always greater than or equal to the probability of A alone or B alone. The picture below is one way to visualize A union B if there is some overlap.

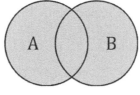

Unlike the "mutually exclusive" situations discussed earlier, union situations often involve overlap (though they need not involve overlap). For example, if event A is having brown hair and event B is having blue eyes, some people have both traits, and that would be "overlap" if we counted those who had A **OR** B. **When events have some overlap we call these "inclusive events."**

If Events A and B are inclusive, then the probability that A or B occurs is the sum of their probabilities minus the probability that both occur; i.e. to find the combined probability of inclusive events, we add the individual probabilities together and subtract the overlap:

$$P(A\text{ or }B) = P(A) + P(B) - P(A\text{ and }B)$$

To help understand the formula, draw a **Venn diagram** to visualize, as in the problem below.

> In in a class of 24 students, if 12 students have blue eyes, the probability of which is denoted by $P(A)$, 8 students have brown hair, the probability of which is denoted by $P(B)$, and 4 students have both, the probability of which is denoted by $P(A\text{ and }B)$, what is the probability that students have brown hair or blue eyes, denoted by $P(A\text{ or }B)$?

First, don't be thrown by all the notation. It's only there to confuse you. I know probability is the number of desired outcomes divided by the possible outcomes. I know I have 24 kids in the class, so that is my denominator. My numerator is the number of blue eyed and brown haired students *inclusive*, i.e. anyone who has either trait or both: those with brown hair and not blue eyes, those with both blue eyes and brown hair, and those with blue eyes and not brown hair. I can start by figuring out each of these cases using the Venn Diagram below. I subtract 4 (overlap) from 8 (number of students who have brown hair) to find the number who have brown hair but not blue eyes and subtract 4 from 12 to find blue-eyed kids without brown hair (8):

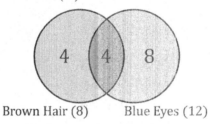

Brown Hair (8)   Blue Eyes (12)

I can now add each segment in my Venn Diagram to find the number of people who have either brown hair, blue eyes, or both ($4+4+8=16$).

I can also find this by taking $8+12-4=16$, per the formula we discussed earlier.

Now I place 16 over 24: $\dfrac{16}{24} = \dfrac{4}{6} = \dfrac{2}{3}$.

Answer: $\dfrac{2}{3}$.

 NOTE: For more on Venn Diagrams see the **Chapter 11: Word Problems** in Book 1.

"INTERSECTION"

Another notation you might see is an upside down "U." We call this an intersection. The probability of A intersection B is denoted by:

$$P(A \cap B)$$

An intersection occurs when **A AND B** are both simultaneously true. It is always less than or equal to the probability of A or B alone.

The picture below depicts what A intersection B looks like; it is the "overlap" of values in both sets:

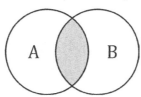

> Let A and B be independent events. Denote $P(A)$ as the probability that event A will occur, $P(A \cup B)$ as the probability that event A or B or both will occur. Which of the following equations *must* be true? (Note: $P(A \cup B) = P(A) + P(B) - P(A)P(B)$.)
>
> **A.** $P(A)P(B) = P(A \cup B)$
>
> **B.** $P(A) > P(A \cup B)$
>
> **C.** $P(B) - P(A) = P(A \cap B) - 2P(A) + P(A)P(B)$
>
> **D.** $P(A \cup B) > P(A) + P(B)$
>
> **E.** $P(A)P(A \cup B) = P(A)^2 - P(A)^2 P(B) + P(A)P(B)$

Let's go through each choice:

A. $P(A)P(B) = P(A \cup B)$

When we multiply two independent probabilities together, we find the chances that *both* occur. Thus this is a calculation of the intersection, NOT the union. A is incorrect.

B. $P(A) > P(A \cup B)$

We know that union essentially adds any items in set B to set A. Thus the probability of A alone cannot be greater than a probability that has at least as many if not more options that make it true. B is also incorrect.

C. $P(B) - P(A) = P(A \cap B) - 2P(A) + P(A)P(B)$

At first, this might look like an algebraic manipulation of the original, given equation. Except it includes the symbol for INTERSECTION not UNION. Be careful! All U's are not the same!

PROBABILITY   SKILLS

D. $P(A \cup B) > P(A) + P(B)$

Again the union is the set of all the items in A plus all the items in B, minus any overlap (if applicable). If there were *no* overlap, the two sides of this expression would be equal. Since that is possible, I know this is not something that MUST be true. In fact, it can't be true. If there *is* overlap, then $P(A) + P(B)$ will overestimate the value of the union (as it is the total before subtraction of the overlapping elements in the sets). Reversing the inequality sign and making it "or equal to" would make this expression true (i.e. $P(A \cup B) \leq P(A) + P(B)$).

E. Looking at the left of the equation, we see that we've simply multiplied the union by $P(A)$. Let's use substitution and expand this expression, using the information in the "Note":

$$P(A)P(A \cup B)$$

We now plug in $P(A) + P(B) - P(A)P(B)$ for $P(A \cup B)$ (this is given in the problem).

$$P(A)\big(P(A) + P(B) - P(A)P(B)\big)$$

Using the distributive property we get:

$$P(A)^2 + P(A)P(B) - P(A)^2 P(B)$$

Now we rearrange using the commutative property:

$$P(A)^2 - P(A)^2 P(B) + P(A)P(B)$$

Answer: **E.**

1. A box contains 10 blue marbles, 30 green marbles, and 24 orange marbles. How many orange marbles must be added so that there is a 60% chance of picking an orange marble at random?

   A. 36
   B. 40
   C. 44
   D. 48
   E. 52

2. A number is chosen from the set $\{1,2,3,4,5,\ldots,24\}$. What is the probability that the number is a factor of 18?

   A. $\dfrac{1}{2}$
   B. $\dfrac{1}{3}$
   C. $\dfrac{1}{4}$
   D. $\dfrac{1}{5}$
   E. $\dfrac{1}{6}$

3. There are 8 children, and you are assigned to line them up from youngest to oldest. You know which child is the oldest, which child is the second youngest, and which child is the youngest, but the other 5 children's ages are unknown. If you randomly sort the middle 5 children, what is the probability that all the children are ordered correctly?

   A. $\dfrac{1}{20}$
   B. $\dfrac{1}{120}$
   C. $\dfrac{1}{760}$
   D. $\dfrac{1}{5}$
   E. $\dfrac{1}{25}$

4. In a set of integers from 1 to 50, inclusive, what is the probability of randomly selecting a prime number?

   A. $\dfrac{1}{10}$
   B. $\dfrac{1}{5}$
   C. $\dfrac{2}{5}$
   D. $\dfrac{3}{10}$
   E. $\dfrac{1}{4}$

5. An integer from 10 through 999, inclusive, is to be chosen at random. What is the probability that the integer has 3 as 2 (not more) of its digits?

   A. $\dfrac{1}{3}$
   B. $\dfrac{1}{33}$
   C. $\dfrac{11}{330}$
   D. $\dfrac{2}{55}$
   E. $\dfrac{3}{110}$

6. There are 200 paper slips in a hat, each numbered from $\sqrt{1}, \sqrt{2}, \ldots, \sqrt{200}$ with no repeats. What is the probability that the number on a slip drawn at random is not irrational?

   A. $\dfrac{7}{100}$
   B. $\dfrac{6}{100}$
   C. $\dfrac{5}{100}$
   D. $\dfrac{93}{100}$
   E. $\dfrac{94}{100}$

7. O'Shea puts 7 blue marbles in a box. He now wants to add enough yellow marbles so that the probability of drawing a blue marble is $\frac{1}{11}$. How many yellow marbles does he need to add?

   A. 50
   B. 60
   C. 70
   D. 80
   E. 90

8. At a party with 100 guests, there is a raffle. Each guest is given a ticket with a number from 00 to 99. There are no repeated numbers. Each guest signs his/her name on the ticket and drops it in a basket. There is a second basket of tickets numbered identically. A guest wins the raffle if his/her ticket is picked from the first basket and a ticket with the same ones digit is picked from the second basket. For example, if a guest has a ticket number 14 and the ticket picked from the first basket is 14 and the ticket from the second basket is 94, the guest wins. If Bernie's ticket number 42 is drawn from the first basket, what is the probability that Bernie will *not* win the raffle?

   A. $\frac{1}{10}$
   B. $\frac{9}{10}$
   C. $\frac{1}{2}$
   D. $\frac{3}{4}$
   E. $\frac{2}{3}$

9. Robin has a fake coin that has a 40% chance of landing on tails and a 60% chance of landing on heads. In 4 coin tosses, what is the probability of getting exactly 3 tails?

   A. .064
   B. .936
   C. .216
   D. .784
   E. .153

10. At a start-up company with a staff of 15 people, 6 people are male and 9 people are female. Two people are randomly chosen to be campus representatives. What is the probability that both representatives are male?

    A. $\frac{6}{15} + \frac{6}{15}$
    B. $\frac{6}{15} \cdot \frac{6}{15}$
    C. $\frac{6}{15} \cdot \frac{5}{14}$
    D. $\frac{9}{15} \cdot \frac{8}{14}$
    E. $\frac{6}{15} \cdot \frac{5}{15}$

11. In a survey conducted at a university, students were asked to write down the number of campus organizations they are involved in. The results are shown below. What are the odds of a student at the university being involved in at least 3 organizations?

Distribution of Student Involvement in Campus Organizations					
# of organizations	0	1	2	3	>3
% of students	14	27	39	16	4

    A. 1:25
    B. 1:5
    C. 1:4
    D. 1:20
    E. 4:25

12. 25% of the dogs at a park are Corgis. There are 28 dogs at the park. How many dogs at the park are not Corgis?

    A. 7
    B. 14
    C. 20
    D. 21
    E. 24

QUESTIONS   PROBABILITY

13. In a Secret Santa gift exchange, there are 3 gift cards, 5 stuffed animals, and 2 articles of clothing. 5 more people want to join the exchange. How many of the 5 people should bring stuffed animals so that the overall probability of getting a stuffed animal gift is 40%?

    A. 1
    B. 2
    C. 3
    D. 4
    E. 5

14. Sam rolls a 6 sided die painted with 3 sides yellow, 2 sides red, and 1 side white. If Sam rolls the die and records the color of the side facing up repeatedly, how many times should Sam expect to record the color red after 180 rolls?

    A. 30
    B. 60
    C. 90
    D. 120
    E. 150

15. A teacher lines 30 students in a single file line and starts passing out candy at the front of the line. The teacher has 15 lollipops, 10 candy canes, and 5 gumdrops. Lisa is 6th in line to get the candy and the students in front of her have received 3 lollipops and 2 candy canes. What is the probability that Lisa will get a lollipop or a gumdrop?

    A. $\frac{12}{25}$
    B. $\frac{17}{30}$
    C. $\frac{2}{3}$
    D. $\frac{17}{25}$
    E. $\frac{4}{5}$

16. In a list of 60 songs, there are 13 songs by artist A, 24 songs by artist B, 13 songs by artist C, and 10 songs by artist D. The first song on the playlist is set to play on random. What is the probability that the first song played is by artist B?

    A. $\frac{1}{5}$
    B. $\frac{13}{60}$
    C. $\frac{2}{5}$
    D. $\frac{3}{5}$
    E. $\frac{4}{5}$

17. In box of 15 pebbles, 4 are white, 6 are black, and 5 are gray. If a blindfolded person is asked to pick one pebble out by random, what is the probability of the person picking a pebble that is not white?

    A. $\frac{11}{15}$
    B. $\frac{9}{15}$
    C. $\frac{6}{15}$
    D. $\frac{5}{15}$
    E. $\frac{4}{15}$

18. Two events are independent if the outcome of one event does not affect the outcome of the other event. One of the following statements does NOT describe independent events. Which one?

    A. An 8 is drawn from a deck of cards, then after replacing the card, an 8 is drawn.
    B. An ace card is pulled from a deck of cards, then, without replacing the card a coin lands tails up.
    C. A coin is flipped and lands heads up, then the same coin is flipped again and lands heads up.
    D. A 4 is drawn from a deck of cards, then after replacing the card, a 3 is drawn.
    E. A 4 is drawn from a deck of cards, then, without replacing the card, a king is drawn.

CHAPTER 10

PROBABILITY QUESTIONS

19. A new version of roulette is played where 2 pockets are green, 9 are red, 9 are black, 9 are blue, and 9 are yellow. If a ball is rolled into one of the pockets at random, what is the probability that it does NOT land in a blue pocket?

   A. $\dfrac{29}{38}$

   B. $\dfrac{27}{38}$

   C. $\dfrac{9}{28}$

   D. $\dfrac{3}{4}$

   E. $\dfrac{1}{4}$

20. A taxi service has 240 taxis in its service. Based on previous data, the company constructed the table below showing the percent of taxis in use and the probabilities of occurring. Based on the probability distribution in the table, to the nearest whole number, what is the expected number of taxis that will be in use any given day?

Taxi Rate Usage	Probability
0.4	0.3
0.6	0.4
0.7	0.2
0.9	0.1

   A. 59
   B. 60
   C. 142
   D. 144
   E. 156

21. For the first 7 possible values of x, the table below gives the probability, $P(x)$, that x inches of rain, to the nearest inch will fall in any given month.

x inches of rain	$P(x)$
0	0.3102
1	0.1020
2	0.1567
3	0.2021
4	0.1166
5	0.0621
6	0.0503

   Which of the following values is closest to the probability that at least 3 inches of rain will fall in any given month?

   A. 0.11
   B. 0.20
   C. 0.40
   D. 0.43
   E. 0.57

22. Let X and Y be independent events. $P(x)$ represents the probability that event X will occur, $P(\sim x)$ represents the probability that event x will not occur, and $P(x \cap y)$ represents the probability that both events X and Y will occur. Which of the following equations *must* be true?

   A. $P(x) = P(y)$
   B. $P(x \cap \sim y) = P(\sim x \cap y)$
   C. $P(x) - P(\sim x) = P(y) - P(\sim y)$
   D. $P(x \cap y) = P(\sim x \cap \sim y)$
   E. $P(x) \geq P(x \cap \sim y)$

23. A lab is testing a new machine to diagnose breast cancer. In 50 trials of 800 individuals, the number of false positives (instances when the machine diagnoses a woman with breast cancer who does not actually have it) were recorded. Based on the distribution below, what is the expected number of false positives that will occur among 50,000 tests?

Number, n, of false positives	Probability that n false positives are produced in a trial of 800 people
0	0.2
1	0.4
2	0.15
3	0.15
4	0.1

A. 1.55
B. 63
C. 97
D. 124
E. 500

24. The probability distribution of the discrete random variable Y is shown in the table below. What is closest to the expected value of Y?

y	$P(Y=y)$
0	$\frac{2}{9}$
1	$\frac{1}{18}$
2	$\frac{5}{18}$
3	$\frac{1}{6}$
4	$\frac{2}{9}$
5	$\frac{1}{18}$

A. 1
B. 2
C. 2.28
D. 3
E. $\frac{2}{9}$

25. The table below shows the results of a survey of 300 people who were asked whether they liked spicy food and whether they liked hiking.

	Like spicy food	Do not like spicy food	Total
Like to hike	75	115	185
Do not like to hike	40	75	115
Total	110	190	300

According to the results, which is closest to the probability that a randomly selected person who was surveyed doesn't like spicy food given that they don't like to hike?

A. 165%
B. 65%
C. 63%
D. 39%
E. 38%

26. The probability that a specific event, E, happens is denoted $P(E)$. The probability that this event does not happen is denoted $P(\text{not } E)$. Which of the following statements is *always* true?

A. $P(\text{not } E) > P(E)$
B. $P(\text{not } E) < P(E)$
C. $P(\text{not } E) = P(E) + 1$
D. $1 - P(E) = P(\text{not } E)$
E. $0 < P(\text{not } E) < P(E)$

27. Suppose that a will be randomly selected from the set $\{-3,-1,0,1,2\}$ and that b will be randomly selected from the set $\{-3,-2,0,1,2,3\}$. What is the probability that $ab < 0$?

A. $\frac{1}{3}$
B. $\frac{3}{20}$
C. $\frac{13}{30}$
D. $\frac{4}{15}$
E. $\frac{1}{15}$

28. Best friends Mylah and Sierra and three other classmates have been instructed to stand in a straight line in a randomly assigned order. What is the probability that Mylah and Sierra will stand next to each other?

   A. $\dfrac{2}{5}$

   B. $\dfrac{1}{5}$

   C. $\dfrac{1}{15}$

   D. $\dfrac{1}{30}$

   E. $\dfrac{1}{120}$

ANSWER KEY

1. A   2. C   3. B   4. D   5. E   6. A   7. C   8. B   9. E   10. C   11. C   12. D   13. A   14. B
15. D   16. C   17. A   18. E   19. A   20. C   21. D   22. E   23. C   24. C   25. B   26. D   27. A   28. A

ANSWER EXPLANATIONS

1. **A.** We have a total of $10+30+24=64$ marbles, 24 of them are orange marbles, and we are looking to find the number of additional orange marbles to add in order to have a 60% probability of picking an orange marble. Let x be the number of additional orange marbles needed. Then, we can say that the probability of picking an orange marble after the addition of the x marbles is $\frac{24+x}{64+x}$, because we'll be adding orange marbles to the orange marbles (numerator) and to the total number of marbles (denominator). We want this to be equal 60%, which equals 0.60 or 6/10, so we set up the equation $\frac{24+x}{64+x}=\frac{6}{10}$. Cross-multiplying and distributing gives us $240+10x=384+6x$. Then, subtracting $6x$ from both sides, we get $240+4x=384$. Subtracting 240 from both sides: $4x=144$. Finally, dividing each side by 4 we find $x=36$. So, we need to add 36 additional orange marbles for there to be a 60% chance of picking an orange marble.

2. **C.** First, we must find all factors of 18 (integers that 18 can be divided by). (See LCM/GCF Chapter for help with factoring using a factor rainbow). They are $1,2,3,6,9$, and 18. We see that 18 has 6 factors and all of these numbers are included in the set $\{1,2,3,4,....,24\}$. The set $\{1,2,3,4,....,24\}$ has 24 numbers, so the probabilty of choosing a factor of 18 from these 24 numbers is $\frac{6}{24}=\frac{1}{4}$.

3. **B.** In order to line the children up from youngest to oldest, we start with the youngest. Since we already know which children are the two youngest and their ages, we know which one child to place in the first spot, and which one child to place in the second spot. The same goes for the last spot since we know which child is the oldest. For the remaining spots $3-7$, we have 5 children left who need to be placed. For spot three there are 5 children who could randomly be placed there. Once one child is randomly placed in the third spot, there are 4 children left who could be placed in the fourth spot, and then 3 children for the fifth spot, 2 choices for the sixth spot, and one remaining child at the end who will take the 7th spot. So, the total possible ways of ordering the middle five children is calculated $1\times 1\times 5\times 4\times 3\times 2\times 1\times 1=120$. Since only one of these line-ups is the correct order, the probability that the children are ordered correctly is $\frac{1}{120}$.

4. **D.** We must first list all the prime numbers between 1 and 50. $2,3,5,7,11,13,17,19,23,29,31,37,41,43,47$. 15 out of the 50 integers from 1 to 50, inclusive, are primes. (Remember, 1 is not prime!) So, the probability of selecting a prime number is $\frac{15}{50}=\frac{3}{10}$. One way to determine if a number is prime is to divide that number by all integers less than or equal to its square root, and if the number cannot be divided by any of these integers other than 1, it is prime. For example, 13 is prime because the square root of 13 is approximately 3.60555, and 13 is not divisible by 3 or 2. See Ch 2 for more on prime numbers.

5. **E.** First, we determine that there are $999-10+1=990$ integers from 10 through 999 (we add 1 because the set is inclusive). (Alternatively, we can reason that we take away numbers $1-9$ from the 999 numbers included in $1-999$ and that leaves 990). Then we count the number of integers that have 3 as exactly two of their digits. If we let x represent any digit from $0-9$ <u>excluding</u> 3 then the numbers we want to count can be represented in the following forms: $x33$, $3x3$, and $33x$. Since there are 9 possible choices for x (0, 1, 2, 4, 5, 6, 7, 8, 9) while the other two digits in our numbers only have one possible choice respectively (3), the number of possible permutations could be calculated for each of the forms. $x33$ has $1\times 9\times 1=9$ possible outcomes, $3x3$ has $1\times 9\times 1=9$ possible outcomes, and $33x$ has $1\times 1\times 9=9$ possible outcomes. This gives us a total of $9+9+9=27$ numbers that satisfy our condition of having 3 as

PROBABILITY ANSWERS

2 of its digits. So, the probability of choosing such an integer out of a total of 990 numbers (calculated earlier) is $\frac{27}{990} = \frac{3}{110}$.

6. **A.** To be "not irrational" is the same as to be rational. Since there are 200 integers from 1 to 200, there are also 200 values from $\sqrt{1}, \sqrt{2}, \ldots, \sqrt{200}$. We must now find the number of values from $\sqrt{1}, \sqrt{2}, \ldots, \sqrt{200}$ that are rational. An rational number is a number that can be expressed as a fraction or in the form $\frac{n}{d}$ where n and d are integers. Numbers of the form $\sqrt{x}$ are rational only if x is a perfect square. So, since there are 14 numbers from 1 to 200 that are perfect squares $(1, 4, 9, 16, 25, 36, 49, 64, 81, 100, 121, 144, 169, \text{ and } 196)$, the probability of drawing one is $\frac{14}{200}$ or $\frac{7}{100}$.

7. **C.** The box initially only has 7 blue marbles. Let x be the number of yellow marbles we want to add in order for the probability of drawing a blue marble to be $\frac{1}{11}$. This means $\frac{7}{7+x} = \frac{1}{11}$. Cross-multiplying this equation, we get $77 = 7 + x$. Subtracting 7 from both sides, we get $70 = x$. Accordingly, we need to add 70 yellow marbles in order for the probability of drawing a blue marble to be equal to $\frac{1}{11}$.

8. **B.** Bernie will not win the raffle if the ticket drawn from the second basket does not end in the same ones digit as 42. So, if the second ticket does not end with a 2, Bernie will not win. The possible ticket numbers that will not allow Bernie to win have 10 possible numbers $(0-9)$ for the first digit and 9 possible numbers $(0-9$ excluding 2) for the second digit (ones place). This gives $10 \times 9 = 90$ possible outcomes out of a total of $99 - 0 + 1 = 100$ tickets to choose from for Bernie to lose. The probability of choosing such a ticket is $\frac{90}{100} = \frac{9}{10}$.

9. **E.** There are 4 orders for which Robin can get 3 tails out of 4 tosses: TTTH, TTHT, THTT, or HTTT. First we find the probability of each possible outcome. Each coin toss is an independent "AND" event, so the probability that she will get, for example, tails in the first toss *AND* tails in the second toss *AND* heads in the third toss *AND* tails in the fourth toss, is found by multiplying the probability of each independent coin toss. The probability of each of these outcomes is, respectively, $(0.4)(0.4)(0.4)(0.6)$, $(0.4)(0.4)(0.6)(0.4)$, $(0.4)(0.6)(0.4)(0.4)$, and $(0.6)(0.4)(0.4)(0.4)$. Now, the probability that she will get one of the four desired outcomes is an independent "OR" event, the probability that she will get TTTH *OR* TTHT *OR* THTT *OR* HTTT, so we must sum the individual probabilities we found before. Thus, the total probability of landing tails 3 out of 4 times is $(0.4)(0.4)(0.4)(0.6) + (0.4)(0.4)(0.6)(0.4) + (0.4)(0.6)(0.4)(0.4) + (0.6)(0.4)(0.4)(0.4)$, or $4 \cdot (0.4)^3 (0.6)$. This simplifies to $4 \times 0.0384 = 0.154$.

10. **C.** When calculating the probability of choosing two males, we are looking for the probability of selecting one male <u>and</u> selecting another male from the <u>remaining</u> staff. Thus these are dependent events. The probability of selecting the first male is $\frac{6}{15}$. Taking out the first selected male, the pool of candidates now consists of 5 males and 9 females, so the probability of selecting a second male from the staff not including the first male is $\frac{5}{14}$. We multiply the two probabilities to find the probability of both occurrences happening, so the probability that both representatives are male is $\frac{6}{15} \times \frac{5}{14}$. A common mistake made is multiplying $\frac{6}{15} \times \frac{6}{15}$. This is the probability of selecting two males with replacement, which means that it is possible to select the same person twice. Since the staff is selecting two different people, they are selecting without replacement.

11. **C.** Most students miss this question because they don't know what "odds" means. Odds are expressed as part to part (ratio) not as a fraction (part of the whole.) The odds of a student being involved in at least 3 organizations is the percentage of the student being in 3 or more clubs against the percentage of the student being in 0, 1, or 2. The percentage of a student being in 3 *or* more than 3 clubs is $16\% + 4\% = 20\%$ and the percentage of students in the remaining tallies are $14\% + 27\% + 39\% = 80\%$. So, the odds are $20:80$, which reduces to $1:4$. Note the word "odds" is rare on the ACT.

12. **D.** Out of the 28 dogs, 25% or $\dfrac{28}{4}=7$ are Corgis. The number of dogs at the park that are NOT Corgis is $28-7=21$.

13. **A.** Initially, there are a total of $3+5+2=10$ gifts, and 5 of these gifts are stuffed animals. With the addition of 5 more gifts, the total number of gifts in the exchange is now $10+5=15$. If we want 40% of those 15 gifts to be stuffed animals, then $0.4\times 15=6$ of the gifts must be stuffed animals. We already know that there are 5 stuffed animal gifts from the original 10 people, so we only need $6-5=1$ out of the 5 people joining to bring a stuffed animal gift.

14. **B.** To find the expected number of rolls that are red side up, we must first determine the probability of rolling a red. Since 2 out of the 6 sides are red, the probability of rolling a red is $\dfrac{2}{6}=\dfrac{1}{3}$. The expected number of rolls that are red side up is then the probability of rolling a red multiplied by the total number of rolls made. Out of the total 180 rolls, $\dfrac{1}{3}$ of them are expected to be red: $180\times\dfrac{1}{3}=60$ rolls.

15. **D.** This is an "OR" situation so we calculate the independent probabilities and add them together. The teacher started out with 15 lollipops, 10 candy canes, and 5 gumdrops. At the time the teacher reaches Lisa, there are $15-3=12$ lollipops left, $10-2=8$ candy canes left, and $5-0=5$ gumdrops remaining. The total number of candies by the time the teacher reaches Lisa is now $30-5=25$. So the probabilities of getting a lollipop, candy cane, and gumdrop are $\dfrac{12}{25}$, $\dfrac{8}{25}$, and $\dfrac{5}{25}$ respectively. The probability of getting a lollipop **or** a gumdrop is $\dfrac{12}{25}+\dfrac{5}{25}=\dfrac{17}{25}$.

16. **C.** There are 24 songs out of 60 that are by artist B, so the probability of the first song to be by artist B is $\dfrac{24}{60}=\dfrac{12}{30}=\dfrac{6}{15}=\dfrac{2}{5}$.

17. **A.** We wish to find the probability of the pebble being NOT white, which is $1-$ (probability of picking a white pebble). The probability of picking a white pebble is $\dfrac{4}{15}$, so the probability of picking a picking a pebble that is *not* white is $1-\dfrac{4}{15}=\dfrac{15}{15}-\dfrac{4}{15}=\dfrac{11}{15}$.

18. **E.** For each answer choice, the events described are independent except for answer choice E because if a 4 is drawn from a deck, and without replacement, a king is drawn, then the probability of drawing the king depends on whether or not the first card drawn from the deck was also a king. All other answer choices describe events that are independent because each draw made from a deck is replaced, making the next draw not dependent on the previous draw. Each coin toss also does not depend on the previous toss.

19. **A.** The probability that an event does not occur is equal to the sum of the probabilities of all other alternative possibilities added together. However, since we may come across a problem where there are too many alternate events to calculate in the time we have, it's better to calculate the probability that an event will not happen as 100%, or 1, minus the probability that the event will happen. In this case, that is $1-\dfrac{9}{38}$, since there are 9 chances for the ball to land in blue out of 38 possibilities. This gives us $\dfrac{38}{38}-\dfrac{9}{38}=\dfrac{29}{38}$.

20. **C.** The expected value is equal to the sum of all possible values, each multiplied by its probability. Our expected taxi usage rate is $0.4(0.3)+0.6(0.4)+0.7(0.2)+0.9(0.1)=0.59$. We expect the taxi service to be using 59% of its taxis at any given time. To find the expected number of taxis, not the rate of taxi use, we multiply the number of taxis, 240, by the rate of use: $240(59\%)=240(0.59)=141.6\approx 142$.

21. **D.** Be careful: we need the probability that *at least* 3 inches of rain will fall. But the chart ONLY lists the **first SEVEN possibilities**. Thus we don't know the probability for more than 6 inches of rain. Our best bet is NOT to try to find the sum of **all** probabilities of $x\geq 3$ in the chart, because the chart leaves off some values. If you tried this method, you would get choice C. Instead, find the probability that this WON'T happen and subtract from one. Doing so will account for the missing chart values. First, take the sum of the probabilities for less than 3 inches of rain. $0.3102+0.1020+0.1567=0.5689$ Now, subtract that value from one: $1-0.5689=0.4311$ The closest value is answer choice (D). Choice B is incorrect as it only is for three inches of rain, not all values equal to or greater than three inches.

PROBABILITY   ANSWERS

22. E. The likelihood of a given event happening is always greater than or equal to the probability of that event happening alongside a second event (and only equal when the probability of the second event is 100%!) Remember, an upside down U shape means intersection, or both events have occured. This is the only choice that MUST be true.

23. C. The expected number of false positives in a group of 800 people will be equal to the sum of each number n of false positives times the probability of each potential outcome in the sample. The expected number in the <u>sample</u> is $0(0.2)+1(0.4)+2(0.15)+3(0.15)+4(0.1) = 0+0.4+0.3+0.45+0.4 = 1.55$. However, we are looking for the number of false positives in a <u>population</u> of 50,000 people, so we can use a proportion to project how many people this would be given the rate of false positives in the sample. We set false positives in the sample over the total in the sample equal to false positive in the population (n) over the total population: $\frac{1.55}{800} = \frac{n}{50,000}$. Cross multiplying, we find $1.55(50,000) = 800n$ or $\frac{1.55(50,000)}{800} = n = 96.875$ which is ≈ 97 false positives.

24. C. The expected value of a variable is the sum of all of its possible values multiplied by their respective probabilities. The expected value of Y equals $0\left(\frac{2}{9}\right)+1\left(\frac{1}{18}\right)+2\left(\frac{5}{18}\right)+3\left(\frac{1}{6}\right)+4\left(\frac{2}{9}\right)+5\left(\frac{1}{18}\right) = \frac{0}{9}+\frac{1}{18}+\frac{10}{18}+\frac{3}{6}+\frac{8}{9}+\frac{5}{18} \approx 2.28$. Just because our inputs are discrete random variables (i.e. integers) doesn't mean round our expected value to an integer. The expected value is a weighted mean or average so should not be rounded to the nearest integer unless specified.

25. B. Since we are given that the person doesn't like to hike, we can get rid of all of the individuals who do like to hike, leaving us with a population of 115. This is our denominator. In this group, 75 people don't like spicy food, so our probability is $\frac{75}{115} \approx 0.65 = 65\%$.

26. D. The probability of an event occurring and the probability of it not occurring must always equal 1, since those are the only two possible outcomes. Thus we know that $1 = P(E)+P(\text{not } E)$, which makes $1-P(E) = P(\text{not } E)$ also true.

27. A. The probability that $ab<0$ is equal to the probability that $a<0$ and $b>0$ plus the probability that $a>0$ and $b<0$. There are 5 possibilities for a, of which 2 are negative and 2 are positive. There are 6 possibilities for b, of which 2 are negative and 3 are positive. The probability that $a<0$ and $b>0 = \frac{2}{5}\times\frac{3}{6} = \frac{6}{30} = \frac{1}{5}$. The probability that $a>0$ and $b<0 = \frac{2}{5}\times\frac{2}{6} = \frac{4}{30} = \frac{2}{15}$. Now we add these two probabilities together: $\frac{1}{5}+\frac{2}{15} = \frac{3}{15}+\frac{2}{15} = \frac{5}{15} = \frac{1}{3}$.

28. A. Here we'll use the principles of arrangements. Remember probability is always:

$$\frac{\text{the number of desired outcomes}}{\text{the number of possible outcomes}}$$

In this line, order matters. Three other classmates means FIVE total. We'll have to manually figure out the numerator but we can use permutations to figure out the denominator. First calculate the number of possible straight line arrangements: 5 people taken 5 at a time ($_5P_5$) or: $5! = 5\times4\times3\times2\times1 = 120$. This is our denominator. Now we need to count the number of desired outcomes for the numerator. Below, 0 denotes other kids and M and S denote Mylah and Sierra:

If Mylah is first and Sierra after her, we have four options: *If Sierra is first and Mylah is second, we have four more:*
     MS000   0MS00   00MS0   000MS   SM000   0SM00   00SM0   000SM

We don't care what position the other kids are in—but we do have to account for them. If there are three other positions to fill, we have $3\times2\times1$ options for arranging those three spots—or in other words each of the above "codes" actually stands for SIX different possibilities. So we need to take the 8 arrangements and multiply each of them by 6—because regardless of where the three open slots are—these three slots represent 6 different orientations. That gives us:

$$\frac{8\times6}{5\times4\times3\times2\times1}$$

We can cancel the 6, and then reduce by dividing out 4:

$$\frac{8\times\cancel{6}}{5\times4\times\cancel{3}\times\cancel{2}\times1} = \frac{8}{20} = \frac{2}{5}$$

This problem is significantly more difficult than the majority of probability problems you'll find on the ACT®. If you can do this, you are set!

PART THREE: GEOMETRY

For our formula cheat sheet covering many formulas in this section go to supertutortv.com/BookOwners

CHAPTER 11

ANGLES AND LINES

> ## SKILLS TO KNOW
> - Straight lines are 180°/Circles of angles sum to 360°
> - Supplementary angles sum to 180° and complementary angles sum to 90° (a right angle)
> - Vertical angles are congruent
> - Parallel lines theorem: the alternate exterior and alternate interior angles of parallel lines intersecting with a transversal are congruent.
> - Triangles: angles sum to 180°, exterior angle theorem, isosceles & equilateral triangles*
> - Quadrilaterals and angles*
> - Angle bisectors
> - Angle "hopping": synthesizing all the rules together.

NOTE: **Angles and Lines** skills also intersect with skills in our chapters on **Triangles, Polygons, and Similar Shapes & Ratios.** Also see these chapters for related skills and problems.

STRAIGHT LINES AND CIRCLES

Straight lines represent 180°. Anytime you see two or more angles popping out to one side of a straight line, these angles must sum to 180°.

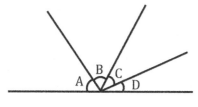
$A+B+C+D=180°$

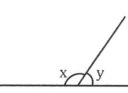
$x+y=180°$

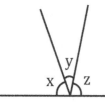
$x+y+z=180°$

The sum of all the angles formed by lines that all converge at a single point (like spokes of a bicycle, or a "circle" of angles) is always 360°. Similarly, the interior angles of a circle always sum to 360°.

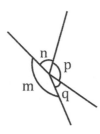
$m+n+p+q=360°$

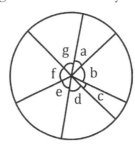
$a+b+c+d+e+f+g=360°$

ANGLES/LINES — SKILLS

SUPPLEMENTARY & COMPLEMENTARY ANGLES

Supplementary angles are any two (and only two) angles that add up to $180°$. An example could look like this:

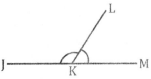

Complementary angles are any two (and only two) angles that add up to $90°$. Below left, angle ABC is complementary to angle CBD. Below right, $\angle a$ is complementary to $\angle b$.

The biggest mistake students make on problems with this vocabulary is confusing these words.

To keep these terms straight, remember: "Supplementary" and "Straight" both start with an "s:" **Supplementary** angles sum to a **straight** angle of 180 degrees. Alternatively, imagine how the "s" in supplementary looks like the **8** in $180°$, and the "c" in complementary looks a little like a backwards "**9**" in "$90°$."

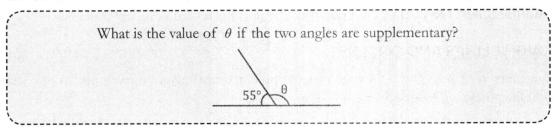

Because we know that the two angles are supplementary, they must add to $180°$. As a result, we can write out the equation $55° + \theta° = 180°$. Solving for $\theta°$, we get $\theta° = 125°$.

Answer: $125°$.

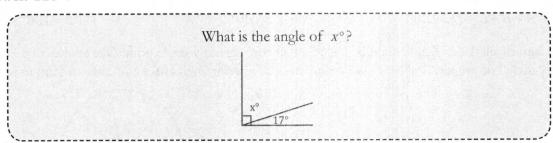

We know that the two angles are complementary and add up to $90°$, so we can write $x° + 17° = 90°$. Solving for $x°$, we get $73°$.

Answer: $73°$.

VERTICAL ANGLES

Here's a diagram of "vertical" angles:

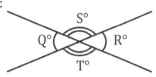

Q and R are vertical angles, as are S and T.

The vertical angles theorem states that vertical angles are congruent; above, angle Q and R equal each other, and angle S and T equal each other. We can see how this relationship holds if we think about the straight angles. We know $Q + S = 180$ because these two angles share a straight line. We also know $S + R = 180$, because these two angles share a straight line. It follows that $Q = 180 - S$ and $R = 180 - S$, and thus $Q = R$. You can see algebraically why this is true.

> In the figure below, line x is perpendicular to line y. Another line intersects the two previous lines at the exact same point. The measure of $\angle 2$ is 57 degrees. What is the difference between $\angle 3$ and $\angle 1$?
>
>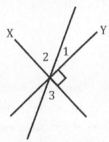

We know that $\angle 2$ and $\angle 1$ must add up to $90°$. This means that if $\angle 2$ is $57°$, $\angle 1$ must be $33°$. Furthermore, $\angle 3$ is the same as $\angle 2$ since they are vertical angles. We thus subtract: $\angle 3 - \angle 1 = 57° - 33° = 24°$.

Answer: $24°$.

PARALLEL LINES THEOREM

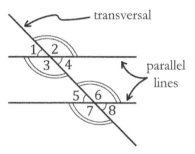

$\angle 1 = \angle 4 = \angle 5 = \angle 8$ (small angles)

$\angle 2 = \angle 3 = \angle 6 = \angle 7$ (big angles)

Any small angle plus any big angle = $180°$

A **transversal** is a line that cuts through any two parallel lines, as shown above. When it does so (and is not perpendicular to the parallel lines), it creates **four equal small angles and four equal big angles**. In other words, **the big angles are all congruent (2, 3, 6, 7)**, and **the small angles are all congruent (1, 4, 5, 8)**. Additionally, **any small angle in this picture added to any big angle will equal 180** degrees. These principles comprise the **parallel lines theorem**.

ANGLES/LINES   SKILLS

Formally, the **parallel lines theorem** says that **alternate exterior angles are equal, alternate interior angles are equal, corresponding angles are equal, corresponding exterior angles are supplementary, and interior angles are supplementary.** Reading that makes my head spin, but you don't really need to worry about all those vocabulary words. As long as you understand the principles on the previous page, you'll be fine.

Lines A and B are parallel in the figure below. A transversal intersects the two lines and three angle measures are given in degrees. What is the value of $x - y$?

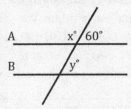

$x°$ is supplementary to 60 degrees so it must be 120 degrees. Furthermore, $y°$ is the same as 60 degrees due to the parallel lines theorem. Therefore, $x - y = 120 - 60 = 60$.

Answer: $60°$.

In the diagram below, lines a and b are parallel with each other and are perpendicular to the horizontal line. Lines d and c are also parallel. Line f is perpendicular to line b. The measure of $\angle \beta$ is $48°$. What is the measure of $\angle \theta$?

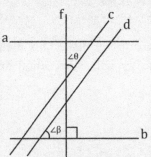

Let's angle hop! First, θ is equal to its **vertical angle**, so we can label that with a θ. Next, we know d and c are parallel, so we know the corresponding angle next to line d is also equal to θ because of the **parallel lines theorem**. Now we have θ and β in the same triangle together.

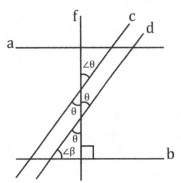

158   CHAPTER 11

Next, we know the other angle in the triangle with θ and β must be 90 degrees, because the problem states line b is perpendicular to line f. A triangle's angles always sum to $180°$ so now we can use algebra to find $\angle 1$:

$$90 + 48 + \theta = 180$$
$$138 + \theta = 180$$
$$\theta = 42$$

TRIANGLE ANGLES

TRIANGLE SUM THEOREM

The sum of the three angles in any triangle must equal $180°$.

Triangle ABC has angles 37 degrees and 61 degrees. What is the third angle in the triangle?

To solve this problem, we can let x equal our unknown angle. We know the sum of all three angles, 37, 61, and x, is 180 so we set up an equation and then simplify:

$$37 + 61 + x = 180$$
$$98 + x = 180$$
$$x = 82$$

EXTERIOR ANGLE THEOREM

The exterior angle of a triangle is equal to the sum of the remote interior angles. $a + b = c$, where c is the exterior angle and a and b are remote interior angles.

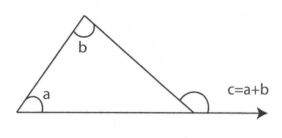

You don't need the **exterior angle theorem**, as any problem it works for can also be solved by "angle hopping" using other rules in this chapter, but it can speed up some geometry calculations. We can prove why it is true. Let's call the unmarked angle in the triangle above x. We know the sum of $a + b + x = 180$, and we also know the sum of $c + x = 180$. Because both equations equal 180, we can set them equal to each other. $a + b + x = c + x$. With algebra, our x's cancel and we get $a + b = c$.

Isosceles triangles are covered in Chapter 12 on Triangles. The **hinge theorem** is covered in Similar Shapes, Chapter 15, page 243. Skim ahead if you need to review.

POLYGON ANGLES

Polygon Angles are discussed further in the **Polygons** chapter. For now: **quadrilaterals** have interior angles that sum to 360. In **parallelograms**, opposite angles are congruent, and interior angles are supplementary.

ANGLES/LINES SKILLS

$A + B = 180$. $A + B + C + D = 360$.

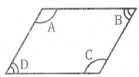

 Also, because **parallelograms** are essentially **pairs of parallel lines**, each line is a transversal of sorts, so we can also imagine a parallelogram in the same way we do parallel lines. **Other polygons may also contain parallel lines.** Remember that any transversals through these shapes (or even sides that intersect two parallel sides) can also be analyzed using rules for parallel lines as described earlier.

TIP: When in doubt, **extend the lines** of complex shapes or drawings with parallel lines so you can better see the transversal and the relationship between the angles. In the **trapezoid** below, angles 1 and 2 are supplementary, and angles 3 and 4 are supplementary. Extending the lines as shown helps us see that relationship.

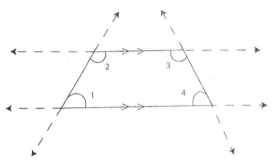

Parallelogram *ABCD* is shown below. If *G* bisects *BD* and *F* bisects *AC*, $\angle BGF = 3x - 19$, $\angle GBA = 6x + 10$, and $\angle AFG = 2y + 14$, what is the value of *y*?

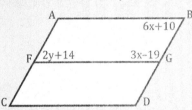

Bisectors in a parallelogram create a line that is also parallel to two lines in the shape. We will discuss why in the chapter on **Similar Shapes**. For now, know segment *FG* is parallel to *AB* and *CD*. Angles *B* and *BGF* are supplementary because of the parallel lines theorem. Thus $6x + 10 + 3x - 19 = 180$. Simplifying, we get $9x - 9 = 180$, or $9x = 189$. So $x = 21$. We also know that angle *ABG* must equal angle *AFG* because opposite angles in a parallelogram are congruent, and side *FG* forms another parallelogram with side *AB*. Thus $6(21) + 10 = 2y + 14$. Simplifying:

$$126 + 10 = 2y + 14$$
$$122 = 2y$$
$$y = 61$$

Answer: **61**.

ANGLE BISECTORS

Lines or rays are said to "bisect" angles when they cut that angle in half.

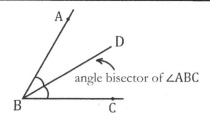

Ray $\overrightarrow{BD}$ bisects $\angle ABC$, which measures $(5x+2)°$. The measure of $\angle ABD$ is $(4x-14)°$. What is the measurement of $\angle DBC$?

We can draw a picture to visualize this problem:

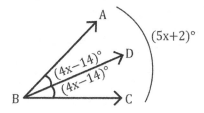

Since ray $\overrightarrow{BD}$ bisects the larger angle, the two smaller angles are equal. We can represent this idea with an equation and then solve for x.

$$2(4x-14) = 5x+2$$
$$8x-28 = 5x+2$$
$$x = 10$$

Next, we plug in our x value into the angle expression for $\angle DBC$.

$$4(10)-14 = 26$$

Answer: 26.

In the figure below, M lies on $\overline{LN}$, $\overline{OM}$ bisects $\angle LMP$, and $\overline{MP}$ bisects $\angle NMO$. What is the measure of $\angle NMO$?

Line $\overline{LN}$ has an angle measure of $180°$. Since $\overline{OM}$ bisects $\angle LMP$ we can label $\angle LMO$ as x and $\angle OMP$ as x as well. Now, when we look at how $\overline{MP}$ bisects $\angle NMO$, we know that $\angle PMN$ must also equal x.

We can now use algebra to solve for x, as we know $3x = 180$ since the three x's are along a straight angle; dividing 180 by 3, we get $60°$ as the value of x. However, x is not what we need. We actually need angle $\angle NMO$, which equals $2x$, or $120°$.

CHAPTER 11

ANGLES/LINES SKILLS

Answer: 120°.

ANGLE HOPPING

Angles and Lines problems often will require you to think creatively and use several of these rules at once. When in doubt:

1. Go through your list of ways you know how angles relate and check to see if any of those conditions apply (vertical angles, straight lines, circle central angles, right angles, parallel lines, triangles, exterior angle theorem, polygons, bisectors, similar shapes, etc.).
2. Draw additional lines to create transversals, additional parallel lines, triangles, or polygons.
3. Extend lines that are parallel to better visualize them.
4. Redraw small portions of the diagram if too many lines exist and you're having trouble seeing the relationship between angles.
5. Whenever your angle problem involves variables, always re-read the question before putting an answer. Often you'll need to plug in the variable to find another value.

Lines m and n are parallel. Given the angles in the diagram below, find the measure of $\angle y$.

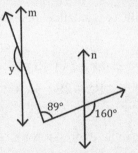

To solve this problem, we'll need to hop from line m to line n. The problem is we don't actually have a complete transversal here. To solve, we thus need to draw more lines, or at least one more line. One way we could do this would be to draw a parallel line through the point that is the vertex of the 89° angle as so:

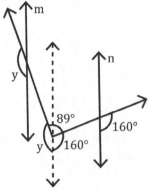

When we do this, we see we get a "spindle" of three angles, one of which is 160°, one of which is y degrees (both because of parallel lines and transversals), and the other is 89°. Because these three angles must sum to 360°, we know:

$$89 + 160 + y = 360$$
$$y = 200 - 89$$
$$y = 111$$

Answer: 111°.

1. In the figure below, *ABCD* is a parallelogram, ∠*DAE* is 64°, ∠*DCB* is 94° and $\overline{BE}$ bisects ∠*ABC*. What is ∠*AEB*?

   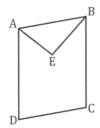

   A. 103°
   B. 105°
   C. 107°
   D. 109°
   E. Cannot be determined from given information

2. In the figure below, *X* is on $\overline{WZ}$ and the angle measures are as given. What is the value of 3*m*?

   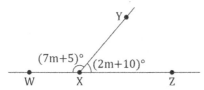

   A. 20°
   B. 55°
   C. 45°
   D. 18.33°
   E. 60°

3. In the figure below, lines $\overline{AD}$ and $\overline{BC}$ are parallel, lines $\overline{EB}$ and $\overline{DC}$ are parallel, ∠*ABE* is 53° and ∠*EAB* is a right angle. What is the measure of ∠*EDC*?

   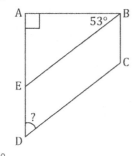

   A. 26.5°
   B. 37°
   C. 53°
   D. 127°
   E. 143°

4. In the figure below, the two vertical lines are parallel. What is the value of *x*?

   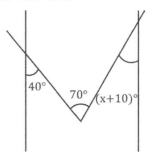

   A. 20
   B. 30
   C. 50
   D. 60
   E. Cannot be determined from the information provided

5. Which of the following statements regarding the figure below is false?

   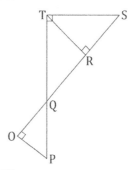

   A. $\overline{OP} \parallel \overline{TR}$
   B. $\overline{RS} \perp \overline{OP}$
   C. $\overline{TR} = \overline{TS}$
   D. ∠*OPQ* = ∠*QTR*
   E. *QR* < *QT*

6. If ∠*ABE* > ∠*EDC*, which of the following statements is always true?

   A. $\overline{AB} \parallel \overline{CD}$
   B. ∠*C* < ∠*A*
   C. ∠*A* < ∠*C*
   D. *AE* > *CE*
   E. *BE* > *ED*

CHAPTER 11   163

7. In the figure below, parallel lines $l1$ and $l2$ intersect parallel lines n, e, and s. If it can be determined, what is the sum of the degree measures of $\angle 1$ and $\angle 2$?

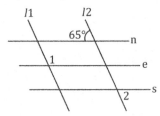

A. 65°
B. 90°
C. 180°
D. 115°
E. It cannot be determined from the information provided

8. In the figure below, $\overrightarrow{BH}$ is parallel to $\overrightarrow{CG}$, I lies on $\overrightarrow{AD}$ and $\overrightarrow{BH}$, E lies on $\overrightarrow{CG}$, the measure of $\angle AIB$ is 120°, and $\angle DIE \cong \angle EIF \cong \angle HIF$. What is the measure of $\angle IED$?

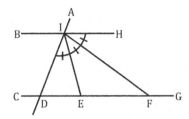

A. 20°
B. 40°
C. 50°
D. 80°
E. 100°

9. In the figure below, lines 1 and 2 are parallel. In terms of y, which of the following is equivalent to the degree measure of $\angle x$?

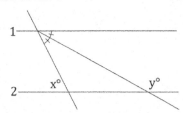

A. $2(180° - y)$
B. $180° - y$
C. $\dfrac{180° - y}{2}$
D. $180° - 2y$
E. $y + 15°$

10. In the figure below, $\overline{AB}$ and $\overline{DE}$ are perpendicular to $\overline{BD}$. Which of the following statements must be true? Note: figure not drawn to scale.

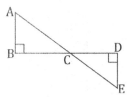

I. $\overline{AB} \cong \overline{DE}$
II. $\angle ACB \cong \angle DCE$
III. $\triangle ABC$ and $\triangle EDC$ are similar triangles

A. I only
B. II only
C. III only
D. I and II
E. II and III

11. In the figure below, $\overline{BD}$ is the longer diagonal of rhombus $ABCD$ and E is on $\overrightarrow{AD}$. The measure of $\angle ABD$ is 30°. What is the measure of $\angle CDE$? Figure not drawn to scale.

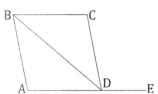

A. 60°
B. 90°
C. 105°
D. 120°
E. 160°

12. A line contains the points A, B, C, D, and E. D is to the right of A and to the left of E. B is to the right of C and to the left of D. C is to the right of A. Which of the following inequalities *must* be true about the length of these segments?

A. $AC < BD$
B. $AB > DB$
C. $AE > BE$
D. $CA > AD$
E. $CE < BD$

13. The following diagram depicts the path of a beam of light inside a modernist dance studio, where 3 of the 4 walls are mirrors. The angle at which light strikes a mirror is equal in measure to the angle at which it is reflected. What is the measure of the indicated angle?

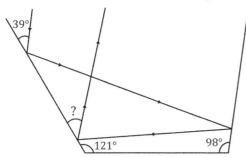

   A. 32°
   B. 43°
   C. 50°
   D. 55°
   E. 63°

14. Suppose the measure of the smaller of 2 supplementary angles is a seventh of the measure of the larger angle. What is the measure of the smaller angler?

   A. 19°
   B. 22.5°
   C. 102.5°
   D. 157.5°
   E. 167°

15. In the figure below $\overline{AE}$ and $\overline{BD}$ intersect at C and the measure of $\angle A$ is four times that of $\angle B$ The measure of $\angle D$ is 75° and the measure of $\angle E$ is 20°. What is the measure of $\angle A$?

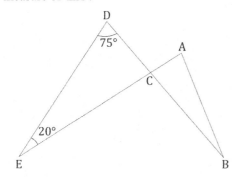

   A. 19°
   B. 72°
   C. 76°
   D. 85°
   E. 95°

16. In the figure below $\overline{AC}$ and $\overline{BE}$ intersect at D. What is the measure of $\angle EAD$?

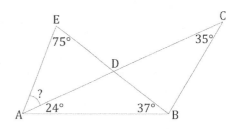

   A. 24°
   B. 35°
   C. 44°
   D. 56°
   E. 75°

17. In the figure below A, C, and E are collinear. $\triangle ABC$ and $\triangle ADC$ are as pictured below, and the angle measures are marked. What is the value of m?

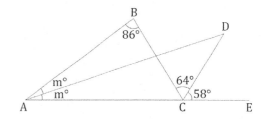

   A. 18°
   B. 23°
   C. 29°
   D. 32°
   E. 36°

18. In the figure below points $A, B, C,$ and D are collinear, points $E, B,$ and F are collinear, and points $F, C,$ and G are collinear. Angle measures are marked below. What is the measure of $\angle AEB$?

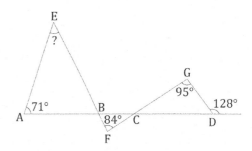

   A. 25°
   B. 46°
   C. 52°
   D. 84°
   E. 85°

19. In △ACD points A, B, and C are collinear. $\overline{AD}$ is perpendicular to $\overline{AC}$, $\overline{DB}$ bisects ∠ADC, and the measure of ∠ACD is 38°. What is the *sum* of ∠ADB and ∠DBC?

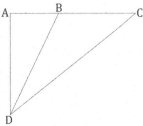

A. 26°
B. 92°
C. 116°
D. 123°
E. 142°

20. Distinct lines *l* and *m* intersect, forming 4 pairs of adjacent angles. Which of the following statement is *not* true about these 4 pairs of angles?

A. The sum of all 4 angles is equal to 360°.
B. The sum of the angle measures in each pair is 180°.
C. The difference of the angle measures in each pair is less than 180°.
D. The sum of one pair of angles is greater than sum of the other remaining pair.
E. The sum of one pair of angles is equal to 360° minus the sum of the other remaining pair.

21. The expression $180 - x$ is the degree measure of a non-zero obtuse angle if and only if?

A. $0 > x > -90$
B. $90 > x > 0$
C. $180 > x > 90$
D. $x < -90$
E. $x > 0$

22. The non-common rays of 2 adjacent angles form a straight angle. The measure of one angle is one fifth the measure of the other angle. What is the measure of the smaller angle?

A. 10°
B. 15°
C. 30°
D. 80°
E. 150°

23. In the diagram below △ABC, is isosceles. The measure of ∠ABC is 7 times the measure of ∠BAC. What is the measure of ∠ABC − ∠EAB?

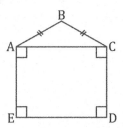

A. 10°
B. 25°
C. 30°
D. 110°
E. 140°

24. In the figure below, lines *l* and *m* are parallel and perpendicular to line $\overline{AB}$. Point C is in line with point E and line *l*. If the measure of ∠ABC is 49°, the measure of ∠BCD is 53°, and the measure of ∠EDC is 108°, what is the measure of the angle that $\overline{ED}$ makes with line *l*?

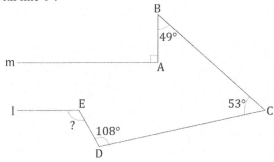

A. 12°
B. 53°
C. 108°
D. 120°
E. 161°

25. In △HIJ, ∠J is a right angle and the measure of ∠I is 63°. Another triangle, △QRS is being constructed such that ∠R is a right angle and the measure of ∠S is one third the measure of ∠I. What is the measure of ∠Q?

A. 21°
B. 22.5°
C. 45°
D. 69°
E. 159°

26. Annie is relaxing in an armchair with the chair back reclined 137°. Looking straight up, she realizes that there is lamp directly above her head, interfering with her rest, so she decides to move to a different armchair. Below is a sketch of the chair and the lamp. At what angle is the lamp with the back of the armchair?

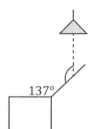

A. 47°
B. 92°
C. 133°
D. 137°
E. 147°

27. In the figure below, B is on $\overline{AC}$ and $\overline{AC} \parallel \overline{DE}$. Which of the following angle congruences holds true?

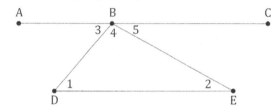

A. $\angle 2 \cong \angle 3$
B. $\angle 2 \cong \angle 4$
C. $\angle 1 \cong \angle 4$
D. $\angle 1 \cong \angle 5$
E. $\angle 1 \cong \angle 3$

28. Given the triangle shown below with exterior angles that measure $a°, b°,$ and $c°$ as shown. What is $a° - b° - c°$, in terms x of and y?

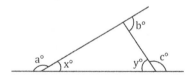

A. $180 - x$
B. $-x - y$
C. $-2x - y$
D. $-2x$
E. $-2x + y$

29. In the figure below X and Z are on line l and Y is on line m. The exterior angle to $\angle a$ at X is 101°. Which of the following statements gives sufficient additional information to find the measure of $\angle c$?

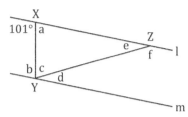

A. Line l is not parallel to line m.
B. Line l is parallel to line m.
C. The measure of $\angle b$ is equal to 79°.
D. The measure of $\angle f$ is 130°.
E. The measure of $\angle d$ is 34°.

30. In the figure below, 3 parallel lines are crossed by two transversals. The points of intersection and some distances, in inches, are labeled. What is the length, in inches, of $\overline{CF}$?

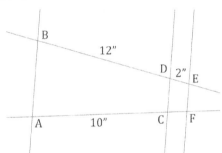

A. $\dfrac{20}{13}$
B. $\dfrac{10}{7}$
C. $\dfrac{5}{4}$
D. $\dfrac{5}{3}$
E. $\dfrac{5}{2}$

31. Lines a, b, c, and d are shown below and c‖d. Which of the following is the set of angles that must be supplementary to ∠x ?

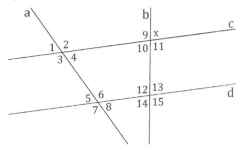

A. {1,4,9,11}
B. {9,11}
C. {1,4,5,8,9,11}
D. {9,11,12,15}
E. {1,4,5,8,9,11,12,15}

32. In the figure below, lines j and k are parallel, and lines l and m are parallel. Given the measure of two angles as shown below, what is the measure of ∠x ?

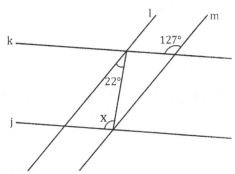

A. 58°
B. 63°
C. 68°
D. 85°
E. 105°

33. In the figure below, $\overline{BG}$, $\overline{CF}$, and $\overline{DE}$ all intersect at point A. The measure of 2 angles are given. What is the measure of ∠DAF ?

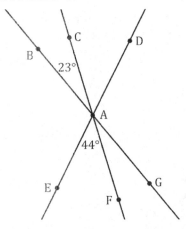

A. 23°
B. 46°
C. 113°
D. 157°
E. 136°

34. In the figure below, lines l and m are parallel. Lines a and b intersect l at the same point. What is the value of x ?

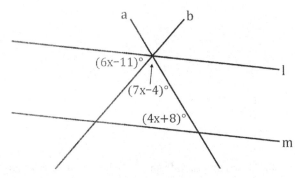

A. 7
B. 11
C. 52
D. 55
E. 73

35. Each of the 3 lines crosses the other 2 lines, as shown below. Which of the following relationships, involving angle measures (in degrees), must be true?

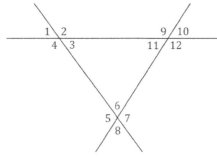

   I.   ∠1 + ∠10 + ∠6 = 180°
   II. ∠3 + ∠11 + ∠6 = 180°
   III. ∠3 + ∠11 + ∠8 = 180°

   A. I only
   B. II only
   C. III only
   D. I and II only
   E. I, II, and III

36. In isosceles trapezoid $ABDC$, $\overline{AB}$ is parallel to $\overline{CD}$. ∠ACB measures 40°, and ∠ADC measures 37°. What is the measure of ∠CAD?

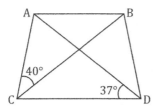

   A. 50°
   B. 66°
   C. 77°
   D. 88°
   E. 114°

37. In the figure below, the measure of ∠BCA is greater than the measure of ∠DCA and $\overline{BC} = \overline{DC}$. Which of the following statements must be true?

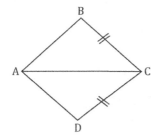

   A. $\overline{BC} < \overline{CD}$
   B. $\overline{AB} > \overline{BC}$
   C. $\overline{AB} > \overline{AD}$
   D. $\overline{AD} > \overline{CD}$
   E. $\overline{AB} = \overline{AD}$

38. In any parallelogram $ABCD$, it is always true that the measures of ∠BCD and ∠CDA:

   A. Are congruent.
   B. Are each 90°.
   C. Are each less than 90°.
   D. Add up to 90°.
   E. Add up to 180°.

ANGLES/LINES ANSWERS

ANSWER KEY

1. C   2. B   3. B   4. A   5. C   6. C   7. C   8. D   9. A   10. E   11. D   12. C   13. E   14. B
15. C   16. C   17. A   18. B   19. E   20. D   21. B   22. C   23. C   24. D   25. D   26. C   27. E   28. D
29. D   30. D   31. D   32. E   33. E   34. B   35. E   36. B   37. C   38. E

ANSWER EXPLANATIONS

1. **C.** Since $ABCD$ is a parallelogram, $\angle DAB$ is congruent to $\angle DCB$, so $\angle DAB = 94°$. We can subtract $\angle DAE (64°)$ from $\angle DAB (94°)$ to find the measure of $\angle EAB$, which we find to be $30°$. Since $\overline{AB} \parallel \overline{DC}$, $\angle ABC$ is supplementary to $\angle DCB$, thus $\angle ABC$ is $86°$. Since $\overline{BE}$ bisects $\angle ABC$, $\angle ABE = 43°$. Since all interior angles of a triangle sum to $180°$, $\angle EAB + \angle ABE + \angle AEB = 180°$. By substitution, $30° + 43° + \angle AEB = 180°$. Simplify to find $\angle AEB = 107°$.

2. **B.** The 2 angles shown are supplementary, since they form a straight line. Thus, $(7m+5) + (2m+10) = 180$. Simplification yields $9m + 15 = 180$. From there, solve for m to find $m = 18.333$, and $3m = 55$.

3. **B.** By parallel lines theorem, $\angle EDC \cong \angle AEB$. Because a triangle's angles must add up to $180°$, we know that $\angle AEB$ measures $180° - \angle EAB - \angle ABE$. By substitution, this becomes $180° - 90° - 53° = 37°$. Thus, $\angle EDC = 37°$.

4. **A.** By extending either of the lines that form the $70°$ angle to create a transversal and then using the parallel lines theorem, we form a triangle made up of a $40°$ angle, the $(x+10)°$ angle, and an angle supplementary to $70°$, which is $110°$. Since the sum of these 3 angles must be $180°$, we can simplify this to $x° + 160° = 180°$. Thus, $x = 20$.

5. **C.** $\overline{TR} = \overline{TS}$ is false because $\overline{TR}$ is a leg of right-triangle $\triangle TRS$ and $\overline{TS}$ is its hypotenuse. The leg of a right triangle cannot be equal to its hypotenuse.

6. **C.** Since the sum of the interior angles of a triangle is $180°$, any angle in a triangle is equal to $180°$ minus the other two angles. Thus, $\angle ABE = 180° - \angle A - \angle AEB$ and $\angle EDC = 180° - \angle C - \angle DEC$ and since $\angle AEB \cong \angle DEC$ by the vertical angles theorem, we can substitute to get $\angle EDC = 180 - \angle C - \angle AEB$. Now substitute these expressions into $\angle ABE > \angle EDC$ to get $180° - \angle A - \angle AEB > 180° - \angle C - \angle AEB$. We can simplify to get $-\angle A > -\angle C$. Dividing by a negative number, in this case negative one, switches the direction of the comparison, so this becomes $\angle A < \angle C$.

7. **C.** Due to the parallel lines, every quadrilateral formed in the diagram is a parallelogram. We can show that $\angle 2$ is the angle in the parallelogram adjacent to $\angle 1$ by using the vertical angles theorem and the parallel lines theorem. Adjacent angles in a parallelogram are supplementary, thus $\angle 1$ and $\angle 2$ sum to $180°$.

8. **D.** $\angle AIB \cong \angle HID$ by vertical angles theorem. $\angle EID$ is one third of $\angle HID$, as shown in the diagram. Thus, $\angle EID$ is one third of $120°$, making it $40°$. $\angle IDE$ is supplementary to $\angle AIB$ by the parallel lines theorem, thus $\angle IDE = 180° - \angle AIB = 180° - 120° = 60°$. Since the sum of the interior angles of a triangle is $180°$, $\angle IED + \angle IDE + \angle EID = 180°$. Substitution renders $\angle IED + 60° + 40° = 180°$. Simplification shows $\angle IED = 80°$.

9. **A.** The parallel lines theorem shows that $\angle x$ is congruent to the sum of the two congruent angles in the upper-left corner of the diagram, thus the marked congruent angles are each equal to $\dfrac{x}{2}$. The angle on the straight line to the right of angle x is equal to $180 - x$. Finally, by the exterior angle theorem, I know the sum of $180 - x$ and $\dfrac{x}{2}$ equals $y: (180 - x) + \dfrac{x}{2} = y$ simplifies to $180 - \dfrac{x}{2} = y \rightarrow \dfrac{x}{2} = 180 - y \rightarrow x = 2(180 - y)$.

10. **E.** Angles $\angle ACB$ and $\angle DCE$ are congruent because of the vertical angles theorem, so II is true. These angles are congruent and $\angle B$ and $\angle E$ are both $90°$ angles. If you take the triangular sum to find angles $\angle A$ and $\angle E$, you see that they are equivalent, congruent, as well. Because all three angles are congruent, $\triangle ABC$ and $\triangle EDC$ are similar triangles, making III true. We have no side lengths at all, so can't confirm choice I must be true. The only answer that includes both II and III is answer E. For more on similar shapes, see chapter 15.

11. **D.** Since the diagonal of a rhombus bisects the angle it is drawn from, $\angle ABC = 2\angle ABD = 2(30°) = 60°$. Since the opposite angles of a rhombus are congruent, $\angle ADC = 60°$. Since $\angle ADC$ and $\angle CDE$ form a straight line, they are supplementary, thus $\angle ADC + \angle CDE = 180°$. By substitution $180° = \angle CDE + 60°$; simplification yields $\angle CDE = 120°$.

12. **C.** This problem builds on spatial arrangement skills we cover at the end of **Counting and Arrangements.** It has little to do with much in this chapter, and probably should be in that chapter, but we ran out of space there and heck it has lines in it. If you need more work on problems like this, head to chapter 9. Drawing a diagram, the first fact, "D is to the right of A and to the left of E" gives us: $A \quad D \quad E$. The second fact, "B is to the right of C and to the left of D" means that $A, B,$ and C can be ordered as $ACB, CAB,$ or CBA to the left of D. The third fact, "C is to the right of A" tells us that they must be ordered ACB. Thus, the final order is $ACBDE$. Now we can look at the answers and tell that the only one that *must* be true is (C), $AE > BE$; since $AE = AB + BE$, AE must necessarily be greater than BE.

13. **E.** This one is a bit tricky to understand. The basic idea you have to get is that the angle of the light against a line "bounces" to create three angles: two that are equal (see below) and a third between them. When the light first bounces off a mirror at $39°$, because the exact degree is reflected, a quadrilateral is formed with inside corners of the studio, with the angles $39°$, $98°$, $121°$, and an unknown angle we can solve for.

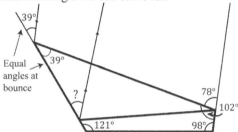

Because a quadrilateral's interior angles always sum to $360°$, the unknown angle is equal to: $360° - 39° - 121° - 98° = 102°$. This angle we just found is supplementary to the next angle that the light is reflected at: $180° - 102° = 78°$. This second angle that is reflected, in turn, forms another angle equal to 78 within a smaller quadrilateral that lies entirely inside the larger one we just looked at, and this smaller quadrilateral's angles are $98°$, $121°$, $78°$, and an unknown angle. Solving for the unknown angle we get $360° - 98° - 121° - 78° = 63°$. The angle we just found, $63°$, when reflected is exactly the angle we are told to find, so our answer is $63°$.

14. **B.** We can translate the verbal expression into an equation. Two supplementary angles means that the two angles summed equal $180°$, and one of the angles is a seventh of the other, so $7x + x = 180$, where x represents the angle measure of the smaller angle. Solving this, we get $8x = 150 \rightarrow x = 22.5$. We could also have created a variable for the larger angle (n in picture below) if that is easier for you to imagine, and then plugged in at the end to find the smaller angle.

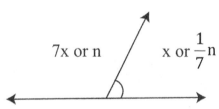

15. **C.** Through the triangle sum theorem (all angles in a triangle sum to $180°$) we find $\angle DCE$ equals $180° - 75° - 20° = 85°$ and by the vertical angles theorem, $\angle ACB$ also equals $85°$. In $\triangle ABC$, $\angle A + \angle B + 85° = 180°$. If we let $\angle A = 4x$ and $\angle B = x$, as the problem tells us, our equation becomes $4x + x + 85° = 180°$. When we solve for x, we find that $x = 19°$. However, we want $\angle A$, not $\angle B$, so we multiply that by 4 to get our final answer, $76°$.

16. **C.** We find $\angle EAD$ through the triangle sum theorem. Looking at $\triangle AEB$, we know that our $\angle ABE = 37°$, $\angle E = 75°$, and $\angle EAB = \angle DAB + 24°$. Thus, $\angle EAD = 180° - 24° - 75° - 37° = 44°$.

17. **A.** You could solve this by first finding $\angle BCA$ because it forms a straight line with $\angle DCE$ and $\angle BCD$, so the sum of the three angles is $180°$, and then using the "triangle sum theorem" on triangle ABC. *Or*, you could use the exterior angle theorem, which states that the exterior angle (which is formed whenever one of the triangle's legs is extended) is equal to the sum of the remote two interior angles. Applying this theorem to $\triangle ABC$, we get: $m° + m° + 86° = 64° + 58° \rightarrow 2m° = 36° \rightarrow m = 18°$.

18. **B.** First find $\angle CDG$ by subtracting it from $180°$ knowing it is supplementary to $128°$ ($180° - 128° = 52°$), and then use the triangle sum theorem to find $\angle DCG$: Alternatively, we could use the exterior angle theorem, which states that the two opposite angles are equal to the outside angle of the third angle, so in this case $95° + \angle DCG = 128°$. Either way, we find that $\angle DCG = 33°$. $\angle DCG$ and $\angle FCB$ are opposite, so by the vertical angles theorem they are congruent. Now that we know

ANGLES/LINES — ANSWERS

both $\angle FCB$ and $\angle BFC$ (given in the diagram), we can find $\angle FBC$ using the triangle sum theorem: $\angle FCB + \angle BFC + \angle FBC = 180° \rightarrow 33° + 84° + \angle FBC = 180° \rightarrow \angle FBC = 63°$. $\angle FBC$ and $\angle ABE$ are opposite angles too, so they are also congruent. Now we know two out of the three angles in $\triangle ABE$, so we can apply the triangle sum theorem a third and final time to find that $\angle AEB = 46°$.

19. **E.** If $\angle ACD$ is $38°$, then by the triangle sum theorem, $\angle ADC$ is $52°$. Since that angle is bisected by $\overline{DB}$, $\angle ADB$ and $\angle BDC$ are both $26°$. If we apply the triangle sum theorem to $\triangle DBC$, we find that $\angle DBC = 116°$. Thus, the sum of $\angle ADB$ and $\angle DBC$ is $26° + 116° = 142°$.

20. **D.** The sum of pairs of adjacent angles on a line will always be $180°$, so one pair will never be greater than or less than the other pair.

21. **B.** For an angle to be obtuse it must be greater than $90°$ but less than $180°$, so the expression given must satisfy those two conditions: $90 < (180 - x) < 180$. Subtract $180°$ from both sides: $-90 < -x < 0$. Multiply all three parts of the expression by -1, and remember to compensate for the negative by changing the direction of the inequality signs: $90 > x > 0$. Alternatively, use logic. If you understand the definition of an obtuse angle, you know you must subtract at least some value because an obtuse angle is less than $180°$, but cannot subtract more than $90°$ or your angle would no longer be obtuse.

22. **C.** First, draw a picture. For our purposes, it doesn't really matter that there are rays. What matters is that we have two adjacent angles that share one side while their two individual sides form a line, so the angles are supplementary.

    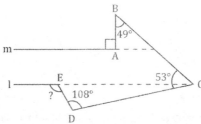 If the larger angle is x, then according to the question the smaller angle is $\dfrac{x}{5}$. Because the angles are supplementary, we know that $x + \dfrac{x}{5} = 180°$. Solving for x, we find that $x = 150°$. However, that is the larger angle. The smaller angle is $\dfrac{x}{5} = \dfrac{150°}{5} = 30°$.

23. **C.** If we let $\angle BAC$ and $\angle BCA$ be x (they are congruent because they are both opposite congruent sides of an isoscoles triangle), then $\angle ABC = 7x$ and $x + x + 7x = 180°$. Simplify to get $9x = 180°$ and divide by 9 on both sides to find x equals $20°$. Because $ABCD$ is a regular quadrilateral, the angles at its corners are equal to $90°$, $\angle EAB = 90° + 20°$ and $\angle ABC = 7x = 7(20°) = 140°$. Using these values, $\angle ABC - \angle EAB = 140° - 110° = 30°$.

24. **D.** If we extend the lines m and l forward until they both reach $\overline{BC}$, we get a two triangles: a right triangle with angle measures $90°$, $49°$, and $41°$ (found using the triangle sum theorem), and an obtuse triangle with only one angle known so far, $108°$.

    $\overline{BC}$ can be treated like a transverse line (transversal), and using the **parallel lines theorem**, we can split $\angle DCB$ into two different angles above and below the dotted line. In the top tiny triangle, using the triangle sum theorem (180-49-90=41), we find the angle formed by BC and the dotted portion of line m is $41°$ so $\angle ECB$, per the parallel lines theorem, is also equal to $41°$, and $\angle DCE = 53° - 41° = 12°$. Now use the **exterior angle theorem** to find our desired exterior angle "?" of triangle DEC by summing opposite interior angles $108° + 12° = 120°$. (Alternatively use the triangle sum theorem with the triangle $\triangle CDE$ to find $\angle DEC$: $\angle DCE + \angle EDC + \angle DEC = 180° \rightarrow 12° + 108° + \angle DEC = 180° \rightarrow \angle DEC = 60°$ $\angle DEC$ is supplementary to our needed angle so $180° - 60° = 120°$).

25. **D.** The measure of $\angle S = \dfrac{1}{3} \angle I = \dfrac{1}{3}(63°) = 21°$. We are given that $\angle R = 90°$. We know that $\angle Q + \angle R + \angle S = 180°$, since the interior angles of a triangle on a plane sum to $180°$. Substitute in the values we have: $\angle Q + 90° + 21° = 180° \rightarrow \angle Q + 111° = 180° \rightarrow \angle Q = 69°$.

26. **C.** One way of looking this problem is to create a right triangle by extending the seat- of the chair and the line of the lamp:

We can find one of the angles of this right triangle by finding the supplemental angle of 137°, which is 43°. Let's say that the angle that we want to find, the angle the lamp makes with the chair back, is x. Using the exterior angle theorem, and setting our desired angle x, as the exterior angle, we see that $x = 43° + 90° = 133°$.

27. **E.** $\overline{AC}$ and $\overline{DE}$ are two parallel lines, and $\overline{BD}$ and $\overline{BE}$ are essentially transversals. By the parallel lines theorem, $\angle 1 \cong \angle 3$ and $\angle 2 \cong \angle 5$. However, only $\angle 1 \cong \angle 3$ is an answer choice.

28. **D.** This question uses the parallel lines theorem and the idea of supplementary angles. Because a and c are the exterior angles of extended triangle sides, $a = 180 - x$ and $c = 180 - y$. By the exterior angle theorem, $b = x + y$. Thus, $a - b - c = (180 - x) - (x + y) - (180 - y) = 180 - x - x - y - 180 + y = -2x$.

29. **D.** Because we are given one of the angles at X, we can use the supplementary angles theorem to get $\angle a$. $180° - 101° = 79°$ Thus, to find $\angle c$, if we knew $\angle e$ we could use the triangle sum theorem to solve for $\angle c$ as $\angle c + \angle e + 79° = 180°$. Alternatively, if we could find $\angle c$ if we knew the sum of $\angle b + \angle d$ as these two angles are supplementary to $\angle c$. Let's first try to find what might help us find $\angle e$: $\angle f$ is supplementary to $\angle e$, so if we can find $\angle f$ we can subtract from 180 to get $\angle e$. Answer choice (A) doesn't tell us any angle values, answer choice (B) would only help if we wanted to find $\angle b$, answer choice (C) tells us that the lines are parallel, but it doesn't give us any numerical values that would help us find $\angle c$ (we'd need $\angle d$ for this fact to be of use to find $\angle e$). Answer (D), however, gives us $\angle f$, which as discussed can help us find $\angle e$. Knowing $\angle e$ and $\angle a$, we can find the value of $\angle c$ with the triangle sum theorem.

30. **D.** Because the lines are parallel, all the corresponding angles are parallel, making any figures created similar; by the **side splitter theorem** (pg 243, Ch 15) we know the segment portions cut by parallel lines are proportional, i.e. $\frac{\overline{DE}}{\overline{BD}} = \frac{\overline{CF}}{\overline{AC}}$. Plugging in we get $\frac{2}{12} = \frac{\overline{CF}}{10}$; now simplify by cross multiplying: $2(10) = 12(\overline{CF}) \to \frac{20}{12} = \overline{CF} \to \frac{5}{3} = \overline{CF}$. See **Chapter 15: Similar Shapes** for more similarity problems.

31. **D.** To start, $\angle x$ is supplementary to the two angles that it is adjacent to, $\angle 9$ and $\angle 11$. Those angles are congruent to their corresponding angles on the parallel line below that intersects with the same transversal, $\angle 12$ and $\angle 15$. Thus, $\angle x$ is supplementary to all four of those angles, so our set is $\{9, 11, 12, 15\}$. Line a and line b are not parallel so no supplementary angles are formed from line a.

32. **E.** The 22° angle and $\angle x$ are in a triangle with a third angle, which by the parallel lines theorem is supplementary to the 127° degree angle. $180° - 127° = 53°$. Knowing this third angle in the triangle, we find that $\angle x = 180° - 22° - 53° = 105°$.

33. **E.** $\angle EAF$ and $\angle DAF$ are supplementary, or sum to 180°. Thus $\angle DAF = 180° - 44° = 136°$.

34. **B.** Ignore line b to see how we have information on a pair of same side interior angles. The top interior angle is formed from $(6x - 11)° + (7x - 4)°$ and the bottom interior angle equals $(4x + 8)°$. The same side interior angles are supplementary, so $(6x - 11) + (7x - 4) + (4x + 8) = 180 \to 17x - 7 = 180 \to 17x = 187 \to x = 11$.

35. **E.** By the triangle sum theorem, $m\angle 3 + m\angle 11 + m\angle 6 = 180°$, so II is true. I is true because angles $\angle 1$ and $\angle 10$ are the vertical angles of (are thus congruent to) two of the interior angles of the triangle, and $\angle 6$ is the third triangle angle. III is true because $\angle 8$ is a vertical angle of $\angle 6$, which when summed with the other two interior triangle angles equals 180°.

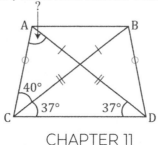

ANGLES/LINES ANSWERS

36. B. The trapezoid is isosceles, so is symmetric about a central axis, meaning that symmetrical pairs of angles in the figure/involving its diagonals are equal: $\angle ACB \cong \angle BDA = 40°$ and $\angle BCD \cong \angle ADC = 37°$. Thus $\angle ACD = \angle ACB + \angle BCD$ so $\angle ACD = 40° + 37° \to \angle ACD = 77°$. The angles in triangle ACD sum to 180, so $\angle CAD = 180° - \angle ACD - \angle ADC = 180° - (77°) - 37° = 66°$.

37. C. One way to solve is by process of elimination. Answer (A) is impossible because the problem literally says that the two are equal. Answers (B), (D), and (E) are wrong because the figure described can be drawn such that $\overline{AB} < \overline{BC}$ and/or $\overline{AD} > \overline{CD}$ and/or $\overline{AB} \neq \overline{AD}$, as shown below, left. We could also solve this faster if we know the **hinge theorem** (Ch 15, pg 243), which states that if you have two congruent sides in two triangles, the size of the angle between those sides dictates how large or small the side opposite that angle is. Thus the larger side is opposite the larger angle, so if $\angle ACB > \angle ACD$ then $\overline{AB} > \overline{AD}$.

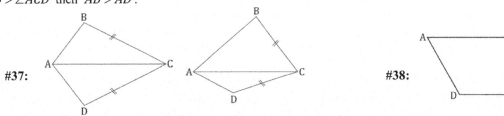

38. E. If you look at a parallelogram and arbitrarily assign vertices, (see above right) you see that $\angle BCD$ and $\angle CDA$ are adjacent. In a parallelogram, opposite angles are congruent and adjacent angles are supplementary, to they must add up to $180°$.

CHAPTER 12

TRIANGLES

> ## SKILLS TO KNOW
> - Triangle types: obtuse, acute, right, equilateral, isosceles, scalene
> - Triangle Area & Perimeter
> - Pythagorean Theorem and its converse
> - Pythagorean Triples
> - Special Triangles (30-60-90, 45-45-90)
> - Triangle Inequality Theorem

NOTE: Similar triangles are covered in Chapter 15: **Similar Shapes**. For more problems that involve triangles, also see this book's chapters on **SOHCAHTOA, Trigonometry, Angles and Lines, Circles,** and **Polygons**.

TRIANGLE BASICS

Triangles are three-sided figures that can be classified according to their angles and/or side lengths. All angles in a triangle always sum to **180 degrees**. Below, you'll find a description of many different classifications of triangles. If you don't remember these basics, review the information below.

Right Triangle	Triangles are right triangles if one angle is **90** degrees. The longest side of a right triangle is called a hypotenuse. The shorter two sides are called legs. Missing pieces of right triangles can be solved using the Pythagorean Theorem and basic Trigonometry (SOHCAHTOA).	
Obtuse Triangle	Triangles are classified as obtuse if one angle is greater than **90** degrees.	
Acute Triangle	Triangles are classified as acute if all angles are less than **90** degrees.	
Scalene Triangle	Triangles are classified as scalene if all sides are unequal. Therefore, all angles will be unequal as well.	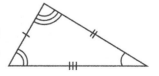

Isosceles Triangle	Triangles are isosceles if at least two sides are congruent. The angles opposite those sides are also congruent.	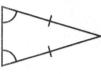
Equilateral Triangle	Triangles are classified as equilateral if all sides are congruent. All angles of an equilateral triangle will be **60** degrees.	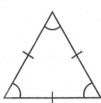

Isosceles Triangles

When two sides of a triangle are congruent, we classify it as an isosceles triangle.

If we know two angles in a triangle are congruent, we also know the sides opposite those angles are congruent. If we know two sides in a triangle are congruent, we also know the angles opposite those sides are congruent. For example, in ΔFGH, if $\overline{FG}$ and $\overline{FH}$ are congruent, then $\angle H$ and $\angle G$ are also congruent. If we knew instead that $\angle H$ and $\angle G$ were congruent, we could also assume that sides $\overline{FG}$ and $\overline{FH}$ are congruent.

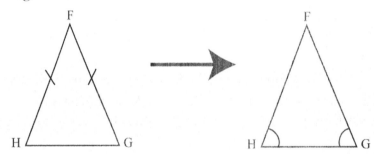

Equal angles in an isosceles triangle are called **base angles** ($\angle H$ and $\angle G$ at right). The non-congruent side in an isosceles triangle (when applicable) is typically called the **base** (side $\overline{HG}$), while the other sides are called the **legs** ($\overline{FH}$ and $\overline{FG}$). The angle included by the legs (angle $\angle F$) is called the **vertex angle**.

The **altitude** to the base of an isosceles triangle is always a **perpendicular bisector** that bisects the vertex angle, forms a perpendicular (right) angle with the base, and cuts that base into two equal pieces.

Likewise, if we have an altitude of a triangle that bisects its vertex angle and is a perpendicular bisector of the opposite side, we also know we have an isosceles triangle. We could prove this rule by similar triangles (HL Theorem), though you can also just memorize this instance. On the ACT®, this rule can help you instantly know more information about a triangle.

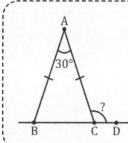

△*ABC* is an isosceles triangle with length $\overline{AB}$ equal to length $\overline{AC}$. ∠*A* has a measurement of 30°. What is the measurement of the ∠*ACD*?

Knowing that △*ABC* is an isosceles triangle, we know ∠*ACB* and ∠*ABC* are congruent. Let's set each of these equal to *x* :

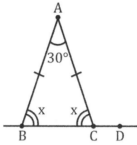

Now to solve, we can use the fact that a triangle has 180° in its interior. First, I create an equation with my variables based on the sum of all angles in the triangle, 180°:

$$180° = 30° + x + x$$
$$150° = 2x$$
$$75° = x$$

∠*C* and ∠*B* are both 75°. We can now find ∠*ACD* with our knowledge that angles along a straight line will sum to 180°. Here I subtract:

$$180° - 75° = 105°$$

Answer: **105°**.

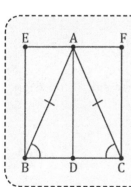

△*ABC* is an isosceles triangle within rectangle *BEFC*. ∠*ABD* measures 50°. Point *D* is the midpoint of $\overline{BC}$. What is the measurement of ∠*BAF*?

$\overline{AD}$ is the perpendicular angle bisector of the inner triangle in this figure, because we know that triangle is isosceles. Since equal sides indicate opposite equal angles, we know that ∠*ABD* and ∠*ACB* are equal, each 50°. We can conclude that ∠*BAC* is 80° by subtracting the other two angles from 180: $180° - (50° + 50°) = 80°$. We can then find ∠*BAD* by cutting 80 degrees in half given the bisector $80° / 2 = 40°$

We also know ∠DAF is 90°, because the bisector of an isosceles triangle meets the base at 90° and we know $\overline{BC}$ is perpendicular to $\overline{AD}$, $\overline{FC}$, and $\overline{EB}$ (given the rectangle information in the question). These three vertical lines are thus parallel and form 90° angles with the horizontal parts of the rectangle. Then we can add ∠BAD and ∠DAF to find our answer:

$$90° + 40° = 130°$$

Answer: 130°.

EQUILATERAL TRIANGLES

An equilateral triangle is a triangle in which all three sides are equal. Equilateral triangles are also equiangular; that is, all three internal angles equal 60° (one-third of 180°).

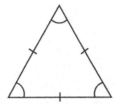

The dimensions of equilateral triangle $\triangle XYZ$ are given in the figure below. What is the value of y in inches?

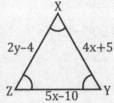

Knowing that $\triangle XYZ$ is an equilateral triangle, we know that all the sides are equal to each other. We can find x by setting $\overline{XY}$ equal to $\overline{YZ}$.

$$4x + 5 = 5x - 10$$
$$5 = x - 10$$
$$15 = x$$

Now we need to find the length of one side. We plug in 15 into either side expression with $x = 4(15) + 5 = 65$. So our side length is 65. Now we can solve for y.

$$2y - 4 = 65$$
$$2y = 69$$
$$y = \frac{69}{2} = 34.5 \text{ inches}$$

Answer: 34.5 in.

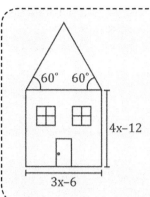

Derek is trying to figure out how much wood he needs to build the frame of his roof. He knows that the front is a square with the side measurements given below. He also knows the angle measurements of the roof as it meets the ceiling of his house. How much wood, in feet, does he need to build the triangular frame?

First, we will find the measurement of the sides of the house by solving for x.

$$3x - 6 = 4x - 12$$
$$-6 = x - 12$$
$$6 = x$$

Now we can plug in x to either one of the sides to find the length.

$$3(6) - 6 = 12$$

Since the problem explains that the front of the house is a square, we can conclude that all sides are 12 feet. This also means that the triangular frame on top will also have sides of 12 feet. Note that the 60° measurements of the angle tell us that the frame is an equilateral triangle. If two angles of a triangle are 60°, so is the third. Thus, we can add up the sides to determine the perimeter and the amount of wood needed.

$$12 + 12 + 12 = 36 \text{ feet}$$

Answer: **36 ft.**

TRIANGLE AREA & PERIMETER

TRIANGLE PERIMETER

The perimeter of a triangle is the sum of its sides. Perimeter $= s_1 + s_2 + s_3$

TRIANGLE AREA

The area of any triangle is:

$$A = \frac{1}{2}bh$$

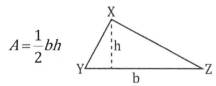

Where b is the measurement of the base of the triangle and h is the measurement of the height of the triangle.

TRIANGLES — SKILLS

Remember ANY side of a triangle can be a base, and the height is the straight line perpendicular distance from the vertex opposite the base to the base or the plane the base rests upon.

> The rectangle below is divided into three triangles, two of which are shaded. What is the total area of the two shaded regions in square inches?
>
> (Rectangle ABDC with E on AB, AE = 5 in, EB labeled, BD = 7 in, CD = 16 in)

$\triangle AEC$ has a height of 7 inches since $\overline{BD}$ and $\overline{AC}$ are the same length. Furthermore, $\overline{EB}$ is 11 inches long, because it is $\overline{CD} - \overline{AE}$. Now we can find the area of both shaded triangles and add them.

$$\frac{1}{2}(5 \times 7) + \frac{1}{2}(7 \times 11) = 56 \text{ square inches}$$

Answer: **56 in.²**

We could also have recognized the unshaded region is half the area of the rectangle that surrounds it, because triangle area is half of the base times the height, while the rectangle area is base times the height. Thus the shaded region is the other half of the rectangle's area, or simply identical to the area of the unshaded region triangle: $\frac{1}{2}(base)(height) = (0.5)(16)(7) = 56$.

PYTHAGOREAN THEOREM/CONVERSE

The Pythagorean theorem relates the lengths of the three sides of a right triangle. If you know two sides of a right triangle, with this theorem, you can always find the third. It states that the square of the hypotenuse equals the sum of the squares of the other two sides.

PYTHAGOREAN THEOREM

For any right triangle with side lengths a, b and c, where c is the longest side (opposite 90 degrees), or hypotenuse:

$$a^2 + b^2 = c^2$$

(Right triangle with legs a, b and hypotenuse c)

We can also work backwards with this formula. If the square of the longest side of a triangle is equal to the sum of the squares of the other two sides, then the triangle is a right triangle. This is called the **converse of the Pythagorean theorem**. With this converse, you can use the Pythagorean Theorem to prove that a triangle is a right triangle, knowing only its side lengths.

Jim wants to buy a ladder so that he can put up lights on his house. The side of Jim's house is perpendicular to the level ground. The ladder must be 10 feet away from the building and must reach 18 feet up the side of the home. Approximately how long does Jim's ladder have to be so that he is able to put up his lights?

When the ladder rests against the side of a house that is perpendicular to the ground, it creates a right triangle with legs of 10 and 18 feet. To find the length of the ladder, we can use Pythagorean Theorem.

$$10^2 + 18^2 = c^2$$
$$424 = c^2$$
$$20.6 \approx y$$

After some basic algebra, we find that the ladder must be around 20.6 ft. long.

ACUTE OR OBTUSE?

We can also extend the Pythagorean theorem to understand whether a triangle with given side lengths is **acute** or **obtuse**. Remember how in a right triangle, $a^2 + b^2 = c^2$? Let's imagine what would happen if a and b stay the same, but c becomes longer. The angle opposite c would expand as sides a and b "hinge" open.

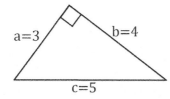 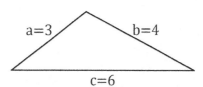

The triangle would thus become obtuse. In other words: if $a^2 + b^2 < c^2$, where c is the longest side, then the triangle is obtuse, with an obtuse angle opposite side c.

Now imagine if instead of "growing" our longest side, we shrank it. In that case, sides a and b would "hinge" tighter together, and c would be smaller.

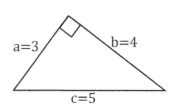 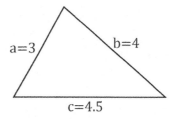

In other words: if $a^2 + b^2 > c^2$, where c is the longest side, then the triangle is acute.

TRIANGLES SKILLS

> The side lengths of a triangle are 3, 6, and 8. Which of the following best describes this triangle?
>
> A. Scalene right
> B. Isosceles obtuse
> C. Scalene obtuse
> D. Isosceles right
> E. Scalene acute

With our knowledge of triangle classification, we can rule out B and D as the side lengths given in the question are all different, indicating that the given triangle is scalene. Next, we must determine whether the triangle is right, obtuse, or acute. Through the Pythagorean theorem, we can rule out A:

$$3^2 + 6^2 = 8^2$$
$$45 \neq 64$$

Note that when we did Pythagorean theorem above, c would have to be $\sqrt{45} \approx 6.708$. This means that the angle opposite of the hypotenuse would have to be larger than $90°$ to make a triangle with a hypotenuse of 8. Thus, the triangle is obtuse. We can imagine this visually also:

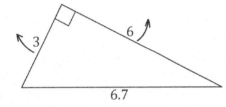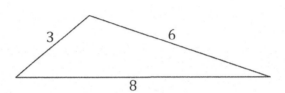

Answer: **C**.

PYTHAGOREAN TRIPLES

A **Pythagorean triple** consists of three positive integers a, b, and c such that $a^2 + b^2 = c^2$. Such a triple is commonly written (a,b,c). Some well-known examples are:

$$(3,4,5), (5,12,13), (8,15,17), \& (7,24,25).$$

 TIMING TIP: MEMORIZE THESE! Knowing these will vastly speed up your math performance!

 BE CAREFUL, though. For the Pythagorean triple to hold, the longest side in the triple pattern must be the hypotenuse. Thus, if I have a triangle marked with sides 5 and 12, and the 12 is the hypotenuse in the picture, the third side is NOT 13! For that, I must run the Pythagorean theorem.

MULTIPLES OF PYTHAGOREAN TRIPLES

If (a,b,c) is a Pythagorean triple, then so is (ka,kb,kc) for any positive integer k

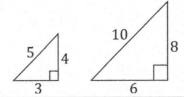

182 CHAPTER 12

△JKL has side lengths, in inches, of 7, 24, and 25. What is the area of the triangle in square inches?

At first this may seem an area problem, and it is. But it's also a Pythagorean triples problem. If you don't know your triples, you won't instantly realize that this is a right triangle and that all you must do to solve is apply the area formula. The legs (at right angles) form the base and height: 7 and 24.

$$A = \frac{1}{2}bh = \frac{1}{2}(24)(7) = 84$$

Answer: 84 in.^2

If you didn't have this Pythagorean triple memorized, you could have run the Pythagorean theorem's converse to check if it is a right triangle, validating that $7^2 + 24^2 = 25^2$.

Otherwise, you could have used **Heron's (or Hero's) formula** if you happened to know it (but learning that is overkill for the ACT, unless you feel like programming it into your calculator).

In right triangle △ADC, $\overline{BE}$ is parallel to $\overline{CD}$ and $\overline{BE}$ is perpendicular to $\overline{AD}$. $\overline{AC}$ has a length of 20 yards. Furthermore, $\overline{BE}$ has a length of 4 yards and $\overline{AE}$ has a length of 3 yards. What is the length, in yards, of $\overline{ED}$?

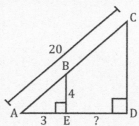

You should instantly recognize that △AEB is a 3−4−5 triangle. This should indicate we can use the 3−4−5 ratio with △ADC to determine the length of the rest of the sides. $\overline{AB} = 5$.

We know because of the similar triangles that $\frac{AB}{AC} = \frac{AE}{AD}$. Since $\overline{AB} = 5$, $\overline{AC} = 20$, and $\overline{AE} = 3$,

$$\frac{5}{20} = \frac{3}{x}$$
$$5x = 60$$
$$x = 12$$

$\overline{AD}$ is 12 yards in length. Now we can subtract $\overline{AE}$ from $\overline{AD}$ to obtain $\overline{ED}$:

$$\overline{AD} - \overline{AE} = \overline{ED}$$
$$12 - 3 = 9$$
$$\overline{ED} = 9 \text{ yards}$$

Answer: $\overline{ED} = 9$ yards. (If this was tough for you, check out **Chapter 15, Similar Shapes**)

SPECIAL TRIANGLES

There are two special triangles worth memorizing.

Special Triangle: 45-45-90

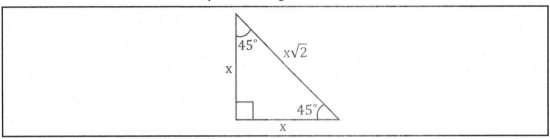

This triangle has particular angle measures that dictate a particular side length ratio. The only isosceles triangle with a right angle, both its base angles are 45 degrees. The hypotenuse of a $45-45-90$ triangle is $x\sqrt{2}$, where x is the length of the congruent sides. Using these variable defined ratios, we can quickly find side lengths in triangles of this type without the Pythagorean theorem. Given sides of this ratio, we can conversely assume that the triangle's angles are $45-45-90$.

> Find the perimeter, in centimeters, of the isosceles right triangle shown below, whose hypotenuse is 16.
>
>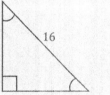

Here, we have a $45-45-90$ triangle (remember any isosceles right triangle fits this pattern). Let's draw it out, but be careful! A big mistake students make is assuming any side you have without the $\sqrt{2}$ is the leg. Here that's not the case! To find the side lengths, set 16 equal to $x\sqrt{2}$.

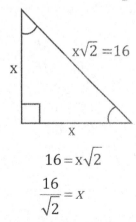

$$16 = x\sqrt{2}$$
$$\frac{16}{\sqrt{2}} = x$$

Now I rationalize the denominator, multiplying by $\sqrt{2}$ on the top and bottom:

$$\frac{16\sqrt{2}}{(\sqrt{2})(\sqrt{2})} = \frac{16\sqrt{2}}{2} = 8\sqrt{2} = x$$

Now I know the lengths:

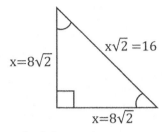

We can add up our sides to find the perimeter:

$$16 + 8\sqrt{2} + 8\sqrt{2} = 16 + 16\sqrt{2}$$

Answer: $16 + 16\sqrt{2}$.

The following figure was made by beginning with square $ABCD$. The midpoints of the four sides of the square were then joined to form another square. The process was repeated to form a third square and finally once more to form the fourth and smallest square in the middle, which has a side length of x. Find the value of x.

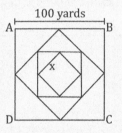

The sides of our outermost square are divided in half by the inner squares. However, what is created is a $45-45-90$ triangle that we can use to determine the length of each inner square. We know the sides of the square one step smaller is equal to $\sqrt{2}$ times the length of half of the outer square's side. Thus we can find x by working our way down.

$$(100 \div 2)\sqrt{2} = 50\sqrt{2} = \text{ side of second largest square}$$
$$(50\sqrt{2} \div 2)\sqrt{2} = 25\sqrt{2}(\sqrt{2}) = 50 = \text{ side of third largest square}$$
$$(50 \div 2)\sqrt{2} = 25\sqrt{2} = \text{ side of fourth largest square}$$
$$25\sqrt{2} = x$$

Answer: $25\sqrt{2}$.

TRIANGLES   SKILLS

Special Triangle: 30-60-90

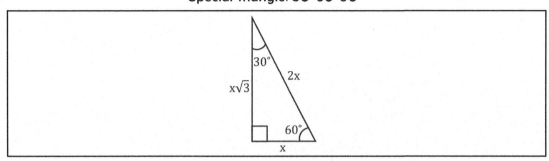

The second special right triangle is the $30-60-90$ triangle. Just like the $45-45-90$ triangle, this triangle is named after the measurements of its angles. Since 30 degrees is the smallest measurement, the side opposite of that angle will be the smallest size, x. The hypotenuse of a $30-60-90$ triangle is opposite the largest angle and equal to $2x$, while the side opposite from the angle of 60 degrees is $x\sqrt{3}$.

Remember that side length size corresponds to opposite angle size, i.e. in this triangle, the longest side is always opposite 90 degrees, and shortest opposite 30 degrees. Here, $x\sqrt{1} < x\sqrt{3} < x\sqrt{4}$ so $1x < x\sqrt{3} < 2x$ and $30° < 60° < 90°$ so x is across from 30 degrees, $x\sqrt{3}$ is opposite 60 degrees, and $2x$ is opposite 90 degrees.

The figure below is a right triangle. The length of $\overline{AC}$ is 8 units and the length of $\overline{CB}$ is 4 units. What is the length of $\overline{AB}$?

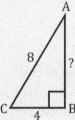

You should notice that the hypotenuse is twice the length of the shortest leg. This is a key indicator that $\triangle ABC$ is a $30-60-90$ triangle. We know that $x = 4$ so we also know that $\overline{AB} = 4\sqrt{3}$.

Answer: $4\sqrt{3}$ units.

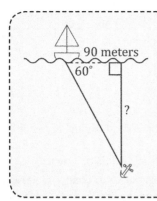
You are on a small boat that has just dropped anchor in the middle of the ocean. After you dropped your anchor, the boat drifted 90 meters away from the original drop site. You measure that the angle of the chain attaching the anchor to your boat is 60 degrees from the water surface as show in the diagram below. How deep, in meters, is the anchor from the surface of the water?

We can use our knowledge of $30-60-90$ triangles to solve this problem. Since we know our shortest side is 90 meters we can determine that the hypotenuse of the triangle is 180 meters and the other leg is $90\sqrt{3}$ meters.

Answer: $90\sqrt{3}$ meters.

TRIANGLE INEQUALITY THEOREM

The Triangle Inequality Theorem can be used to determine if the sides of a triangle are long enough to fully create a triangle. The rule for triangle lengths is that the sum of any two sides must be greater than the third side.

$$a+b>c$$
$$a+c>b$$
$$b+c>a$$

For example, the side lengths 5, 3, and 7 can create a triangle because $3+5>7$. However, side lengths $3, 4,$ and 7 cannot create a triangle as $3+4=7$. Also, the side lengths of 3, 3, and 7 are not able to create a triangle because $3+3$ is not greater than 7.

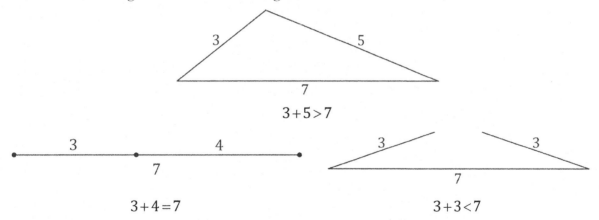

To visualize this, line up your short sides above your longest side. If the sum is equal, you have a two straight lines on top of a straight line. If you try to "pop" up the short legs at an angle, you get a drawbridge not a triangle. If the sum is less than the long side, you don't even have a drawbridge, you have two helpless legs that can't touch each other! However, if the two shortest sides are longer than the third they can "pop up" to form a triangle.

TRIANGLES — QUESTIONS

1. In △ABC below, $\overline{AC} = \overline{CB}$ and ∠A measures 70°. What is the measure of ∠C?

   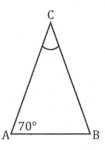

   A. 35°
   B. 40°
   C. 45°
   D. 50°
   E. 55°

2. In isosceles triangle △RST, $\overline{RS}$ is congruent to $\overline{ST}$ and the measure of one base angle ∠R is 67.5°. What is the measure of vertex angle ∠S?

   A. 13.5°
   B. 45°
   C. 67.5°
   D. 85°
   E. 135°

3. What is the length, in meters, of the hypotenuse of a right triangle with legs that are 3 meters long and 7 meters long, respectively?

   A. $\sqrt{10}$
   B. $\sqrt{58}$
   C. 10
   D. 14
   E. 21

4. Which of the following sets of 3 numbers could be the side lengths, in feet, of a right triangle?

   A. 1,1,1
   B. 2,6,8
   C. 3,7,10
   D. 5,12,13
   E. 7,10,29

5. What is the area, in square meters, of a right triangle with sides of length 7 meters, 24 meters, and 25 meters?

   A. 84
   B. 87.5
   C. 168
   D. 300
   E. 336.5

6. In △XYZ, $\overline{XY} \cong \overline{XZ}$, and the measure of ∠X is 44°. What is the measure of the sum of ∠X and ∠Y, in degrees?

   A. 44°
   B. 68°
   C. 112°
   D. 136°
   E. Cannot be determined from the given information.

7. Triangles △EFG and △HFG, shown below, are isosceles with base $\overline{FG}$. Segments $\overline{FH}$ and $\overline{GH}$ bisect ∠EFG and ∠EGF, respectively. If ∠EFH is 32°, what is the measure, in degrees, of ∠FGH?

   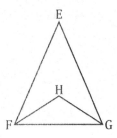

   A. 24°
   B. 28°
   C. 32°
   D. 36°
   E. 64°

8. In △DEF, ∠D and ∠F are congruent, and the measure of ∠E is 56°. What is the measure of ∠D?

   A. 28°
   B. 56°
   C. 62°
   D. 100°
   E. 124°

9. In △ABC, the measure of ∠B is twice the measure of ∠A, and ∠C is three times the measure of ∠A. What is the measure, in degrees, of the sum of ∠B and ∠C?

   A. 30°
   B. 60°
   C. 90°
   D. 150°
   E. 180°

10. In △XYZ, the measure of ∠X is 57° and the measure of ∠Y is 32°. Which of the following inequalities involving the lengths of the sides of △XYZ is FALSE?

   A. $\overline{YZ} > \overline{XZ}$
   B. $\overline{XY} > \overline{YZ}$
   C. $\overline{XY} > \overline{XZ}$
   D. $\overline{XZ} > \overline{YZ}$
   E. Not enough information

11. In isosceles triangle △DEF, base angles ∠D and ∠F each measure 36°. Points D, E, and G are collinear points, with E between D and G. What is the measure of ∠GEF?

   A. 18°
   B. 24°
   C. 36°
   D. 68°
   E. 72°

12. Two triangles are presented in the diagram below, each with sides of lengths a and b. If the area of the top triangle is 30 square centimeters, what is the area of the bottom triangle?

   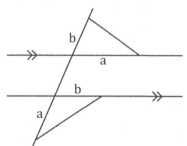

   A. 15
   B. 19
   C. 25
   D. 30
   E. 33

13. In the figure below, $\overline{YA}$ is an altitude of equilateral triangle △XYZ. If $\overline{YX}$ is 8 units long, how many units long is $\overline{YA}$?

   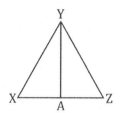

   A. 4
   B. $4\sqrt{3}$
   C. 8
   D. $8\sqrt{3}$
   E. 16

14. The measure of a vertex angle of an isosceles triangle is $(x-12)°$. The base angles each measure $(3x+40)°$. What is the measure in degrees of the vertex angle?

   A. 4°
   B. 13°
   C. 16°
   D. 35°
   E. 88°

15. In △ADE below, B lies on $\overline{AE}$; C lies on $\overline{AD}$; and w, x, y, and z are angle measures, in degrees. The measure of ∠A is 50°. Which of the following must be true?

   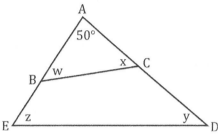

   A. $w + x < 50°$
   B. $x + y > 50°$
   C. $z = w$
   D. $z + y > w + x$
   E. $w + x = z + y$

16. The area of △ABC below is 16 square inches and the area of △ABD is 10 square inches. If $\overline{AC}$ is 8 inches long, how long is length $\overline{AD}$, in inches?

   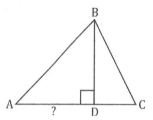

   A. 3
   B. 4
   C. 5
   D. 6
   E. 7

17. Harry is examining the shape of a spear for his history class. In the figure below, the arrowhead is represented by $\triangle XYZ$. Base $\overline{YZ}$ is 20 inches long, and sides $\overline{XY}$ and $\overline{XZ}$ are each 26 inches long. The shaft, represented by $\overline{XA}$, is perpendicular to the base and extends 12 inches below the bottom of the arrowhead. How many inches long is the shaft?

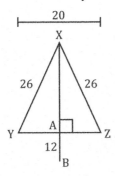

- A. 24
- B. 36
- C. 43
- D. 130
- E. 142

18. Shown below are right triangles $\triangle ACD$ and $\triangle ABE$ with lengths given in feet. What is the length, in feet, of $\overline{AE}$?

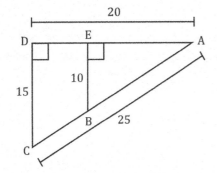

- A. $10\frac{1}{5}$
- B. $10\frac{1}{4}$
- C. $10\frac{3}{4}$
- D. $13\frac{1}{3}$
- E. $37\frac{1}{2}$

19. The lengths, in meters, of all three sides of a triangle are positive integers, and one side is 9 inches long. If the area of this triangle is positive, what is the smallest possible perimeter for this triangle, in meters?

- A. 11
- B. 12
- C. 18
- D. 19
- E. 27

20. Mark is shopping to replace his bicycle's wheels. He places two wheels, one with a radius of 60 centimeters and one with a radius of 40 centimeters, parallel with each other as if they were the two front tires of a car on level ground. As he does so, he notices that in the late afternoon light their shadows end in the same below, as shown in the diagram below. How far apart did he place the wheels, measuring from the wheel's left edges? (Note: the wheels' widths are negligent.)

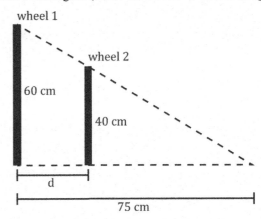

- A. 25
- B. 37.5
- C. 40
- D. 50
- E. 112.5

21. In order to paint the window of her house, Shelley needs to find a ladder of appropriate length. The side of her house is perpendicular to the level ground so that the base of the ladder is 15 feet away from the base of the building. How long does the ladder need to be to reach 20 feet up the house?

- A. 5
- B. 15
- C. 17
- D. 25
- E. 35

22. In the 30°–60°–90° right triangle below, $AB = 10$ cm. What is the length, in centimeters, of $\overline{BC}$?

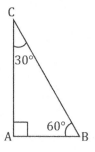

A. 20
B. $10\sqrt{3}$
C. 10
D. $\dfrac{10\sqrt{3}}{3}$
E. 5

23. In right triangle $\triangle LMF$ below, distances are shown in yards. How many yards long is $\overline{LO}$?

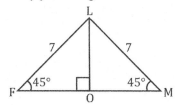

A. $\dfrac{14\sqrt{2}}{2}$
B. $14\sqrt{2}$
C. 7
D. $7\sqrt{2}$
E. $\dfrac{7\sqrt{2}}{2}$

TRIANGLES ANSWERS

ANSWER KEY

1. B	2. B	3. B	4. D	5. A	6. C	7. C	8. C	9. D	10. D	11. E	12. D	13. B	14. A
15. E	16. C	17. B	18. D	19. D	20. A	21. D	22. A	23. E					

ANSWER EXPLANATIONS

1. **B.** Since $\overline{AC} \cong \overline{CB}$, $\angle A$ and $\angle B$ are also congruent. Since all angles in a triangle must equal $180°$, and the sum of $\angle A$ and $\angle B$ is $140°$, $\angle C$ must be $40°$.

2. **B.** In an isosceles triangle, the two base angles are congruent. Since $\angle R$ is given to us to be $67.5°$, $\angle T$, the other base angle, must also be $67.5°$. All angles in a triangle must sum to $180°$, so it follows that $\angle S$ must be $45°$.

3. **B.** This question depends on our knowledge of the Pythagorean Theorem: $a^2 + b^2 = c^2$. Since a and b represent the legs of the right triangle and c represents the hypotenuse: $3^2 + 7^2 = c^2 \rightarrow 9 + 49 = c^2 \rightarrow 58 = c^2$. Thus we find c to be $\sqrt{58}$.

4. **D.** If we know what to look for and recognize the Pythagorean triple in answer D, then we've found the correct answer and can move on. But if we don't recognize the Pythagorean triple, we can find the answer another way. In any triangle, the sum of two side lengths must always be greater than the third side length, which eliminates answer choices B and C. Now look for which set of side lengths is a right triangle, which means that the lengths satisfy the Pythagorean Theorem. Answer A is an equilateral triangle, whose angles by definition are all $60°$, while answer E doesn't satisfy the Pythagorean Theorem. Answer D, (5, 12, 13) does.

5. **A.** In a right triangle, the hypotenuse will always have the greatest length. Therefore, 7 and 24 must be the legs of the triangle, which in a right triangle are also our base and height. Thus, we can plug these numbers into the triangle area formula, $A = \left(\dfrac{1}{2}\right)(7)(24)$, and find the area to be 84 square meters.

6. **C.** It always helps to sketch a quick diagram. We can determine that $\angle X$ is the vertex angle by looking at the graph or because it appears in both the congruent line segments. Since $\angle X$ is $44°$, and all angles in a triangle must sum to $180°$, the sum of base angles $\angle Y$ and $\angle Z$ must be $136°$. Because the triangle is isosceles, we can divide $136°$ by 2 in order to find the measure of $\angle Y$, or $68°$. Thus the addition of $\angle X$ and $\angle Y$ is $112°$.

7. **C.** Since $\triangle EFG$ is an isosceles triangle, $\angle EFG$ and $\angle EGF$ are congruent to each other. Secondly, since segments $\overline{FH}$ and $\overline{GH}$ bisect these angles, the four angles created ($\angle EFH$, $\angle HFG$, $\angle EGH$, $\angle HGF$) are all congruent. Thus, $\angle EFH$ is congruent to $\angle FGH$, making the answer $32°$.

8. **C.** The sum of all angles in a triangle is $180°$. If we subtract the measure of $\angle E$ from the total degrees, we see that the sum of angle $\angle D$ and $\angle F$ is $124°$. Because these two angles are congruent, we can divide $124°$ by 2 to find the measure of $\angle D$.

9. **D.** We can start by assigning variables to the various angles. Let $\angle A = x$, because the other angles are described in terms of $\angle A$. Because $\angle B$ is twice the value of $\angle A$ and $\angle C$ is three times that value, we can let $\angle B = 2x$ and $\angle C = 3x$. Since all angles in a triangle must sum to $180°$ we can set the sum of our assigned variables equal to $180°$: $x + 2x + 3x = 180$. Solving this, we find that x, or $\angle A$, equals $30°$. Thus, $\angle B$ is $60°$ and $\angle C$ is $90°$. The question asks for the sum of $\angle B$ and $\angle C$, so the answer is D, $150°$.

10. **D.** If $\angle X$ is $57°$ and $\angle Y$ is $32°$, and all angles in a triangle sum to 180, then $\angle Z$ must be $91°$. Since the side opposite the largest angle is the largest side, and the side opposite the smallest angle is the smallest side, $XY > YZ > XZ$. Only choice D, stating that $XZ > YZ$, is not true.

11. **E.** This problem is testing for knowledge of geometrical theorems. We must remember that an exterior angle in a triangle is equal to the sum of the two remote interior angles. The interior angles remote to exterior angle $\angle GEF$ would be base angles $\angle D$ and $\angle F$. The sum of these base angles is $72°$. See Angles & Lines chapter to review the exterior angle theorem.

12. **D.** Because of the parallel lines, we know the exterior angle of the triangle at the top of the diagram equals the included angle between a and b in the lower triangle. Thus, we can redraw the lower triangle such that the side b is matched with the side b in the upper triangle, and the base becomes a. We thus have a picture of both triangles on the same flat plane, each with a base of length a, and with sides b touching/overlapping. From this picture we can tell both triangles also share the same height. Two triangles with the same base and height will have the same area, so the area of the other triangle is also 30. Alternatively we could think of this using trigonometry: since the angles of the triangles formed by the transversal across the parallel lines are supplementary, the values of the sine of the angle between sides *a* and *b* are the same. Since the area of a triangle is equal to one half the sine of an angle times the lengths of the adjacent sides to that angle, and the sides of the two triangles are equal, the areas of the triangles are equal. If you're interested in this 2nd method, Google formulas for sine & triangle area. It's beyond what I think most students need, but it's fun to know.

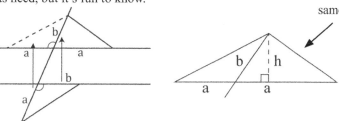

same height, same base, same area

13. **B.** This question is concerned with the $30-60-90$ special right triangle. This theorem says that the side length opposite the $30°$ angle is x units, the side length opposite the $60°$ angle is $x\sqrt{3}$ units, and the side length opposite the $90°$ angle is $2x$ units. Since $\triangle XYZ$ is an equilateral triangle, $\angle YXZ$ must be $60°$, bisected angle $\angle XYA$ must be $30°$, and because $\overline{YA}$ is the altitude, by the definition of an altitude $\angle YAX$ must be $90°$. Therefore, side $\overline{YX}$, which is 8 units, constitutes as side $2x$. That tells us that $x=4$, so it follows that $\overline{YA}$ is $x\sqrt{3}$, in our case $4\sqrt{3}$.

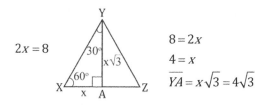

14. **A.** First we must find the value of unknown variable x. Since all angles in a triangle sum to $180°$, we can set the sum of the given expressions to the that as well: $(x-12)+(3x+40)+(3x+40)=180 \rightarrow 7x+68=180 \rightarrow 7x=112 \rightarrow x=16$. Since the question is asking for the vertex angle, we plug x into the vertex angle expression $(x-12)$ and find that the vertex angle is $4°$.

15. **E.** Both the angles in $\triangle ADE$ and the angles in $\triangle ABC$ must sum to $180°$. The two triangles share one common angle, $\angle A$, so for both triangles the sum of the remaining angles must be equal to $180-\angle A$. $180-\angle A=w+x=z+y$, so Choice E is correct.

16. **C.** We can first use the length of $\overline{AC}$ and the area of $\triangle ABC$ in order to find the length of altitude $\overline{DB}$. Using the formula for the area of a triangle: $16=\left(\frac{1}{2}\right)(8)h=\left(\frac{1}{2}\right)(8)(\overline{DB})$, we can find the height $\overline{DB}$ to be 4. Next we can use the area of $\triangle ABD$ and its height to find length $\overline{AD}$. Using the formula for the area of the triangle: $10=\left(\frac{1}{2}\right)b(4)=\left(\frac{1}{2}\right)(\overline{AD})(4)$, we find that the base length $\overline{AD}$ is 5.

17. **B.** Since $\triangle XYA$ and $\triangle XZA$ have two sides of common length (their hypotenuses and $\overline{XA}$), we can deduce that their third sides, $\overline{YA}$ and $\overline{AZ}$ must be congruent, each 10 inches long. Now we can see that both triangles have a leg of 10 and hypotenuse of 26. If we can recognize that the doubled Pythagorean triple of $5-12-13$, we can save time by finding the length of

$\overline{XA}$ to be 24. If not, we can use the Pythagorean theorem to solve for $\overline{XA}$. Either way, we must add $\overline{XA}$, 24 inches, to $\overline{AB}$, given as 12 inches, in order to find the total length of the shaft, which is 36 inches.

18. **D.** $\angle A$ in $\triangle ABE$ and $\angle A$ in $\triangle ACD$ are congruent to each other. $\angle BEA$ in $\triangle ABE$ and $\angle CDA$ in $\triangle ACD$ are congruent to each other. Therefore, since the two angles are congruent, the third must also be congruent, and the triangles are similar. Thus, the corresponding sides must be proportional to each other. We can solve this by setting up a proportional equation, placing the smaller triangle's sides over the larger triangle's corresponding side: $\frac{\overline{EB}}{\overline{DC}} = \frac{\overline{AE}}{\overline{AD}} \rightarrow \frac{10}{15} = \frac{\overline{AE}}{20}$. Solve and we find that $\overline{AE} = 13\frac{1}{3}$.

19. **D.** The sum of the lengths of any two sides of a triangle must be greater than the third side. Therefore, if one side of the triangle is 9 inches, then the sum of the two other sides must be at the least 10 inches since the sides must be integers. Thus, the smallest possible perimeter is 19 inches.

20. **A.** If two angles in a triangle are congruent to two corresponding angles in a second triangle, then these triangles are similar. The triangles formed by Block A and Block B and their shadows each have a right angle and a shared angle. Thus, the two triangles are similar. In two similar triangles, the corresponding sides are proportional, so we can solve by setting up a proportional equation relating these corresponding sides. Let d equal the distance between the bikes. Relating the corresponding side of the two triangles, our equation is $\frac{40}{60} = \frac{75-d}{75}$. Solve and we find that $d = 25$. Note that it doesn't matter which value is in the numerator or denominator, so long as the relationship is consistent. For example, we could have set our equation as $\frac{40}{75-d} = \frac{60}{75}$, and the answer would have been the same. For more similar triangles, see Ch 15.

21. **D.** When the ladder rests against the side of a house that is perpendicular to the ground, it creates a right triangle with legs of 15 and 20 feet. To find the length of the ladder, we can use Pythagorean Theorem, $15^2 + 20^2 = c^2$. After algebraic calculation, we can see that c, or the length of the ladder, is equal to 25.

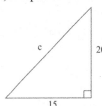

22. **A.** Remember that the corresponding ratio of the sides of a $30°-60°-90°$ triangle is $1:\sqrt{3}:2$, respectively. Since 10 is opposite of $30°$, and the side we are looking for is opposite $90°$, the ratio of 10 to the side we're solving for is equal to $1:2$. $\frac{10}{n} = \frac{1}{2}$, where n is the side we are seeking. Simple algebra shows that $n = 20$.

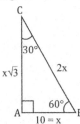

23. **E.** One way to solve this problem is to recognize that the diagram shown depicts half of a square, cut across the diagonal. Each triangle is a 45-45-90 special triangle. The diagonal of a 45-45-90 triangle is equal to the side times $\sqrt{2}$. Thus we can reason that $\angle FLM$ is 90 degrees and so $\overline{FM} = 7\sqrt{2}$. Since $\overline{LO}$ will be equal to $\overline{FO}$ or $\overline{OM}$, as this is an isosceles right triangle, at this point we need only divide $\overline{FM}$ by 2 to get the answer: $\frac{7\sqrt{2}}{2}$.

CHAPTER

13

CIRCLES

SKILLS TO KNOW

- Formulas for circle area, diameter, circumference
- Sectors of circles (area and arc length)
- Problem solving & circles: tangents, radii, & drawing more lines
- Circles, arc measures, and angles
- Pie charts: circle angles & fractions/percents/probability
- Area subtraction ("donut" problems)

NOTE: Circle equations are covered in **book 1, Chapter 21 Conic Sections**.

BASIC CIRCLE FORMULAS

Be sure you have all of these memorized, and if you're prone to careless errors, always double check which element you need when you read the question! Remember the **diameter** is the span across an entire circle. The **radius** is half the length of the diameter, stretching from the center point to any point on the perimeter of the circle. The **circumference** is the perimeter distance around a circle.

> Area: $A = \pi r^2$
> Diameter: $d = 2r$
> Circumference: $C = \pi d$ or $C = 2\pi r$
> Where r is the radius of the circle, d is the diameter of the circle, and
> pi (π) is approximately equal to 3.14.

If the circumference of a circle is 120π, what is the area of the circle?

A. 60   B. 60π C. 120   D. 3600   E. 3600π

For this problem, we'll need to first use what we know to find the radius, and then use that information to find the area. Remember if you find the radius, you can always use that to find any basic circle parameter such as area or circumference.

First, because $C = 2\pi r$, we set:
$$120\pi = 2\pi r.$$

Now we solve for r, dividing both sides by 2π:
$$\frac{120\pi}{2\pi} = \frac{2\pi r}{2\pi}$$

$$\frac{120\pi}{2\pi} = \frac{\cancel{2}\pi r}{\cancel{2}\pi}$$

Simplifying, the π's cancel on both sides, the 2's cancel on the right, leaving:

$$\frac{120}{2} = r$$
$$60 = r$$

But we're not done! The question asks for the **area**. Plugging in: $A = \pi r^2 = \pi(60^2)$ or 3600π.

Answer: **E**.

CIRCLE SECTORS

Sectors are "slices" of a circle (think like a pizza or pie slice). Arcs are the length of the circumference that the sector involves.

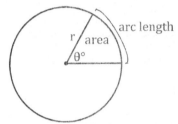

Problems involving sectors and arcs often ask you to find the area of the sector, the length of the arc, or the angle of the sector. All of the formulas we use are based on the basic idea that the angle measure of a sector and its respective arc length and area are proportional, or $\frac{\theta°}{360°} = \frac{\text{PART}}{\text{WHOLE}}$.

Nearly all circle sector problems on the ACT are in **degrees**, so we'll focus on those equations. If you're also taking the SAT, you should learn the sector formulas for radians as well.

AREA OF A SECTOR FORMULA

$$\frac{\text{SECTOR AREA}}{\text{TOTAL AREA}} = \frac{\theta°}{360°}$$

$$\frac{S}{\pi r^2} = \frac{\theta°}{360°} \quad \text{or} \quad S = \frac{\theta°}{360°}\pi r^2 \quad \text{or} \quad S = \frac{\theta°}{360°}A$$

Where A is the area of the whole circle, S is the area of the sector, θ is the measure of the sector angle in degrees, and r is the radius.

ARC LENGTH FORMULA

$$\frac{\text{ARC LENGTH}}{\text{TOTAL CIRCUMFERENCE}} = \frac{\theta°}{360°}$$

$$\frac{L}{2\pi r} = \frac{\theta°}{360°} \quad \text{or} \quad L = \frac{\theta°}{360°}2\pi r \quad \text{or} \quad L = \frac{\theta°}{360°}C$$

Where L is the arc length, C is the circumference of the circle, θ is the measure of the sector angle in degrees, and r is the radius.

If $\angle ACB = 80°$, and $AC = 5$ cm, what is the arc length of $\overset{\frown}{AB}$?

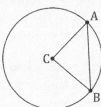

A. $\dfrac{10}{9}\pi$ cm **B.** $\dfrac{20}{9}\pi$ cm **C.** 5π cm **D.** $5\sin 40°$ cm **E.** $10\sin 40°$ cm

We are given the length of $\overline{AC}$, which is the radius. Let's label it **5**. With this, we can find the circumference using the equation $C = 2\pi r$.

Plugging in r, we get:

$$C = 2\pi 5 \text{ or } 10\pi$$

Now we use our arc length formula. $\angle ACB$ is $80°$, or θ in our equation.

$$\frac{L}{C} = \frac{\theta°}{360°} \text{ so plugging in we get } \frac{L}{10\pi} = \frac{80°}{360°}$$

Cross multiplying we get:

$$(10\pi)(80) = 360L$$

Now we isolate the L and simplify:

$$\frac{800\pi}{360} = L \quad \text{which simplifies to} \quad \frac{20\pi}{9} = L$$

Answer: **B.**

PROBLEM SOLVING & CIRCLES: TANGENTS, RADII, & DRAWING MORE LINES

Circle problems can sometimes overwhelm students. Where do you start? What do you know?

One element that gives you additional information in a circle problem is a **tangent line**. By definition, **tangent lines** that touch circles create **right angles with the radius of the circle**.

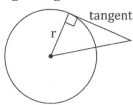

When you see the word TANGENT in a problem involving a circle:
- **Draw a radius** from the center to the tangent point(s).
- **Mark the angle** formed by a radius and tangent line as 90 degrees.
- **Look for right triangles** or draw any necessary additional lines to create them.
- **Use the Pythagorean Theorem, SOH CAH TOA, or other rules** you know as appropriate to solve.

CIRCLES   SKILLS

RADII are your FRIENDS!

Another element to keep in mind is the **RADIUS** of the circle. Often, general circle problems require that you **draw more radii** than are already drawn for you. Over half the time when my students are stuck on circle problems, they failed to recognize or draw radii.

When you see a circle with lines or intersecting/inscribed shapes:
- **Draw all the radii you can** to any points given on the circle (and particularly to any tangents)
- **Look for triangles** in circles or draw lines to make triangles (particularly right triangles or isosceles ones).
- **Radii are always equal**; mark them as such.
- **Mark triangles formed by two radii as isosceles.** The opposite angles will be equal and the altitude will always be a perpendicular bisector to the other side.

When in doubt, DRAW MORE LINES!
- **Draw additional lines or chords** to form right or isosceles triangles, squares, rectangles or other shapes formed by existing points, if possible.
- **Cut complex shapes into more manageable pieces**, particularly with area or sector length problems.

Let's take a look at an example and apply these ideas!

What is the length of the chord $\overline{AB}$ if the radius of the inner circle is 5 and the radius of the larger circle is 8? (Chord $\overline{AB}$ is tangent to the inner circle.)

Remember our bullet points: because radii are our friends, we draw radii to the points marked, A and B. Because we see the word tangent, we draw a radius to the point of tangency, and mark the angle formed as 90 degrees. These radii create right triangles:

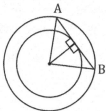

Because we know the radii of the 2 circles, we can solve for the triangle.

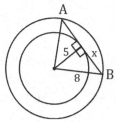

$$8^2 = 5^2 + x^2$$
$$64 = 25 + x^2$$
$$39 = x^2$$
$$x = \sqrt{39}$$

But we need the length of the whole chord. We know our drawing is symmetric. We know the other triangle has the same features and is identical. The two triangles have sides $5, \sqrt{39}, 8$. To find the whole chord, we simply multiply the half chord by two. . .

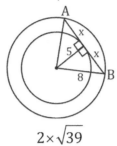

Answer: $2\sqrt{39}$.

CIRCLE ARC MEASURES AND ANGLES

Angles

Central angles (angles formed by two radii that touch the circle in two places) in a circle are always the same degree measure as their corresponding arcs. For example, in the following picture, the arc is $47°$ and so is the central angle:

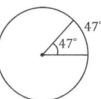

Inscribed angles (formed by two chords) are **half the corresponding arc measure**. For example, a 50 degree angle always opens to a 100 degree arc. A 90 degree angle always opens to a 180 degree arc. In fact, this is the example I use to remember this rule if I forget it, because it's so common to see in problems.

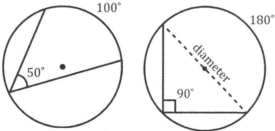

It's also good to note that **any inscribed angles that measure 90 degrees always open to the diameter of the circle**. Thus if a triangle is inscribed in a circle, and one side is the diameter, it's a right triangle. When you know an inscribed angle measures 90 degrees, it intercepts the diameter.

CIRCLES SKILLS

When a square or rectangle is inscribed in a circle, its angles are inscribed right angles, so its diagonals are diameters, equal in length, bisecting each other at the radius.

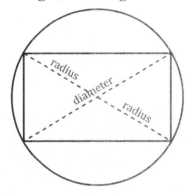

Pie Charts: Circle Angles & Fractions/Percents/Probability

Some ACT® problems will ask you for the **central angle** for a segment of a pie chart or circle graph. Remember that pie charts are composed of segments that proportionally represent data. Also remember that a circle's central angle encompasses 360°, so you want to figure out a proportional amount **of** 360°. Remember **of** means **multiply**! To find the central angle of an element in a pie chart, first figure out the fraction of the whole, the percent over 100, or the probability of the element in question, which equals **the part divided by the whole**. Then multiply this fraction by 360 degrees to find the degree measure for the pie chart segment.

DEGREE MEASURE OF PIE CHART SEGMENT

$$\frac{Part}{Whole} \times 360° = \text{Degree Measure of Slice in a Pie Chart}$$

You can also create a proportion to solve these:

PROPORTION OF PIE CHART SEGMENT

$$\frac{Part}{Whole} = \frac{\text{Degree Measure of Pie Piece}}{360°}$$

The idea of proportionality also extends to the **area of a pie piece**. For example, whenever a circle represents **probability**, the area of any segment divided by the area of the whole circle equals the probability of the event that segment represents. When a circle represents data, likewise, **the percent or fraction** that represents a part of the data is equal to the area of the pie piece over the area of the whole pie chart.

PROPORTIONALITY USING AREA

$$\frac{Part}{Whole} = \frac{\text{Area of Pie Piece}}{\text{Area of Whole Pie Chart}(\pi r^2)}$$

$$\frac{Part}{Whole} \times \text{Area of Whole Pie Chart}(\pi r^2) = \text{Area of Pie Piece}$$

> David wants to draw a circle graph showing all the favorite desserts of his friends. When he polled his friends, $\frac{2}{5}$ said ice cream, $\frac{1}{5}$ said cake, $\frac{3}{10}$ said cookies, $\frac{1}{20}$ said candy, and the remaining friends said other desserts. What is the angle measure of the segment of friends who preferred other desserts?

First, we need to know what fraction of friends said other desserts (our "part over whole"). We find this by subtracting all the other fractions from 1 (all these fractions must sum to 1, or one whole "pie", so to find the remaining fraction, we subtract all the parts we know from the whole).

CALCULATOR TIP: If it's easier, convert these to fractions in your calculator to decimals. Alternatively, use lots of parentheses, subtracting one fraction at a time, and let your calculator do the work.

$$1 - \left(\frac{2}{5}\right) - \left(\frac{1}{5}\right) - \left(\frac{3}{10}\right) - \left(\frac{1}{20}\right) = \frac{1}{20}$$

Thus $\frac{1}{20}$ of his friends said other desserts. Now, we multiply $\frac{1}{20} \cdot 360$ to find the proportional amount of the central angle that will create an appropriate sized sector for these respondents.

$$\frac{360}{20} = 18°$$

Answer: $18°$.

DONUT AND COMPLEX CIRCLE AREA PROBLEMS

Sometimes, you'll be asked to find the area of something that isn't quite a circle or a circle sector, but a part of a circle or a shape that's made from overlapping circles. Whenever you have a funky shaped area to solve for, try to figure out what "shapes" are subtracted from larger shapes to create this resultant shape. Then subtract the area of the elements that are "cut out" from the larger shape's area.

You can apply a similar technique to problems involving circumference or perimeter: divide the funky shape into segments, solve for each, and add the results together. Alternatively, cut up the shape into pieces and rearrange the position of elements to find an easier way to solve for the needed value.

CIRCLES — SKILLS

The figure below shows a face of a small circular washer. If the inner circle has a radius of 2 centimeters while the width of the washer is 3 cm, what is the total area, in square centimeters, of the shaded area of this washer?

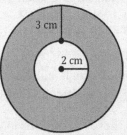

To find the answer, we need the area of the biggest circle (a circle of total radius 5, which we find by adding 2 and 3). Then we'll subtract the area of the smaller circle (with radius 2):

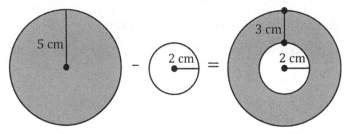

$$\textit{Area of Large Circle} - \textit{Area of Small Circle} = 5^2\pi - 2^2\pi$$
$$= 25\pi - 4\pi$$
$$= 21\pi$$

Answer: 21π.

In the figure below, four congruent circles are equally spaced within a single larger circle. Two of the smaller circles are tangent to each other and the sides of the larger circle. The other two smaller circles are also tangent to each other and the larger circle. The largest circle's circumference is 64π. What is the area of the shaded region?

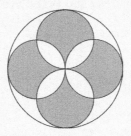

There are a few ways to solve this. The fastest is hard to see (skip to the end if you think you know it), so we'll start with a more involved version. Though this problem has a shortcut, not all similar problems do. For this problem, we can figure out the area of one small circle and then subtract two "petals" from that area. Then we'll multiply that funky shape by four.

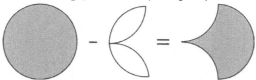

First, let's find the area of the small circle. We know the large circle's circumference is 64π. Thus the diameter must be 64. Thus the diameter of one smaller circle is 32, and its radius is 16.

Now the area of a circle is πr^2 so for one small circle, $16^2 \pi = \text{area} = 256\pi$.

Now we must find the area of one petal. Because we know the shape is even, we can assume each intersecting arc forms a $90°$ span. There are $360°$ in the circles in total, and the "petals" are evenly spaced about that around the very center of the picture, each at $90°$. Thus the petal is formed by two identical curved "parts" of a quarter circle sector:

We can find the area of half of this "petal" by first finding the area of $\frac{1}{4}$ of one of these smaller circles, and then subtracting off the triangular part:

$$\frac{256\pi}{4}=64\pi$$

So the $\frac{1}{4}$ circle area is 64π. Now I want to subtract off the triangle in this "pie piece" or sector. I know the triangle has two legs that are part of a quarter circle, so this triangle is a right isosceles triangle (two sides are identical radii; the other side is the hypotenuse opposite $90°$).

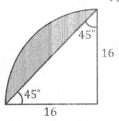

I can find the triangle's area by multiplying the radius times itself and dividing by 2. Since the radius is 16, I take $16^2 = 256$ and divide by 2 to get 128.

Now we subtract: $64\pi - 132 =$ area of half a petal.

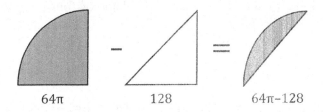

$64\pi \qquad\qquad 128 \qquad\qquad 64\pi-128$

Now we multiply by 2 to get a whole petal area: $2(64\pi - 128) = 128\pi - 256$.

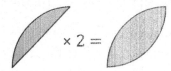

Now we subtract two whole petals from one smaller circle to get that funky shape we have four of:

$256\pi \qquad -2(128\pi-256)$

I distribute the negative two to get:

$$256\pi - 256\pi + 512 \text{ or } 512$$

Finally, we multiply the area of this "funky shape" by 4:

$$4(512) = 2048$$

That was complicated! And there is an easier way (but easy can be hard to see...)

We could also rearrange the two petals as cutouts on the small circles to form a square in the middle and realize the shaded region is equivalent to the area of an inscribed square.

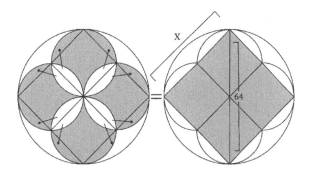

Then we could find the side length of the large square, using what we know about the circle diameters and **45-45-90** triangles, and find the area of the large square. This way may be faster, but it's less intuitive to "see."

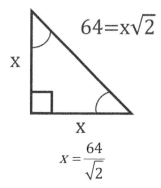

We would need x^2 or $\left(\dfrac{64}{\sqrt{2}}\right)^2 = \dfrac{64^2}{2} = \dfrac{4096}{2} = 2048$

CIRCLES — QUESTIONS

1. In the figure shown below, two congruent semicircles are adjacent to one another inside a bigger semicircle. The diameter of each congruent semicircle is 6 cm. What is the sum of the lengths, in centimeters, of the three arcs of these semicircles $\overset{\frown}{AB}$, $\overset{\frown}{BC}$, and $\overset{\frown}{AC}$?

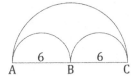

   A. 6π
   B. 9π
   C. 12π
   D. 15π
   E. 18π

2. In the figure below, the corners of the rectangle with sides of length 8 and 15 is inscribed in a circle. What is the area of the circle?

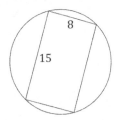

   A. 289π
   B. $\dfrac{289\pi}{4}$
   C. $\dfrac{289\pi}{2}$
   D. $2\sqrt{161}\pi$
   E. $\dfrac{161\pi}{4}$

3. In the figure below, square $WXYZ$ with side lengths $= 4$ is inscribed in a circle with center C. What is the area of the circle?

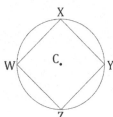

   A. 4π
   B. 8π
   C. 12π
   D. 16π
   E. 64π

4. In the circle shown below, chords $\overline{AC}$ and $\overline{BD}$ intersect at E, which is the center of the circle, and the measure of minor arc $\overset{\frown}{CD}$ is $20°$. What is the measure of $\angle BCE$?

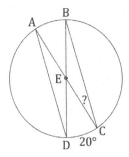

   A. $10°$
   B. $15°$
   C. $20°$
   D. $25°$
   E. $30°$

5. In the figure below, central angle $\angle EGF$ is $120°$ and the arc $\overset{\frown}{EF}$ is 3π. What is the circle's diameter?

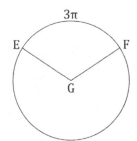

   A. 3
   B. 3π
   C. 9
   D. 9π
   E. 5π

6. In the circle below centered at D with radius 5, the arc length of the obtuse sector is 8π. What is the value of angle $\angle EDF$ in degrees?

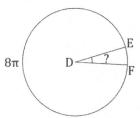

   A. $12°$
   B. $25°$
   C. $36°$
   D. $72°$
   E. $144°$

7. In the circle below with diameter 8, the central angle intercepting arc $\overset{\frown}{CD}$ is 52°. What is the length of arc $\overset{\frown}{CD}$?

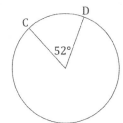

A. 52π

B. $\dfrac{52\pi}{45}$

C. $\dfrac{7\pi}{6}$

D. $\dfrac{10\pi}{9}$

E. $\dfrac{104\pi}{45}$

8. The diagram below shows a quarter of each of 2 circles both having point R as their center. Point S lies of $\overline{TR}$ and point Q lies on $\overline{PR}$. The length of $\overline{PQ}$ is $x-2$ centimeters and the length of $\overline{QR}$ is $x+1$ centimeters. What is the area, in square centimeters, of the shaded portion of the entire circle?

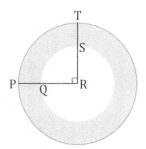

A. $(3x^2 - 2x)\pi$

B. $(3x^2 - 6x)\pi$

C. $(3x^2 - 6x + 2)\pi$

D. $(3x^2 - 2x + 2)\pi$

E. $(3x^2 + 4x + 2)\pi$

9. Becca is making a bet with her friend that she can throw a dart and hit the bull's eye (innermost circle). If the outermost circle has a diameter of 20 inches and there is a 2-inch difference in radius for each inner circle, what are the chances that Becca will win the bet?

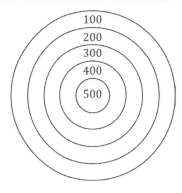

A. $\dfrac{49}{100}$

B. $\dfrac{21}{50}$

C. $\dfrac{1}{25}$

D. $\dfrac{4}{25}$

E. $\dfrac{9}{25}$

10. To make one of a pair of googly eyes, Alex pastes a dark circular piece of felt onto a larger circular piece of plastic, as shown below. The radius of the larger circle is 5 centimeters. If the smaller circle's area is $\dfrac{1}{3}$ the of the uncovered area of the white circle, what is the circumference, in centimeters, of the smaller, dark circle?

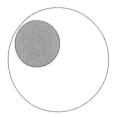

A. $\dfrac{5\pi}{2}$

B. $\dfrac{10\pi}{3}$

C. $\dfrac{10\pi}{\sqrt{3}}$

D. $\dfrac{25\pi}{4}$

E. 5π

CIRCLES QUESTIONS

11. A square with side length 4 is inscribed in a half circle with diameter 10. The area of the space in the half circle to the left of the square is equal to 3π. What is the area of the shaded region?

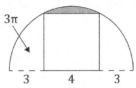

A. $94\pi - 16$
B. $19\pi - 16$
C. $41\pi - 16$
D. $\dfrac{13\pi}{2} - 16$
E. $13\pi - 16$

12. In the figure below, the square has side length of 16 inches. The circles within the square are congruent, and each circle is tangent to 2 of the other circles. The region that is interior to the square and exterior to all 4 circles is shaded. What is the perimeter, to the nearest inch, of the shaded region?

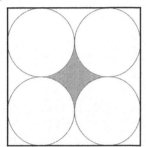

A. 16
B. 32
C. 4π
D. 8π
E. 16π

13. In the figure below, each of the congruent circles are tangent to another circle or the edge of the square with side length 18. There are 16 circles. What is the circumference of one of these circles?

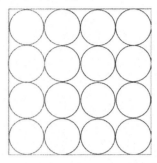

A. 9π
B. $\dfrac{9\pi}{2}$
C. $\dfrac{9\pi}{4}$
D. $\dfrac{81\pi}{4}$
E. $\dfrac{81\pi}{16}$

14. In the figure below, chord $\overline{PR}$ intersects chord $\overline{SQ}$ at point T. If segment ST is equal to 8, what is the length of segment TQ? (Note: The product of segments of intersecting chords are equal: $PT \cdot TR = QT \cdot TS$)

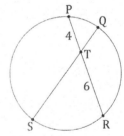

A. 1
B. 2
C. 3
D. $\dfrac{3}{2}$
E. 4

208 CHAPTER 13

15. In the figure below, three tangent circles with centers $X, Y,$ and Z and with respective radii $x, y,$ and z are shown. The perimeter of the triangle $\triangle XYZ$ is 30. What is $x + y + z$?

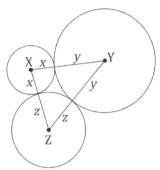

- A. 30
- B. 15
- C. 10
- D. $\dfrac{15}{2}$
- E. 15π

16. In the figure below, the circles centered at R and U intersect at points Q and V. The points R, S, T and U are collinear. If the lengths of $\overline{ST}$, $\overline{RV}$, and $\overline{VU}$ are 1, 6, and 4 respectively, what is the length of the segment $\overline{RU}$?

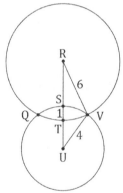

- A. 9
- B. 10
- C. 8
- D. 7
- E. $\sqrt{20}$

17. In the figure below, the points $A, B,$ and C lie on the circle centered at point O. If angle $\angle ABO = 8°$, what is $\angle OCA$?

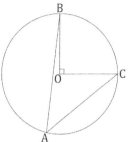

- A. 37
- B. 42
- C. 45
- D. 47
- E. 50

18. In the figure below, a circle is inscribed in a square with side s, what is the length of segment $\overline{OD}$ in terms of s?

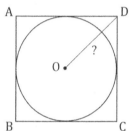

- A. s
- B. $2s$
- C. $s\sqrt{2}$
- D. $s\sqrt{3}$
- E. $\dfrac{s\sqrt{2}}{2}$

19. In the figure below, the chord $\overline{XW}$ of length 20 is parallel to the diameter $\overline{UV}$ of length 30. If the circle is centered at point Y, what is the length of the segment $\overline{YZ}$?

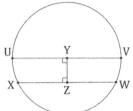

A. 5
B. $5\sqrt{5}$
C. $10\sqrt{5}$
D. $\dfrac{5\sqrt{5}}{2}$
E. $\sqrt{5}$

20. A circle with radius 6 has a sector with central angle of 48°. What is the area of the sector?

A. $\dfrac{12}{5}\pi$
B. $\dfrac{4}{5}\pi$
C. $\dfrac{3}{4}\pi$
D. $\dfrac{24}{5}\pi$
E. $\dfrac{48}{5}\pi$

21. The Ferris wheel at the amusement park has a radius of 50 feet, rotates at a constant speed, and completes 1 rotation in 5 minutes. How many feet does a Ferris wheel passenger car travel along the circular path in 30 seconds?

A. 12π
B. 10π
C. 20π
D. 10
E. 20

22. Jenny is sewing a tablecloth for a circular table with a radius of 2 feet for a Girl Scouts merit badge. The finished tablecloth must hang down 8 inches over the edge of the table. The tablecloth's edges must be hemmed to be considered finished, which means she will have to fold over 1 inch of the material at the edge. Jenny wants to use only one piece of fabric, a rectangular piece that is 6 feet wide. What is the shortest length of fabric, in inches, Jenny could use without needing to use a second piece of fabric?

A. 16
B. 40
C. 42
D. 64
E. 66

23. Given two circles of different diameters, what is the maximum number of points of intersection that the two circles can have?

A. 0 only
B. 0 or 1 only
C. 2 only
D. 0, 1, or 2
E. Infinitely many

24. Stacy is designing a new logo for Candy Cane Cable Company. Her design consists of two concentric circles. The bigger circle has a radius of 8 feet. The smaller circle is one fourth the area of the bigger circle. Which of the following is an expression for the area, in square feet, of the smaller circle?

A. $\dfrac{8}{4}\pi$
B. $\left(\dfrac{8}{4}\right)^2 \pi$
C. $\dfrac{64}{4}\pi$
D. 4π
E. 2π

25. What is the area of a circle having the points $(5,3)$ and $(-7,-13)$ as endpoints of a diameter?

    A. 10π
    B. 100π
    C. 400π
    D. 20π
    E. 200π

CIRCLES   ANSWERS

ANSWER KEY

1. C   2. B   3. B   4. A   5. C   6. D   7. B   8. B   9. C   10. E   11. D   12. D   13. B   14. C
15. B   16. A   17. A   18. E   19. B   20. D   21. B   22. E   23. D   24. C   25. B

ANSWER EXPLANATIONS

1. **C.** The two smaller semi circles have diameter $=6$ while the larger semi-circle has diameter $=6+6=12$. The length of an arc is equal to half the length of the full circle's circumference. So, using the formula $C = \pi d$, the sum of arc-lengths $\overarc{AB}$, $\overarc{BC}$, and $\overarc{AC}$ is equal to $\frac{1}{2}\pi(6) + \frac{1}{2}\pi(6) + \frac{1}{2}\pi(12) = 3\pi + 3\pi + 6\pi \rightarrow 12\pi$.

2. **B.** Since the rectangle is inscribed in the circle, the diagonal of the rectangle is equal to the circle's diameter. We find the length of the rectangle's diagonal using the Pythagorean Theorem $a^2 + b^2 = c^2$. Plugging in $a = 8$ and $b = 15$, we get $8^2 + 15^2 = c^2 \rightarrow c = \sqrt{8^2 + 15^2} \rightarrow c = \sqrt{64 + 225} \rightarrow c = \sqrt{289} = 17$. Now, we know that the diameter of the circle is equal to 17, so the radius of the circle is $\frac{17}{2}$. We now plug in $\frac{17}{2}$ for the radius of the circle to find the area of the circle using the formula $A = \pi r^2$. $A = \left(\frac{17}{2}\right)^2 \pi \rightarrow \frac{289\pi}{4}$.

3. **B.** We are given the side lengths of square $WXYZ$: 4. We can draw out a triangle using the points XCW. We know that $XC = CW$ because they are both radii of the circle and that they are perpendicular because they lie on the diagonal of the square (diagonals of squares and kites are perpendicular). Thus, using the Pythagorean theorem, we can set up the equation $r^2 + r^2 = 4^2 \rightarrow 2r^2 = 16 \rightarrow r^2 = 8 \rightarrow r = \sqrt{8}$. Using the formula to find the area of a circle ($A = \pi r^2$), we can solve for the area: $A = \pi\left(\sqrt{8}\right)^2 \rightarrow A = 8\pi$.

4. **A.** Since $\angle DBC$ is the inscribed angle of $\overarc{DC}$, it measures half of $\overarc{DC}$'s central angle, which makes $\angle DBC = 10°$. $\overline{BE}$ and $\overline{CE}$ are equal since they are both radii of the circle, which makes the angles opposite them in the triangle equal. Thus, $\angle BCE = \angle DBC = 10°$.

5. **C.** We wish to first find the circumference by finding the arc length that would be proportional to $360°$ as 3π is proportional to the central angle of $\angle EGF = 120°$. To solve this, we set up the proportions $\frac{3\pi}{120°} = \frac{x}{360°}$ and solve for x. Cross multiplying the equation, we get $3\pi(360) = 120x$. Dividing by 120 on both sides, we get 9π is the circumference of the circle. Since we know that $C = \pi d$, we can solve for the diameter of the circle by solving $9\pi = \pi d$. Dividing π on both sides, we get $d = 9$.

6. **D.** Using the formula $C = 2\pi r$, we get $C = 2(5)\pi = 10\pi$. Since we know the arc length of the obtuse sector is 8π, we know the arc length of the acute sector is $10\pi - 8\pi = 2\pi$. We wish to first find the angle that would be proportional to 2π as the circumference 10π is proportional to $360°$. To solve this, we set up the proportions $\frac{2\pi}{x°} = \frac{10\pi}{360°}$ and solve for x. Cross-multiplying the equation, we get $2\pi(360) = 10\pi x$. Dividing by 10π on both sides, we get $72° = x$.

7. **B.** Using the formula $C = \pi d$, we find $C = 8\pi$. We wish to first find the arc length that would be proportional to $52°$ as the circumference 8π is proportional to $360°$. Remembering that part over whole equals part over whole, set up the proportion $\frac{x}{8\pi} = \frac{52°}{360°}$ and solve for x. Cross multiplying the equation, we get $360x = 8(52)\pi$. Dividing both sides by 360, we get $x = \frac{8(52)\pi}{360} \rightarrow \frac{52\pi}{45}$.

8. **B.** The area of the shaded region is equal to the area of the big circle minus the area of the small circle. The big circle has radius $PR = x - 2 + x + 1 = 2x - 1$. The small circle has radius $QR = x + 1$. So, the area of the big circle is $A = \pi r^2 \rightarrow (2x-1)^2 \pi \rightarrow (4x^2 - 4x + 1)\pi$. The area of the small circle is $A = \pi r^2 \rightarrow (x+1)^2 \pi \rightarrow (x^2 + 2x + 1)\pi$. Now, we find the shaded area by subtracting the two areas we just found:

$(4x^2-4x+1)\pi-(x^2+2x+1)\pi=(4x^2-4x+1-x^2-2x-1)\pi \to (4x^2-x^2-4x-2x+1-1)\pi \to (3x^2-6x)\pi$.

9. **C.** The probability that Becca will hit the bulls eye can be calculated by the area of the innermost circle divided by the area of the outermost circle. We know that the diameter of the outermost circle is 20 inches, which means the radius is $r=\frac{1}{2}d=\frac{1}{2}(20)\to 10$. Since each inner circle has a radius 2 inches smaller than the next bigger circle, we know that the second outermost circle has radius $=10-2=8$ inches, the third outermost circle has radius $=8-2=6$ inches, the fourth outermost circle has radius $=6-2=4$ inches, and the innermost circle has radius $=4-2=2$ inches. Using the formula $A=\pi r^2$, we calculate the area of the innermost circle to be $A=\pi 2^2 \to 4\pi$ and the area of the outermost circle to be $A=\pi 10^2 \to 100\pi$. The probability can then be calculated as $\frac{4\pi}{100\pi}=\frac{4}{100}\to \frac{1}{25}$.

10. **E.** Knowing $A=\pi r^2$, we calculate the area of the big circle to be $\pi 5^2 = 25\pi$. Let x be the area of the smaller circle. We are told that the area of the smaller circle is equal to $\frac{1}{3}$ the uncovered area of the white circle, which means it is equal to $\frac{1}{3}$ (area of big circle – area of small circle). Plugging the values above into this expression, we get $x=\frac{1}{3}(25\pi-x)$. Multiplying by 3 on both sides gives us $3x=25\pi-x$. Adding x on both sides gives us $4x=25\pi$. Finally, dividing by 4 on both sides gives us $x=\frac{25\pi}{4}$. Using the formula $A=\pi r^2$ again, we find the radius of the small circle is $\frac{25\pi}{4}=\pi r^2 \to \frac{25}{4}=r^2 \to r=\frac{5}{2}$. Using the formula $C=2\pi r$, we find $C=2\pi\left(\frac{5}{2}\right)\to 5\pi$.

11. **D.** To solve, we'll find the area of the half circle and subtract off all the areas except the shaded portion. We first calculate the area of the half circle by using the formula $A=\frac{\pi r^2}{2}$. We are given that the diameter of the half circle is 10, so the radius is $\frac{10}{2}=5$. Plugging in $r=5$, we get $A=\frac{\pi(5)^2}{2}\to \frac{25\pi}{2}$. Now, we can calculate the area of the shaded region as the area of the half circle minus the areas of the non-shaded areas in the half circle. We are given that the space to the left of the square is 3π and the side of the square is 4, so the non-shaded areas add up to equal $3\pi+4(4)+3\pi=6\pi+16$. Subtracting this from the area of the half circle, we get $\frac{25\pi}{2}-(6\pi+16)=\frac{25\pi}{2}-6\pi-16 \to \frac{25\pi}{2}-\frac{12\pi}{2}-16 \to \frac{13\pi}{2}-16$.

12. **D.** The perimeter of the shaded region is equal to $4\left(\frac{1}{4}\right)C$ where $C=\pi d$. We are given that the side of the square has length of 16 inches, and since there are 2 circles aligned with each side of the square, the diameter of one circle is $\frac{16}{2}=8$ inches. Plugging in $d=8$, we get $C=\pi d=8\pi$. Plugging in $C=8\pi$, we get $4\left(\frac{1}{4}\right)C=4\left(\frac{1}{4}\right)(8\pi)\to 8\pi$.

13. **B.** The side of the square is 18, and there are 4 circles aligned with each side of the square, so the diameter of each of the circles is $\frac{18}{4}=\frac{9}{2}$. Using the formula $C=\pi d$ and plugging in $d=\frac{9}{2}$, we get $C=\frac{9}{2}\pi$.

14. **C.** According to the given formula, when two chords of a circle intersect within the circle, the product of the two segments of one chord is equal to the product of the other two segments of the other chord. So, the product of the segments of chord $\overline{PR}$ is equal to $4(6)=24$. Setting the product of the segments of chord $\overline{SQ}$ also equal to 24, we get $8x=24$ where $x=$ the length of segment $\overline{TQ}$. Thus, $\overline{TQ}=\frac{24}{8}\to 3$.

15. **B.** The perimeter of the triangle is equal to $\overline{XY}+\overline{YZ}+\overline{ZX}=x+y+y+z+z+x\to 2x+2y+2z \to 2(x+y+z)$. We are given this is equal to 30, so $2(x+y+z)=30 \to x+y+z=\frac{30}{2}\to 15$.

CHAPTER 13

CIRCLES ANSWERS

16. **A.** Since $\overline{RV}$ of length 6 connects the center of circle R to the side of the circle, it is the radius of circle R. We can observe that segment $\overline{RT}$ is also a radius of circle R and thus must also have length 6. Similarly, since $\overline{VU}$ of length 4 connects the center of circle U to the side of the circle, it is the radius of circle U. We can observe that segment $\overline{US}$ is also a radius of circle U and thus must also have length 4. The length of segment $\overline{RU}$ is $RT + US - ST$ because the middle segment $\overline{ST}$ is counted twice (once as part of $\overline{RT}$ and again as part of $\overline{US}$). Plugging in the values for the segments, we get $RU = RT + US - ST \rightarrow 6 + 4 - 1 = 9$. Note that we cannot use the Pythagorean theorem to solve for the third side of the triangle in this case because it is not stated that the triangle is a right triangle.

17. **A.** Since angle $\angle BOC$ is at the midpoint of circle O, any angle inscribed in circle O and tangent to points B and C and any other point on the circle between the larger arc $\overset{\frown}{BC}$ is equal to half of angle $\angle BOC$. So, angle $\angle BAC$ is equal to $\frac{90}{2} = 45°$. The sum of the angles in any 4-sided polygon is $360°$, so plugging in the angles in polygon $ABOC$ that we know, we get $45 + 8 + (360 - 90) + \angle OCA = 360 \rightarrow 45 + 8 + 270 + \angle OCA = 360 \rightarrow \angle OCA = 37$.

18. **E.** We can calculate the length of segment $\overline{OD}$ by imagining a right triangle formed by the midpoint of segment $\overline{AD}$ and points O and D. This triangle has sides of length $\frac{s}{2}, \frac{s}{2}$ and the hypotenuse $\overline{OD}$. Using the Pythagorean theorem or the 45-45-90 triangle rule, we can calculate the hypotenuse $\overline{OD}$ is equal to $\sqrt{\left(\frac{s}{2}\right)^2 + \left(\frac{s}{2}\right)^2} = \sqrt{\frac{2s^2}{4}} \rightarrow \frac{s\sqrt{2}}{2}$.

19. **B.** Since points Y and Z are midpoints of the segments $\overline{UV}$ and $\overline{XW}$ respectively, the length of $\overline{YV} = \frac{UV}{2} \rightarrow \frac{30}{2} = 15$ and the length of $\overline{ZW} = \frac{XW}{2} \rightarrow \frac{20}{2} = 10$. Since $\overline{UV}$ is the diameter of the circle, we know that $\overline{YV}$ is the radius. If we draw a segment connecting points Y and W, the segment $\overline{YW}$ will also be a radius and have length 15. Now, we can imagine a triangle formed by the points $Y, Z,$ and W. Using the Pythagorean theorem, we can solve for segment $\overline{YZ}$ by plugging in $a = 10$ and $c = 15$ and solving for b. We get $10^2 + b^2 = 15^2 \rightarrow 100 + b^2 = 225 \rightarrow b^2 = 125 \rightarrow b = 5\sqrt{5}$.

20. **D.** Using the formula $A = \pi r^2$, we can first find the area of the entire circle to be $A = \pi(6)^2 \rightarrow 36\pi$. This area corresponds to $360°$ of the circle. We now want to find the sector that corresponds to $48°$. Remembering that part over whole equals part over whole, we set up the proportion $\frac{48°}{360°} = \frac{x}{36\pi}$, where $x =$ the area of the sector. Cross multiplying, we get $48(36\pi) = 360x \rightarrow 48\pi = 10x \rightarrow x = \frac{48\pi}{10} = \frac{24\pi}{5}$.

21. **B.** The Ferris wheel has a circumference of 100π. If the Ferris wheel completes 1 rotation in 5 minutes, we can convert the five minutes to seconds using dimensional analysis: $5 \text{ min} \times \frac{60 \text{ sec}}{1 \text{ min}} = 300 \text{ seconds}$. We can now express the rate at which the ride travels as total rotation distance (circumference) per second, remembering that per means divide: $\frac{100\pi}{300 \text{ seconds}}$. If you divide top and bottom by 10, you find how much it rotates in 30 seconds: $\frac{100\pi/10}{300/10 \text{ seconds}} = \frac{10\pi}{30 \text{ seconds}}$. Thus, our answer is 10π.

22. **E.** The table has a 2 foot radius, meaning it is 4 feet, or 48 inches, across. In addition, the tablecloth must stretch down 9 inches on either side, 8 inches that hang and 1 inch for the hem. As such, the total width of the fabric must at least be $48 + 2(9) = 66$ inches.

23. **D.** The circles have different radii, which means they cannot be congruent and thus they cannot have infinitely many

points of intersection. Additionally, nowhere is it said that the two circle share the same center point. One must be larger, and one must be smaller, so there are three possible scenarios: that the smaller circle is completely inside or outside the larger circle, meaning there are 0 points of intersection, that the smaller circle and larger circle are tangent to one another, thus intersecting at one point only, and finally that the smaller circle is cut across by the large circle, meaning it has 2 points of intersection.

24. **C.** The area of the bigger circle can be calculated by the formula $A = \pi r^2 = \pi 8^2 = 64\pi$. The smaller circle's area is $\frac{1}{4}$ of the big circle's area, so it is $\frac{64\pi}{4}$.

25. **B.** Because we are given the points of the circle's diameter, using the distance formula, we can calculate the distance of the diameter: $\sqrt{(5-(-7))^2 + (3-(-13))^2} \to \sqrt{12^2 + 16^2} \to \sqrt{144 + 256} \to \sqrt{400} = 20$. Now that we know the diameter is 20, we also know that the radius is 10. Using the formula to find the area of a circle, $A = \pi 10^2 \to 100\pi$.

CHAPTER

14

POLYGONS

SKILLS TO KNOW

- Regular polygons
- Names and area formulas for various polygons
- Picture frames and border word problems

WHAT IS A POLYGON?

A polygon is an enclosed, two-dimensional shape that is made of entirely straight lines. All lines are connected, making vertices. Polygons contain no curved sides.

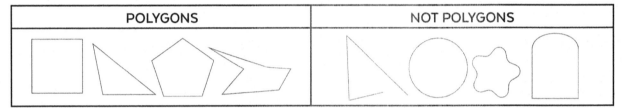

For any given polygon, the number of angles will always be equal to the number of sides. For example, a polygon with 5 sides will also have 5 angles.

We largely classify polygons by the number of sides they have. A **triangle**, for example has 3 sides (tri- being the prefix for three). Other shapes to know include the **quadrilateral** (4 sides), **pentagon** (5 sides), **hexagon** (6 sides), **heptagon** (7 sides, rare), and **octagon** (8 sides).

Perimeter of Polygons

One of the easiest elements of a polygon to solve for is its perimeter, equal to the sum of all the side lengths. One type of perimeter problem that occurs on the ACT®, which involves irregular polygons with lots of right angles, looks something like this:

Jamie received a new floor plan request for a house that she is designing. Assuming all walls meet at ninety degree angles, what is the perimeter of the new floor plan?

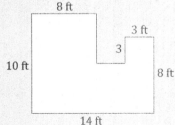

This kind of problem presents a drawing that gives many side lengths, but may not give all of them. However, because we know all the angles are $90°$, we can infer the other side lengths by looking at the "span" of parallel sides.

To find the perimeter, we need to find the sides labeled a and b below. To find b, we know that the horizontal lines all span a width of 14 ft: because the top lines on the figure are all parallel to that 14 ft side, if I imagine all the vertical lines disappearing, and collapse all the horizontal lines down to meet the bottom, they would span the same length as the total length along the bottom. Thus:

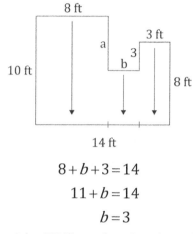

$$8 + b + 3 = 14$$
$$11 + b = 14$$
$$b = 3$$

Now the vertical line is a bit more tricky. We'll need to "section off" the 3 by 3 area on the right to see that side a begins 5 feet vertically higher than the base of the figure. We get this by subtracting 3, the height of the little "square" pop up on the right from the 8 on the far right of the figure.

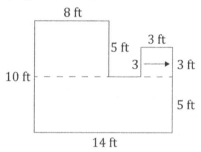

To find a, we now subtract this length of 5 from 10, the total span of the vertical figure given by the line on the left:

$$10 - 5 = 5$$

So $a = 5$.

Now we add all the sides together:

$$10 + 8 + 5 + 3 + 3 + 3 + 8 + 14 = 54$$

Answer: 54.

REGULAR POLYGONS

As seen below, **regular polygons** have equal sides and angles all around the shape.

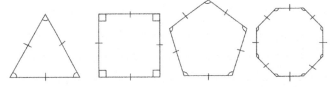

In contrast, **irregular polygons** do not have all equal sides and angles:

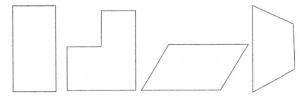

We can find the area of any regular polygon using a single formula:

AREA OF A REGULAR POLYGON FORMULA

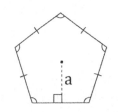

$$\text{Area for Regular Polygons} = \frac{apothem \times perimeter}{2}$$

Where the apothem (a) is the length of the distance from the center of the shape to a side at a right angle.

A simpler way to think of this formula is to cut your shape into multiple tiny isosceles triangles, which all converge at the center of the shape:

We can think of the area of each of these whole regular polygons as the sum of the areas of the smaller triangles. To find the area of any one of these little triangles, we calculate $\frac{1}{2}$ base (b in diagram below) times the height (a in the diagram below, aka the apothem, or distance to the center at a right angle). We then multiply the area of one little triangle by the number of sides.

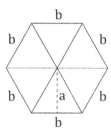

Area of the little triangle: $\frac{1}{2}ba$

Area of total hexagon (6 triangles): $=6\left(\frac{1}{2}ba\right)$

However, we could also simply add all the "bases" together first to find the perimeter before multiplying by the height of the little triangle (The apothem of the shape) and multiplying by $\frac{1}{2}$. The perimeter would be $6b$. We can factor out the b to visualize this algebraically:

$$(6b)(a)(1/2)$$

That is where this area formula comes from.

In any case, if asked to find the area of a regular polygon, you can either:
1. Memorize the above formula
   OR
2. Divide the polygon into isosceles triangles that converge at the center, find the area of one triangle, and multiply by the number of sides to find the total area.

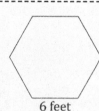

6 feet

A regular hexagon with side lengths of 6 feet is given above. What is the total area of the hexagon?

Let's use the triangle method to solve:

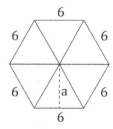

$b=6$. We can find a by figuring out the angles in the little triangle. I see the center of the shape cuts $360°$ into 6 equal slices.

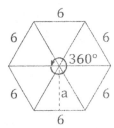

Thus, I can find the top angle of the little triangle by dividing 360 by 6 = 60 degrees. I know the little triangle is isosceles, so I can find the base angles by taking 180 − 60 = 120 degrees left for the bottom two equal angles. I divide **120** by **2** to get **60** degrees. Aha! This is an equilateral triangle (some of you may have had that memorized already; if so, bravo!). I don't even need SOHCAHTOA or law of sines to find this height, which in an isosceles (or equilateral) triangle is always a perpendicular bisector.

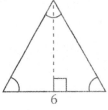

I can see above how the altitude cuts this little triangle into two 30-60-90 special triangles, and use this fact to solve for a. (If you don't know special triangles, see the Triangles chapter.)

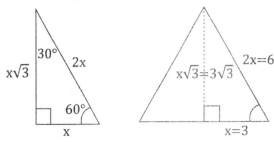

If the "hypotenuse" of the 30-60-90 triangle is **6**, the little base, x, (half of the base of our cut out triangle) is **3**, then the side a, $x\sqrt{3}$, is opposite **60** degrees and equal to $3\sqrt{3}$.

Now I can find the area of the little triangle: $\frac{1}{2}bh = \frac{1}{2}(6)(3\sqrt{3}) = 9\sqrt{3}$

I multiply by **6** (the number of little triangles I drew) to get the total area:

$$(6)(9\sqrt{3}) = 54\sqrt{3}$$

Answer: $54\sqrt{3}$.

True, finding the "apothem" is difficult in some circumstances, and slicing things into triangles can take time. Also, we don't always have regular polygons, and will need the area of non-regular shapes, too. We also use some of the more specific formulas delineated below to solve for polygon areas.

PARALLELOGRAMS

A parallelogram is a quadrilateral in which each set of opposite sides is both parallel and congruent (equal) with one another. The length may be different than the width, but both widths will be equal and both lengths will be equal.

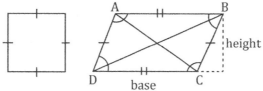

Parallelograms have equal opposite angles (the angles on the diagonals): in the diagram above, $\angle A = \angle C$ and $\angle B = \angle D$ The adjacent angles are supplementary (meaning any two angles that touch the same side of parallelogram will add up to **180** degrees). In the diagram above, $\angle A + \angle B = 180$, $\angle C + \angle B = 180$, $\angle C + \angle D = 180$, $\angle D + \angle A = 180$.

AREA OF A PARALLELOGRAM

$$Area = base \times height$$

 FUN FACT: the diagonals of a parallelogram bisect each other.

Rhombus

A rhombus is a parallelogram in which all four sides are equal.

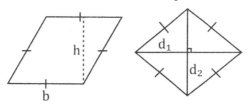

AREA OF A RHOMBUS

$$Area = bh \text{ or } \frac{d_1 \times d_2}{2}$$

Where b is the base, h is the height and d_1 and d_2 are the diagonals.

 FUN FACT: The diagonals of a rhombus form a **90°** angle. A parallelogram is a rhombus if and only if its diagonals form a right angle. All quadrilaterals with diagonals that form right angles are called **kites**. A rhombus is a special kind of **kite. The area of a kite is half the product of its diagonals.**

> What is the area of a rhombus with diagonals of length 5 and 8?

We can calculate the area of a rhombus as the product of its diagonals divided by two:

$$5 \times 8 = 40$$

$$\frac{40}{2} = 20$$

CHAPTER 14

We are essentially finding the area of the "rectangle" we could draw around the rhombus that has a height and width equal to the diagonals. We divide by two, because if we quartered the rhombus, we would see four triangles, each half of a smaller rectangle as shown below:

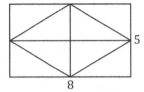

Answer: 20.

Rectangle

A rectangle is a special kind of parallelogram in which each angle is 90 degrees.

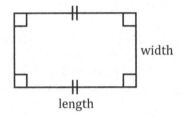

AREA OF A RECTANGLE

$$Area = length \times width$$

The **diagonals of a rectangle** are **congruent** and **bisect** each other.

Square

If a rectangle has an equal length and width, it is called a square. This means that a square is a type of rectangle AND a type of rhombus. But NOT all rectangles are squares, nor are all rhombuses.

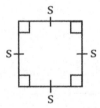

AREA AND PERIMETER OF A SQUARE

$$Area = side\ length\ squared\ \ or\ \ s^2$$
$$Perimeter = 4s$$

Trapezoid

A trapezoid is a quadrilateral with <u>only</u> one set of parallel sides. The other two sides are non-parallel. We find the area of a trapezoid by averaging the two bases and multiplying that by the height.

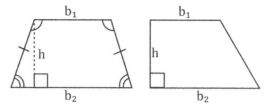

AREA OF A TRAPEZOID

$$Area = h\left(\frac{b_1 + b_2}{2}\right)$$

Where h is the height, and b_1 and b_2 are the base lengths.

 TIP: If you forget the formula for a trapezoid (or another polygon), cut it into a rectangle and a couple of triangles and you can find the area by adding together the areas of these smaller pieces.

We call a trapezoid with equal side lengths and base angles (see above, left) an "isosceles trapezoid." These trapezoids are symmetric about the line between the midpoints of the top and bottom bases, and have equal diagonals.

 TIP: The median of a trapezoid is parallel to the bases as well as half of the sum of the length of the bases as shown below.

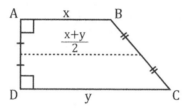

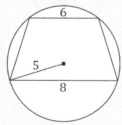

An isosceles trapezoid whose bases have lengths 6 and 8 is inscribed in a circle of radius 5 as shown below. The center of the circle lies in the interior of the trapezoid. What is the area of the trapezoid?

To do this problem, you'll also need to know about circles. One of our biggest tips in that chapter is to **always draw radii** whenever you have any points that don't have radii drawn to them. In this diagram, we'll draw radii from the center to the vertices of the trapezoid.

Remember that isosceles trapezoids are symmetric and have parallel sides. That means we can cut the trapezoid in half and create right angles with this midline. We also know this midline bisects the trapezoid. Again we can do this because it is isosceles.

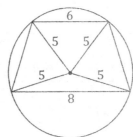

Here, once we draw radii, look for triangles with right angles.

We can now use the Pythagorean theorem to solve for the missing lengths, **4** and **3**.

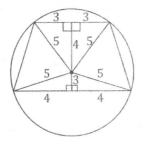

Now we know the height of the trapezoid is **7**, and we use the formula:

$$\frac{1}{2}\big(sum\:of\:bases\big)\cdot height = \frac{1}{2}\big(6+8\big)\times 7 = 49$$

Answer: **49**.

Polygon Angles

Whether your polygon is regular or irregular, the sum of its interior degrees will always have a sum determined by the number of sides on the polygon.

For example, the interior angles of a quadrilateral—whether kite, square, trapezoid, or other—will always add up to 360 degrees.

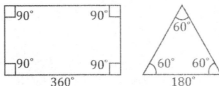

By that same notion, the interior angles of a triangle sum to 180 degrees, whether the triangle is equilateral (a regular polygon), isosceles, acute, or obtuse.

We determine the sum of the interior angles in one of two ways:
1. Draw all the diagonals possible from a single vertex, count the number of distinct "triangles" you've created, and multiply that number of "triangles" by 180 (as triangles have an interior angle sum of 180):

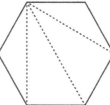

In the hexagon above, we've created four triangles from this method, so we calculate $4 \times 180 = 720°$. Thus the hexagon's interior angles sum to $720°$.

2. Use the formula for the sum of the interior angles.

SUM OF INTERIOR ANGLES

> The sum of the interior angles in a polygon $= (n-2)180$, where n is the number of sides on the polygon.

Remember, if you know the sum of the interior angles, you can find each angle of a regular polygon simply by dividing by the number of angles/sides. For example, a regular hexagon would have individual interior angles that measure $\frac{720}{6}$ or 120 degrees each.

Exterior Angles

Exterior angles for EVERY polygon always sum to 360 degrees. Exterior angles are the angles that are supplementary to interior angles. I like to teach this idea because it is SO EASY to remember, even easier than the sum of the interior angles. To draw exterior angles, extend the lines of a polygon to make it look like a pinwheel. In the polygon below, $\angle x$ is an exterior angle:

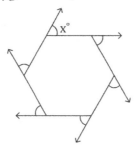

EXTERIOR ANGLE OF REGULAR POLYGONS

> The measure of a single exterior angle of **regular polygon** $= \frac{360}{n}$, where the shape has n-sides.

Remember, if you know an exterior angle, you can easily calculate the associated interior angle as well by subtracting from 180.

POLYGONS — SKILLS

In heptagon *ABCDEFG* below, exterior angle at *A* of $\angle\theta = 112°$. What is the sum of the remaining exterior angles?

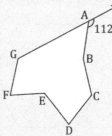

In this problem, we essentially have one exterior angle and we need the rest. All we need to do is subtract $112°$ from $360°$, because we know the sum of the exterior angles of ANY polygon is always $360°$: $360° - 112° = 248°$.

I realize this is a concave polygon. In truth, finding EACH exterior angle would involve some negative angles (yes that's confusing). But if you stick to the rules and keep it simple, this is not a hard problem.

Answer: 248.

Number of Diagonals

Sometimes the ACT® may ask the number of distinct diagonals in a polygon. Again, you can find this information using the formula or by drawing it out (or a combination of the two). You can also solve these using combinations (8 vertices taken 2 at a time, or $_8C_2$).

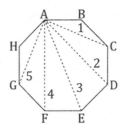

To draw this out, first calculate the number of diagonals you can form from a single vertex. In this octagon, that's five.

Then multiply by 8 (the number of vertices) and you'll have the total number of diagonals you could draw BUT your calculation involves a lot of repeats. I.e., you drew a line from A to C but also from C to A. To divide out these repeats, simply divide the product you got by 2 (how many times you've counted each combination of two letters). (See Arrangements Chapter for more on this topic.)

$$\frac{8 \times 5}{2} = 20 \text{ diagonals}$$

Alternatively, memorize this formula or program it into your calculator to solve:

NUMBER OF DIAGONALS OF A POLYGON

$$\text{Number of Diagonals in a polygon of } n-\text{sides} = \frac{n(n-3)}{2}$$

$$\frac{8(8-3)}{2} = 20$$

Answer: 20.

AREA WORD PROBLEMS

Word problems involving area can be tackled in the same way that traditional word problems are: take things one step at a time, make up variables if necessary, make up numbers as you go to decipher relationships, and convert your units as needed. Often, these problems are culmination of many skills, including polygons, units, and percents. At other times, you'll have weird shapes. **Remember you can cut shapes into easier to manage pieces and add the areas together, or subtract out cutouts to solve for areas of shapes with missing pieces or holes.**

> Marina is tiling the area around her bathtub in a herringbone pattern using rectangular subway tile. She has to cover two wall segments that each measure 2.5×6 feet and another that measures 5×6 feet. She wants to order at least 15% overage in tile to ensure she has enough given that she will be cutting tiles on a diagonal to achieve this pattern. Each tile sits flush to each other tile, and measures 2 in by 8 in. If tiles come in a box of 100, how many boxes of tile should she order at minimum?

To solve this problem, we first need the area in total of the walls. Let's convert EVERYTHING to inches first. When you do conversions with area, you'll need to account for a different conversion factor than if you were in a single dimension. I.e. if I find the area of each wall first and THEN convert to inches, there are 144 square inches in one square foot, not 12 square inches in one square foot. We square the ratio of the two "single dimension" measures when working with area, so a $1:12$ ratio turns into a $1:144$ ratio. I realize that can be confusing. Translating everything first in this problem will help me avoid this complication. I use dimensional analysis to keep my units straight.

$$5\text{ft} = 5 \text{ feet} \times \frac{12 \text{ inches}}{1 \text{ foot}} = 60 \text{ inches}$$

$$6\text{ft} = 6 \text{ feet} \times \frac{12 \text{ inches}}{1 \text{ foot}} = 72 \text{ inches}$$

$$2.5\text{ft} = 2.5 \text{ feet} \times \frac{12 \text{ inches}}{1 \text{ foot}} = 30 \text{ inches}$$

Now we need to find the total area to cover with tile. Be sure to get ALL THREE walls!

Short walls: $2.5\text{ft.} \times 6\text{ft.} = 30\text{ in.} \times 72\text{ in.} = 2160 \text{ sq. in.}$ <u>PER WALL</u> (there are two!!)
Now I multiply the short wall amount by 2: $2 \times 2160 = \mathbf{4320}$
Long walls: $60 \text{ in.} \times 72 \text{ in.} = 4320 \text{ sq. in.}$
I add the <u>TWO</u> short walls to the long wall: $4320 + 4320 = 8640 \text{ sq. in.}$

CHAPTER 14

Cool. Now we can take this total square footage and translate it into tiles. To do so, we need the rate of area per tile. That rate will help us jump from area to number of tiles.

$$\text{Each tile is } 2 \times 8 = 16 \text{ sq. in. per tile}$$

We have inches squared in terms of total square footage, so we flip our rate of area per tile upside down so the labels cancel and we multiply:

$$8640 \text{ in.}^2 \times \frac{1 \text{ tile}}{16 \text{ in.}^2} = \frac{8640}{16} \times 1 \text{ tile} = 540 \text{ tiles}$$

Now, I need to calculate my 15% overage. To do this, I multiply 540 times 1.15, which is a shortcut to give me 115% of 540 or 540 plus 15% of 540. That gives me 621 tiles.

With 100 tiles per box, Marina will need to order 7 boxes (700 tiles total) to give her a minimum 15% of overage.

Answer: 7.

> Pamela is building a pool in her backyard, which measures 14×20 feet. Her backyard will be covered in grass except for the area taken up by the pool. If the pool is a kidney bean shape, has a uniform depth of 10 feet, and a volume of 1000 cubic feet, what is the area of her grass lawn, to the nearest square foot?

Remember, ANY 3-D shape with uniform depth is like a prism, and the volume of a prism is the area of the base times the height. Here, the pool's surface area, or "footprint," equals the area of the base of such a "prism."

Here, we can subtract the surface area of the pool, which we find by dividing its volume by the depth. $\frac{1000}{10} = 100$ (see our chapter on "Solids" if you need help with this step). Now we know the surface area of the pool is 100 sq ft, so we can subtract this from the area of her yard: 14×20 or 280 square feet. $280 - 100 = 180$ sq. ft.

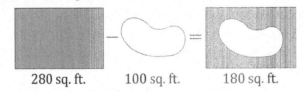

280 sq. ft.   100 sq. ft.   180 sq. ft.

Thus her lawn will take up 180 square feet.

Answer: 180.

QUESTIONS   POLYGONS

1. A polygon has the dimensions below. Each pair of intersecting line segments meet at right angles, and all the dimensions given are in inches. What is the area, in square inches, of the polygon?

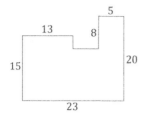

A. 86
B. 295
C. 355
D. 370
E. 460

2. A polygon has the dimensions below. Each pair of intersecting line segments meet at right angles. What is the area, in square feet, of the figure below?

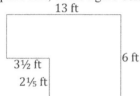

A. 38
B. $70\frac{3}{10}$
C. $71\frac{9}{10}$
D. $72\frac{9}{10}$
E. 78

3. As shown below, rectangle $WXYZ$ is divided into 2 big squares (labeled B) each x inches on a side, 16 small squares (labeled S) each y inches on a side, and 6 rectangles (labeled R) each x inches by y inches. What is the total area, in square inches of $WXYZ$?

A. $2x + 6xy + 16y$
B. $4x^2 + 16y^2$
C. $2x^2 + 34y^2$
D. $2x^2 + 6xy + 16y^2$
E. $2x^2 + 6x^2y^2 + 16y^2$

4. In the figure below, the vertices of rectangle $STUV$ lie on the circle and the vertices of rhombus $WXYZ$ are the midpoints of the sides of rectangle $STUV$. Point R is at the center of the circle, rectangle, and rhombus. The radius of the circle is 15 units. What is the perimeter of the rhombus, in units?

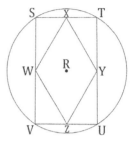

A. 15
B. 30
C. 40
D. 60
E. 80

5. What is the maximum number of distinct diagonals that can be drawn in the pentagram shown below?

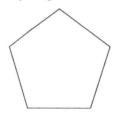

A. 2
B. 5
C. 7
D. 10
E. 15

6. The heptagon shown below has 7 sides of equal length. What is the sum of the measures of the interior angles in the heptagon?

A. 360°
B. 540°
C. 720°
D. 900°
E. 1080°

CHAPTER 14   229

7. Inscribed in the circle below is a regular octagon ABCDEFGH with some diagonals shown. What is the measure of ∠DAE ?

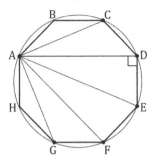

A. 11.25°
B. 15°
C. 17.5°
D. 22.5°
E. 25°

8. The area of a rectangular card is 544 square inches. A rectangular photograph that is twice as wide as it is tall is glued to the card as shown below so that there are 5 inches of clearance above and below it and 11 inches of clearance to either side. How many inches tall is the photograph?

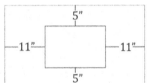

A. 5
B. 6
C. 8
D. 12
E. 27

9. Depicted here is a painting in its frame. The frame has a uniform width of $3\frac{2}{5}$ inches. What is the area, in square inches, of the visible part of the painting?

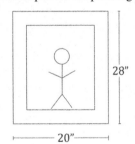

A. 233.6
B. 279.84
C. 326.08
D. 408.36
E. 560

10. What is the area, in square centimeters, of the parallelogram below?

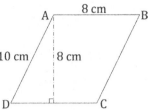

A. 60
B. 64
C. 76
D. 80
E. 104

11. In the parallelogram below, lengths are given in centimeters. What is the area of the parallelogram, in square centimeters?

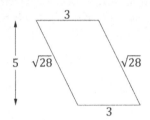

A. 13.1
B. 15
C. 16
D. 16.3
E. Cannot be determined.

12. In the figure below, each pair of intersecting line segments meets at a right angle, and all the lengths are given in inches. What is the perimeter, in inches, of the figure?

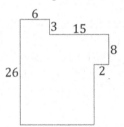

A. 60
B. 81
C. 94
D. 99
E. 107

13. All line segments in the figure below are horizontal or vertical, and units are given in inches. What is the perimeter, in inches, of the figure below?

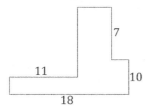

A. 46
B. 53
C. 63
D. 70
E. 77

14. In rhombus $ABCD$, shown below, $\overline{BD}$ is 20 inches long and $\overline{AC}$ is 14 inches long. What is the area, in square inches, of the rhombus?

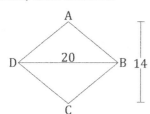

A. $2\sqrt{149}$
B. $4\sqrt{149}$
C. 70
D. 140
E. 280

15. The area of rhombus $RHOM$ is 230 square inches. If $RO = 5$ inches, what is the length of $\overline{HM}$ in inches?

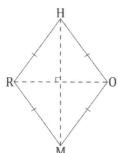

A. 18.4
B. 23
C. 25
D. 46
E. 92

16. A right triangle, with legs of length 5 and 12 inches, was cut off of a square with 13 inch sides, resulting in the shaded portion of the figure below. What is the approximate perimeter, in inches, of the shaded portion?

A. 39
B. 43
C. 47
D. 48
E. 52

17. In the figure below, the area of the larger square is 162 square centimeters, and the area of the smaller square is 72 square centimeters. What is s, in centimeters?

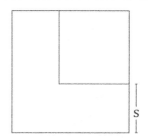

A. 3
B. $3\sqrt{2}$
C. $3\sqrt{10}$
D. $6\sqrt{2}$
E. 45

18. In the figure below, $QRST$ is a square and $W, X, Y,$ and Z are the midpoints of the square's sides. If $RS = 16$ inches, what is the perimeter of $WXYZ$, in inches?

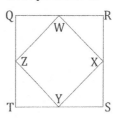

A. 16
B. $16\sqrt{2}$
C. 32
D. $32\sqrt{2}$
E. $64\sqrt{2}$

19. In trapezoid *STUV* below, ∠*U* is 30° and ∠*S* = ∠*V* = 90°. Side lengths are given in meters. Which of the following values is closest to the area, in square meters, of *STUV*? (Note: drawing not to scale)

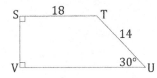

   A. 83.6
   B. 103.4
   C. 126
   D. 150.5
   E. 168.5

20. The trapezoid below is formed by two shaded triangles and an unshaded square. Lengths are given in feet. What is the combined area, in square feet, of the two shaded triangles?

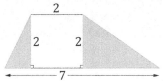

   A. 4
   B. 5
   C. 6
   D. 7
   E. 10

21. The area of the trapezoid below is 24 square inches. The altitude is 3 inches. One base measures 11 inches. What is the length of the other base, *b*, in inches?

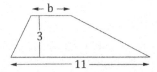

   A. 4
   B. 5
   C. 6
   D. 7
   E. 8

22. A regular hexagon is shown below. What is the measure of the designated angle? (The measure of each interior angle of a regular *n*-sided polygon is $\frac{(n-2)180°}{n}$).

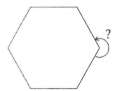

   A. 120°
   B. 270°
   C. 225°
   D. 240°
   E. 250°

23. In octagon *ABCDEFGH*, shown below, ∠*A* measures 25°. What is the total measure of all 8 interior angles?

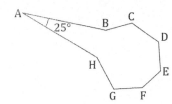

   A. 360°
   B. 135°
   C. 1260°
   D. 1440°
   E. 1080°

24. If the area of a rectangle that is twice as wide as it is long is 128 square feet, which of the following could be the perimeter of the rectangle?

   A. 8
   B. 16
   C. 48
   D. 128
   E. 256

25. A rectangle has a diagonal of 13 units and a side of 5 units. What is the perimeter, in units, of the rectangle?

   A. 34
   B. 36
   C. 50
   D. 20
   E. 48

26. A rectangular animal pen is 3 times as wide as it is long. If the area of the pen is 867 square meters, what is the perimeter of the pen?

    A. 51 meters
    B. 117.6 meters
    C. 136 meters
    D. 140 meters
    E. 143.6 meters

27. The area of a large wall tapestry is 126 square feet. The length of the tapestry is 4 more than twice the width. What is the perimeter of the tapestry?

    A. 252 feet
    B. 73 feet
    C. 62 feet
    D. 56.1 feet
    E. 50 feet

28. George is putting tile flooring in his living room. The living room is rectangular and measures 10 feet by 25 feet. The tiles measure 6 inches by 10 inches. What is the minimum number of tiles that George will need?

    A. 417
    B. 500
    C. 570
    D. 600
    E. 630

29. A rectangular poster has a width that is 5 centimeters longer than its length and an area of 500 square centimeters. What is the length of the poster in centimeters?

    A. 17.5
    B. 20
    C. 25
    D. 40
    E. 100

30. A parallelogram has a perimeter of 68 centimeters and a side that measure 22 centimeters. If it can be determined, what are the lengths of the other 3 sides?

    A. 22, 3.1, 3.1
    B. 22, 12, 12
    C. 22, 14.5, 14.5
    D. 22, 15, 15
    E. Cannot be determined from the given information

31. Jimmy has 240 feet of fence to build an enclosure around his garden. If the enclosure is to be rectangular with a length 20 feet longer than its width, what will be the approximate dimensions of the enclosure?

    A. 50 feet long, 30 feet wide
    B. 50 feet long, 70 feet wide
    C. 70 feet long, 50 feet wide
    D. 75 feet long, 55 feet wide
    E. 90 feet long, 70 feet wide

32. A rhombus has sides that measure 9 inches and two of its interior angles measure 120°. How many square inches is the area of the rhombus?

    A. $\dfrac{81}{4}$
    B. $\dfrac{81}{2}$
    C. $\dfrac{81\sqrt{3}}{2}$
    D. 81
    E. $81\sqrt{3}$

33. What is the area, in square meters, of a trapezoid with parallel bases of 10 feet and 13 feet and a height of 6 feet?

    A. 60
    B. 69
    C. 72
    D. 78
    E. 138

34. A trapezoid has an area of 77 square inches and parallel bases that measure 9 inches and 5 inches. What is the height in inches of the trapezoid?

    A. 7
    B. 8.6
    C. 11
    D. 15.4
    E. 16

35. Lorenzo has a lawn in his backyard that forms a rectangle that is 25 meters wide and 30 meters long. Around the entirety of the lawn is a walkway with a width of 3 meters. What is the total area of the lawn and the walkway in square meters?

   A. 750
   B. 915
   C. 924
   D. 1008
   E. 1116

36. A rectangular wall that measures 20 feet by 34 feet is going to be painted. In the middle of the wall, there is a circular mirror with a diameter of 8 feet. If the wall will not be painted under the mirror, what is the approximate area of the wall that will be painted in square feet?

   A. 479
   B. 590
   C. 615
   D. 630
   E. 655

37. A quadrilateral has sides that are all n inches long. Which of the following cannot be this quadrilateral?

   A. Rhombus
   B. Square
   C. Trapezoid
   D. Parallelogram
   E. Rectangle

38. All of the following quadrilaterals have diagonals that are not always congruent except:

   A. Rhombus
   B. Rectangle
   C. Parallelogram
   D. Trapezoid
   E. Kite

ANSWERS   POLYGONS

ANSWER KEY

1. C	2. B	3. D	4. D	5. B	6. D	7. D	8. B	9. B	10. B	11. B	12. C	13. D	14. D
15. E	16. D	17. B	18. D	19. E	20. B	21. B	22. D	23. E	24. C	25. A	26. C	27. E	28. D
29. B	30. B	31. C	32. C	33. B	34. C	35. E	36. D	37. C	38. B				

ANSWER EXPLANATIONS

1. **C.** We divide the polygon into 3 rectangles by drawing a line down from the right end of the 13 unit segment and another down from the left end of the 5 unit segment. The leftmost rectangle has a width of 13 and a height of 15, giving it an area of 195. The rightmost rectangle has a width of 5 and a height of 20, giving it an area of 100. The center rectangle is the width of the total width of the polygon, 23, minus the other horizontal line segments, 13 and 5: $23-(13+5)=5$. We can find the height of the center rectangle by starting with 20 from the segment to the right and subtracting 8 due to the segment that stretches from the top of the tallest part to the top of the center rectangle: $20-8=12$. The area of the center rectangle is $12\times 5=60$. Summing the areas of the rectangles gives us $195+100+60=355$.

2. **B.** In contrast to the answer to question 1, we can also find the area by subtracting the area of what is cut off from a larger polygon. In this case, we can subtract the $3\frac{1}{2}$ by $2\frac{1}{5}$ feet rectangle from the 13 by 6 feet rectangle: $(13\times 6)-\left(3\frac{1}{2}\times 2\frac{1}{5}\right)=78-7.7=70.3$ or $70\frac{3}{10}$.

3. **D.** The area of each of the big squares is the side length squared: x^2. The area of each of the small squares is the respective side length squared: y^2. The area of each of the rectangles is x times y: xy. There are 2 big squares, which together have an area of $2x^2$. There are 6 rectangles, which together have an area of 6xy. There are 16 small squares, which together have an area of $16y^2$. To get the total area, we sum the total areas of each type of shape: $2x^2+6xy+16y^2$.

4. **D.** Draw a line from X to Z and from W to Y, dividing the rectangle into 4 smaller rectangles. We can see that each of the sides of the rhombus is a diagonal of one of the smaller rectangles. The diagonals of a rectangle are congruent. If we draw the other matching diagonals that are not pictured but are nevertheless congruent, we see that they stretch from the center to points on the circle, and thus equal the radius, 15. Since the 4 sides of the rhombus are congruent to the 4 diagonals, the 4 sides of the rhombus added together, the perimeter, are equal to 4 times the radius. Therefore the perimeter of the rhombus equals $4\cdot 15=60$.

5. **B.** One way to solve is to calculate the number of diagonals from a single vertex (2) and then multiply by the number of vertices (5) to get 10 diagonals. However, this method "double counts" each diagonal, so we must divide by 2 to get the number of unique diagonals: $\frac{10}{2}=5$. We could alternatively draw the distinct diagonals in the pentagon form a pentagram, a 5-pointed star. It is made up of 5 lines, thus 5 distinct diagonals. Finally, we could also use the general equation for the number of distinct diagonals in a polygon $\frac{n(n-3)}{2}$, plugging in 5 for n to get $\left(\frac{5(2)}{2}\right)=5$.

6. **D.** The sum of the measures of the interior angles in a polygon equals $(n-2)\times 180°$ where n is the number of sides of the polygon. For a heptagon, that's $(7-2)\times 180°=5\times 180°=900°$.

7. **D.** We're going to find the measure of $\angle DAE$ by finding the measure of $\angle DEA$ and using the triangle sum theorem (which states that the measure of the interior angles of a triangle sums to $180°$). We know that the total sum of the interior angles of any octagon is $(8-2)\times 180°=1080°$, and that if all the angles are congruent (as they are in a regular octagon), each interior angle equals $\frac{1080°}{8}=135°$. Now we proceed by cutting the octagon half along $\overline{AE}$. In cutting the octagon in half, the angle $\angle BAH$ and $\angle DEF$ are both bisected, so the resultant angles created are congruent: $\angle BAE \cong \angle DEA$. $\angle DEF$ originally equaled $135°$, so $\angle DEA=\frac{135°}{2}=67.5°$. $\angle ADE$ is a right angle, by substitution $\angle DAE+90°+67.5°=180°$. Simplifying, $\angle DAE=22.5°$.

CHAPTER 14

8. **B.** Let x be the height of the photograph. Consequently, the width of the photograph can be expressed as $2x$. The height of the card is $x+2(5)=x+10$. The width of the card is $2x+2(11)=2x+22$. The area of the card is the product of the width and height: $(x+10)(2x+22)=544$. Multiply the binomials to get $2x^2+42x+220=544$. This simplifies to $x^2+21x-162=0$. Factoring this, we get $(x-6)(x+27)=0$. We solve to find that the positive root is $x=6$. We must use the positive root because it is the measure of a line segment, which must have a positive length. Thus, the photograph is 6 inches in height.

9. **B.** The height of the painting is the height of the frame minus two times the width of the frame, since the frame runs along the top and the bottom: $28-2\left(3\frac{2}{5}\right)=21\frac{1}{5}$. The width of the painting is found by applying the same concept to the width of the rectangle formed by the frame: $20-2\left(3\frac{2}{5}\right)=13\frac{1}{5}$. The area of the visible portion of the painting is thus $21\frac{1}{5}\cdot 13\frac{1}{5}=279.84$.

10. **B.** The area of a parallelogram is the base times the height. Here, the base is 8 cm and the height is 8 cm, so the area is $8\cdot 8=64$.

11. **B.** The area of a parallelogram is the base times the height. Here, the base is 3 cm and the height is 5 cm, as indicated on the left, so the area is $3\cdot 5=15$.

12. **C.** The perimeter of a polygon like this is equal to 2 times the total horizontal length plus 2 times the total vertical length. It is easier to solve using this method instead of finding the length of every line segment and adding them together. The total vertical length is 26, seen in the long vertical line segment on the left. The total horizontal length is 21, seen in the 6 and 15 inch long segments on the top that together span the entire horizontal length (or that segments of 19 and 2 along the bottom). Thus, the perimeter is $2(26)+2(21)=94$.

13. **D.** As we did for #12, we will find the total horizontal and vertical lengths of the polygon to find the perimeter. The total horizontal length is 18 inches, from the horizontal line segment that spans the entire horizontal length along the bottom. The total vertical length is 17, from the 7 and 10 inch segments that span the entire vertical length on the right. The perimeter of the figure is thus $2(18)+2(17)=70$.

14. **D.** All rhombuses are kites, and the area of a kite is equal to half the product of its diagonals. Thus, the area of the rhombus is $\frac{20\cdot 14}{2}=\frac{280}{2}=140$. If you forget this formula, break the figure into triangles and use the triangle area formula.

15. **E.** The area of a rhombus, like any kite, is equal to half the product of its diagonals. We set up our equation as $\frac{5\cdot HM}{2}=230$. Simplifying, $HM=92$. If you forget this formula, you can also break the figure down into triangles to solve.

16. **D.** We can find two of the five sides of the shaded portion easily as they are the sides of the square, so they are both 13 inches. Another two sides can be found as the sides of the square minus the legs of the triangle that are cut off: 13 minus 5 and 12, respectively, giving sides of lengths 8 and 1. The final side is the hypotenuse of the right triangle, which we find, using Pythagorean triples, is 13 (note: remember that 5, 12, and 13 is a Pythagorean triple; alternatively, use the Pythagorean theorem). The total perimeter of the shaded portion is $13+13+8+1+13=48$.

17. **B.** If the area of the smaller square is equal to 72, then the side length of the smaller square is $\sqrt{72}=6\sqrt{2}$. Since the area of the larger square is equal to 162, the side length of the larger square is $\sqrt{162}=9\sqrt{2}$. s is the difference between the larger side and the smaller side: $9\sqrt{2}-6\sqrt{2}=3\sqrt{2}$.

18. **D.** Each of the sides of the interior square are the hypotenuse of an isosceles right triangle with legs that are each equal to half of the side length of the larger square. The leg of each 45-45-90 triangle formed is 8 (since the legs are bisected segments of the outer square's length). Using our knowledge of the ratio of sides of a 45-45-90 triangle, we can conclude that the hypotenuse of our triangles, which is the side of the smaller square is $8\sqrt{2}$. Since there are 4 hypotenuses/interior square sides, the total perimeter of the smaller square is four times $8\sqrt{2}$ or $32\sqrt{2}$.

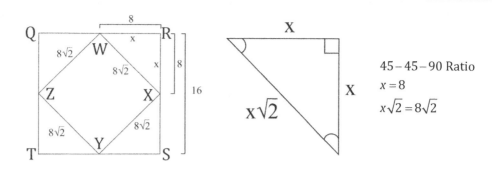

19. E. Drawing an altitude down from T, we divide the figure into a rectangle and a 30-60-90 triangle. For reference, I now draw out a 30-60-90 triangle with the appropriate ratios as a reference, orienting that triangle in the same way the triangle is oriented in my problem. We can find the height of both using the ratio of sides of a 30-60-90 triangle. Since a segment of length 14 is directly across from the 90° angle, and corresponds to the $2x$ side of the 30-60-90 triangle, the side directly across from the 30° x must be half this length 7; the base of the triangle (across from the 60° angle) must correspond to $x\sqrt{3}$ which is $7\sqrt{3}$. Thus, the area of the triangle is $\frac{1}{2} \cdot 7 \cdot 7\sqrt{3} \approx 42.4$. We know that the area of the rectangle is $7 \cdot 18 = 126$, so adding the areas together, the area of the entire figure is $126 + 42.4 = 168.4$, which is closest to 168.5.

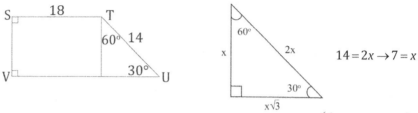

20. B. The combined area of the shaded triangles is equal to the total area of the trapezoid $\left(\left(\frac{\text{sum of bases}}{2}\right) \times \text{height}\right)$ minus the area of the unshaded square. The area of the trapezoid is $\frac{2+7}{2} \cdot 2 = 9$. The area of the unshaded square is $2 \cdot 2 = 4$. Thus the area of the shaded triangles is $9 - 4 = 5$.

21. B. We solve this problem by plugging the values given into the equation for the area of a trapezoid $\left(\left(\frac{\text{sum of bases}}{2}\right) \times \text{height}\right)$ and solving for b: $\frac{b+11}{2} \times 3 = 24 \rightarrow b+11 = 16 \rightarrow b = 5$.

22. D. We are given the formula for finding the interior angle of a regular n-sided polygon, so we plug in 6 for a hexagon: $\frac{(6-2)180°}{6} = 120°$. The exterior angle and the interior angle must sum to 360°, so the exterior angle is $360° - 120° = 240°$.

23. E. The sum of the measures of the interior angles of any polygon is only dependent on the number of sides, not on how funky the angles look. Because this is an octagon, we can use the formula $(n-2)180$ for the sum of the measures of the interior angles: $(8-2)180° = 1080°$.

24. C. Let n be the length of the rectangle. $2n$ is the width. The area given is equal to the product of the length and the width: $2n \cdot n = 128$. This simplifies to $n^2 = 64$, so $n = 8$. The perimeter is $2(2 \cdot 8) + 2(8) = 48$.

25. A. The diagonal of the rectangle is the hypotenuse of a right triangle, where the side is one of its legs. The other leg, and so the other side, can be found as $\sqrt{13^2 - 5^2} = 12$ (note: it could also be found by remembering that 5, 12, and 13 is a Pythagorean triple). The perimeter is thus $2(5) + 2(12) = 34$.

26. C. Let l be the length of the pen. The width is $3l$. The relationship between l and the area is $l * 3l = 867$. From this, we can isolate l to get $l = 17$. The perimeter is thus $2(17) + 2(3 \cdot 17) = 136$.

27. E. Let w be the width of the tapestry. Per the description, the length of the tapestry is $2w + 4$. The relationship to the area is $w(2w+4)=126$. We distribute to get $2w^2+4w=126$, which becomes $2w^2+4w-126=0$. We simplify to $w^2+2w-63=0$. Factoring gives us $(w-7)(w+9)=0$. We take the positive root, $w=7$, since w is a physical length and thus must be positive. The perimeter of the tapestry is thus $2w+2(2w+4)=2(7)+2(2\cdot 7+4)=14+36=50$.

28. **D.** First, we convert our measurements into the same units. To avoid working in fractions, let us convert dimensions given in feet into inches. The living room measures 10 feet by 25 feet, which is 120 inches by 3000 inches. The total area, in inches squared, is $120\cdot 300=36000$. An individual tiles covers $6\cdot 10=60$ square inches. Thus, we need at least $\frac{36000\,in^2}{60\,in^2}=600$ tiles.

29. **B.** Let l be the length of the poster. The width is $l+5$. The area is $l(l+5)=500$. We distribute to get $l^2+5l=500$, which rearranged becomes $l^2+5l-500=0$. Factor and we find that $(l-20)(l+25)=0$, meaning $l=20$ or $l=-25$. l cannot be negative, so l must be 20.

30. **B.** We know that one of the other sides must be 22, since the opposite sides of a parallelogram are equal. The other 2 sides plus the sides with length 22 must sum to 68. We express this as $2x+44=68$, where x is the length of the other sides. Solving, we find that $x=12$. Thus, 2 of the other sides have length 12. The 3 sides we are not given measure 22, 12, and 12, respectively.

31. **C.** Let w be the width of the enclosure. The length is $w+20$. The total perimeter is 240, so $2(w)+2(w+20)=240$. This becomes $4w+40=240$. Solving, we determine that $w=50$, and so $l=70$. The dimensions are 50 feet wide and 70 feet long.

32. **C. When in doubt, draw it out!** Because two of the interior angles are 120°, and a rhombus is a type of parallelogram with all sides parallel, we know the other two small interior angles must each be supplementary to each of these big angles, and sum to 180. This is the same relationship of angles in this orientation across a transversal (see Lines & Angles chapter). $180-120=60$, so the smaller angles in the rhombus must equal 60°. If we sketch the rhombus and draw an altitude down from one of the 120° vertices, we see that it forms a right triangle. Because we know the small angle in the rhombus is one angle in this triangle, this angle is equal to 60°. The other angle must be 30°, which we can find either by subtracting 90 from the 120 degree angle or by subtracting the two angles we know in the triangle to find the third: $180-60-90=30$, given that triangle angles sum to 180. Since we now have a 30-60-90 triangle, we can use our knowledge of the ratios of the sides of a 30-60-90 triangle to find the height of the rhombus, which is the altitude we drew. The 9-inch segment is directly across from the 90° angle, so setting that equal to its corresponding "ratio" $9=2x$, or $x=4.5$. Thus the height, the segment across from the 60° angle, is $x\sqrt{3}$ or $4.5\sqrt{3}$. The area of the rhombus or any parallelogram is base times height: $9\cdot 4.5\sqrt{3}$, or $9\cdot\frac{9}{2}\sqrt{3}=\frac{81\sqrt{3}}{2}$.

33. **B.** The area of a trapezoid is half the sum of the bases times the height: $\frac{10+13}{2}\cdot 6=69$.

34. **C.** We fill in as much of the trapezoid area formula as we can: $\frac{9+5}{2}h=77$. By isolating h we get $h=\frac{154}{14}=11$.

35. **E.** To get the length and width of the combined lawn and walkway, we add 2 times the width of the walkway to the length and width of the lawn, respectively. The length is $30+2(3)=36$ m long. The width is $25+2(3)=31$ m. The area of the combined lawn and walkway is $36\cdot 31=1116$ m².

36. **D.** The area that is to be painted is equal to the total area of the wall minus the area of the mirror: $20\cdot 34-\pi(4)^2$ (note that the radius is 4, the question only indicates the diameter). This is equal to $680-16\pi\approx 630.$.

37. **C.** The quadrilateral cannot be a trapezoid because trapezoids by definition have 2 bases that are of unequal length.

38. **B.** The diagonals of a rectangle are already congruent. There are many proofs of this; one is that the diagonals are hypotenuses of right triangles with congruent legs, and so must themselves be congruent.

CHAPTER 15

SIMILAR SHAPES

> ## SKILLS TO KNOW
> - Ratios
> - Similar shapes
> - Similar triangles
> - Triangle congruency
> - Hinge theorem
> - Parallel lines theorem
> - Similar solids
> - Surface area and volume of similar solids

RATIOS

A **ratio** compares two different values. In the case of geometry, it may compare two areas or even two side lengths of two shapes. We cover how to solve problems using basic ratios in chapter 9 on **Ratios, Rates and Units in Book 1**.

SIMILAR SHAPES

Imagine blowing up a picture of a triangle using a copy machine or even zooming in on a square on your computer screen. That's essentially what similar shapes are: zoomed in or zoomed out versions of the same shape.

Similar shapes can be different sizes but must have the same proportionate shape. All corresponding angles of similar figures are equal. For example, all regular polygons and circles are similar since they share the same shape. Here are some examples:

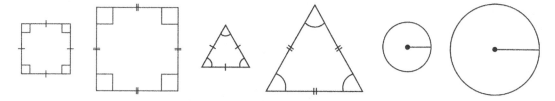

On the other hand, these rectangles are not similar. Their sides are not in equal ratios, i.e. 7:5 is not equal to 13:1.

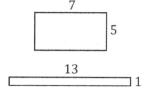

SIMILAR SHAPES — SKILLS

These rectangles, on the other hand, are similar: $5:10 = 1:2$, just as $3:6 = 1:2$. The sides of the right figure are twice the corresponding sides on the left figure.

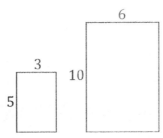

SIMILAR TRIANGLES

Triangles that have the same shape but different sizes are similar. This means they will share the same corresponding angles and the corresponding side lengths will be proportional.

Similar triangles can show up in different ways on the test. Common orientations are shown below: one inside the other, sitting on top of each other via parallel lines, or side by side.

 BE CAREFUL: the order of the letters in your similarity statement MATTERS.

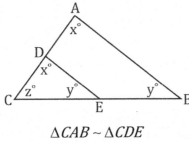

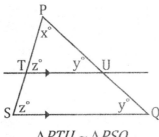

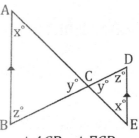

$\triangle CAB \sim \triangle CDE$ $\triangle PTU \sim \triangle PSQ$ $\triangle ACB \sim \triangle ECD$

For instance, if $\triangle ACB \sim \triangle ECD$ then $\overline{AC}$ is similar to $\overline{EC}$ and $\overline{CB}$ is similar to $\overline{CD}$ as they are in the same corresponding positions in the name of each triangle. Confusing which sides are similar is a common careless error on similar shape questions.

Ways you can prove similarity

AA: If two angles are congruent to two angles of another triangle, then the triangles must be similar.

SSS: If all the corresponding sides of two triangles are in the same proportion, then the two triangles must be similar.

SAS: If one angle of a triangle is congruent to the corresponding angle of another triangle and the lengths of the sides adjacent to the angle are proportional, then the triangles must be similar.

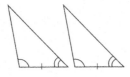

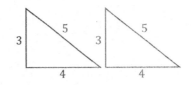

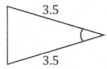

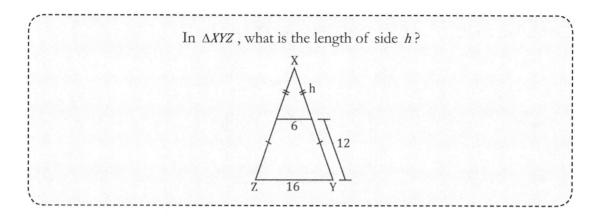

In △XYZ, what is the length of side h?

We can set up a ratio because we can see congruent sides of the similar triangle *ZXY* and the shorter triangle formed with a base of 6. We set the short leg on the right, h, proportional to the long leg on the right, h plus 12 and the short base on the upper, smaller triangle proportional to the larger base, 16: $\frac{h}{h+12}=\frac{6}{16}$.

Be careful! Students will often think $\frac{h}{12}=\frac{6}{16}$. However, we're dealing with triangle bases, so we must consider the WHOLE length of the triangle side. Cross-multiplying, we have $16h=72+6h$, $10h=7.2$, which gives us $h=7.2$ units.

Answer: $h=7.2$.

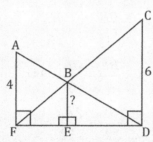

Lines $\overline{AF}$, $\overline{BE}$, and $\overline{CD}$ are all parallel with one another in the diagram above, which shows two right triangles intersecting one another. What is the length of $\overline{BE}$?

First, we know that there are sets of similar triangles within the diagram. Since △*CFD* and △*BFE* share the same angle at F, we can conclude that the two triangles are similar. We can use the same logic to conclude that △*ADF* is similar to △*BDE*. With this knowledge, we can start setting up proportions to find the missing sides to the triangles.

SIMILAR SHAPES   SKILLS

Let's call $\overline{FD}$ the variable y, $\overline{FE}$ the variable x and $\overline{ED}$ $y-x$.

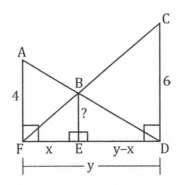

Now let's set up a proportion for $\triangle CFD$ compared to $\triangle BFE$:

$$\frac{BE}{6} = \frac{x}{y}$$

And the same for $\triangle ADF$ to $\triangle BDE$.

$$\frac{BE}{4} = \frac{y-x}{y}$$

For the latter proportion, we can simplify it down to this:

$$\frac{BE}{4} = 1 - \frac{x}{y}$$

Notice now that we can substitute the proportion for $\triangle CFD$ and $\triangle BFE$ and solve.

$$\frac{BE}{4} = 1 - \frac{BE}{6}$$

$$\frac{BE}{4} + \frac{BE}{6} = 1$$

$$\frac{3BE}{12} + \frac{2BE}{12} = 1$$

$$\frac{5BE}{12} = 1$$

$$BE = \frac{12}{5}$$

TRIANGLE CONGRUENCY

For reasons unknown, the ACT has many more questions on similarity than on congruency. However, I'll briefly review the three means of proving congruency below in case they appear. For most of you this should be review.

Remember that **Corresponding Parts of Congruent Triangles are Congruent! (CPCTC!)**. If you prove congruency, EVERYTHING that corresponds in the shapes is equal.

ASA: If two angles are congruent to two angles of another triangle, and the side between them is congruent, then the triangles must be congruent.

SSS: If all the corresponding sides of two triangles are congruent, then the two triangles must be congruent.

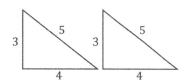

SAS: If one angle of a triangle is congruent to the corresponding angle of another triangle and the lengths of the sides adjacent to the angle are congruent, then the triangles must be congruent.

HINGE THEOREM

The hinge theorem states that if two triangles have two congruent sides (see picture below) joined by different angles, then the triangle with the larger angle between those sides will have a longer third side (below, $y > x$). This also gives way to the converse of the hinge theorem, which states that if two triangles have two congruent sides, then the triangle with the longer third side will have a larger angle opposite that third side. Below, y must be greater than x because $47° > 32°$:

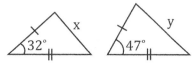

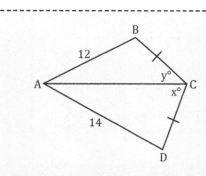

In the diagram shown above, what can be concluded about angles x and y?

A. $x > y$ **B.** $x < y$ **C.** $x = y$ **D.** $x + y = 90$ **E.** No conclusion can be made.

We see that $\overline{AC}$ is a shared side by both triangles and $\overline{BC} = \overline{CD}$. Thus, we can use the hinge theorem to assume that because $14 > 12$ it follows that the angles opposite these side lengths will hold the same comparative relationship. In other words, $x > y$.

Answer: **A**.

SIMILAR SHAPES — SKILLS

PARALLEL LINES THEOREM & SIDE SPLITTER THEOREM

Parallel lines to the triangle's base always divide triangles into two similar triangles. Below, you can see how $\triangle ADE \sim \triangle ACB$.

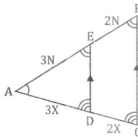

The **side splitter theorem** states that if a line is parallel to a side of a triangle and intersects <u>the other</u> two sides, then this line divides those two sides proportionally. Above, you can see how if $AD:DC$ is in a ratio of $3:2$ then $\overline{AE}$ to $\overline{EB}$ must also be in a ratio of $3:2$.

This rule also extends to trapezoids—because we can extend the lines of trapezoids, we can see how this proportionality happens when we have trapezoids with parallel "cuts" horizontally, too.

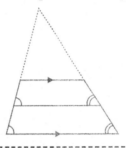

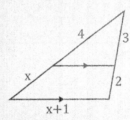

In the diagram above, a scalene triangle is split by a line that is parallel to the base of the triangle. What is the perimeter of the triangle?

Given the parallel lines, we can conclude that the two triangles are similar. With our knowledge of the side splitter theorem, we know if the right side is split in a ratio of $\frac{3}{2}$, then so is the left. Thus:

$$\frac{4}{x} = \frac{3}{2}$$

$$3x = 8$$

$$x = \frac{8}{3}$$

We could also have set up our proportion as follows, using similar triangles:

$$\frac{3}{4} = \frac{5}{4+x}$$

Now solve for x via cross multiplication.

$$12+3x=20$$
$$3x=8$$
$$x=\frac{8}{3}$$

Now we add all the sides together to find the perimeter.

$$4+3+2+1+\frac{8}{3}+\frac{8}{3}=15\frac{1}{3}$$

SIMILAR SOLIDS

Similar solids also operate much like similar shapes, but with an extra dimension.

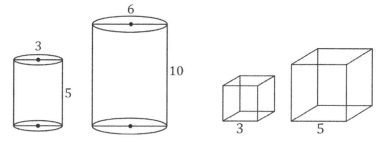

SURFACE AREA AND VOLUME OF SIMILAR SOLIDS

NOTE: **Solids** are covered in the next chapter. Feel free to skip ahead to that chapter and come back to this section later if necessary if you want to review volume and surface area concepts.

Two shapes are similar if all their <u>corresponding angles</u> are congruent and all their corresponding sides are proportional. **Two solids are similar** if they are the same type of solid and their corresponding radii, heights, base lengths, widths, etc. are proportional.

Surface Areas of Similar Solids

SURFACE AREA RATIO

If two solids are similar with sides, heights, or other one dimensional attributes in a ratio of $a:b$, (i.e. the scale factor is $a:b$) then the surface areas are in a ratio of $\left(\dfrac{a}{b}\right)^2$.

The dimensions of the rectangular prism below are tripled. What is the surface area of the new solid?

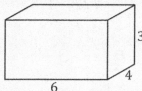

First, let's find the surface area of the smaller prism.

$$2(4\times 6)+2(6\times 3)+2(3\times 4)=96\,units^2$$

If the dimensions are tripled, the ratio of the side lengths of the prism would be $1:3$. This means that the surface area would have a ratio of $1:9$. Now we set up a proportion to find the surface area of the bigger solid, using our ratio, $1:9$.

$$\frac{96}{x}=\frac{1}{9}$$

$$x=864\,units^2$$

Alternatively, we could first upsize each side length: 3 would become 9, 4 would become 12 and 6 would become 18. The surface area would then be:

$$2(9\times 12+9\times 18+18\times 12)=864$$

Answer: 864.

Volumes of Similar Solids

VOLUME RATIO

> If two solids are similar with single dimension lengths (& thus scale factor) in the ratio of $a:b$, then the volumes are in a ratio of $\left(\frac{a}{b}\right)^3$.

> If the volumes of two spheres are in a ratio of $8:27$, what is the ratio of their diameters?

To solve this problem, we know that the three-dimensional ratio (i.e. ratio of the volumes) is the single dimension ratio (or diameter ratio) cubed, i.e.:

$$\frac{8}{27}=\left(\frac{x}{y}\right)^3$$

Where $\frac{x}{y}$ is the ratio of the diameters (or of any specific single dimension part to another single dimension part).

Thus, we solve by taking the cube root of $\frac{8}{27}$:

$$\sqrt[3]{\frac{8}{27}}=\frac{2}{3}$$

Answer: $2:3$.

1. The volume, V, of the right circular cylinder below is given by the formula $V = \pi r^2 h$, where r is the radius of the base and h is the height of the cylinder shown below. If r is tripled and h is halved, the cylinder's new volume would be:

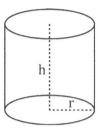

   A. $\frac{3}{4}V$
   B. $\frac{2}{3}V$
   C. $\frac{3}{2}V$
   D. $\frac{9}{4}V$
   E. $\frac{9}{2}V$

2. The formula for the surface area (S) of a rectangular solid (shown below) is $S = 2lw + 2lh + 2wh$, where l represents the length, w the width, and h the height of the solid. Tripling each of the dimensions (l, w, and h) will increase the surface area to how many times its original size?

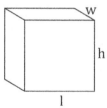

   A. 3
   B. 6
   C. 9
   D. 12
   E. Impossible to determine without knowing the original measurements.

3. Isosceles trapezoid $PQRS$ below has side lengths as marked. Its diagonals intersect at T. What is the ratio of the length of $\overline{PR}$ to $\overline{TQ}$?

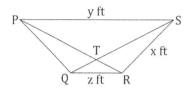

   A. $\dfrac{z}{y}$
   B. $\dfrac{y}{y+z}$
   C. $\dfrac{y+1}{z}$
   D. $\dfrac{y}{z}$
   E. $\dfrac{y+z}{z}$

4. In the figure below, $X, Y,$ and Z are the midpoints of the sides of $\triangle ABC$, and $D, E,$ and F are the midpoints of the sides of $\triangle XYZ$. The interiors of $\triangle AXY$, $\triangle CXZ$, $\triangle BYZ$, and $\triangle DEF$ are dotted. What percent of the interior of $\triangle ABC$ is *not* dotted?

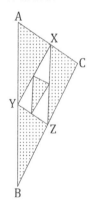

   A. 18.75%
   B. 22%
   C. 37.5%
   D. 56.25%
   E. Impossible to determine from the given information.

5. In the figure below, U lies $\frac{2}{3}$ of the way from R to S on the rectangle $QRST$. The area of $\triangle QUT$ is what fraction of the area of rectangle $QRST$?

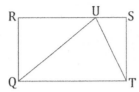

- A. $\frac{1}{3}$
- B. $\frac{1}{2}$
- C. $\frac{2}{3}$
- D. $\frac{3}{4}$
- E. $\frac{4}{5}$

6. Rectangle $ABCD$ consists of 4 congruent rectangles as shown in the figure below. Which of the following is the ratio of the length of $\overline{EF}$ to the length of $\overline{AD}$?

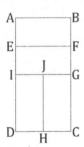

- A. $1:4$
- B. $1:3$
- C. $1:2$
- D. $2:3$
- E. $3:4$

7. For the triangles in the figure below, which of the following ratios of side lengths is equivalent to the ratio of the perimeter of $\triangle ABD$ to $\triangle ABC$?

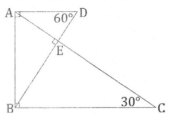

- A. $\overline{AB}:\overline{BC}$
- B. $\overline{AB}:\overline{AC}$
- C. $\overline{AD}:\overline{BC}$
- D. $\overline{AD}:\overline{AC}$
- E. $\overline{DB}:\overline{CB}$

8. Triangles $\triangle TUV$ and $\triangle XYZ$, shown below, are similar with $\angle T \cong \angle X$ and $\angle U \cong \angle Y$. The given lengths are in meters. What is the length, in meters, of $\overline{YZ}$?

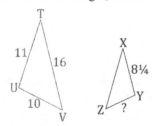

- A. $5\frac{5}{32}$
- B. $7\frac{1}{4}$
- C. $7\frac{1}{2}$
- D. $9\frac{3}{40}$
- E. $13\frac{1}{5}$

9. Lines $L1$ and $L2$ intersect each other and 3 parallel lines, $L3$, $L4$, and $L5$, as shown below. The ratio of the perimeter of $\triangle HKL$ to $\triangle HMN$ is $5:8$. The ratio of IJ to MN is $3:10$. What is the ratio of KL to IJ?

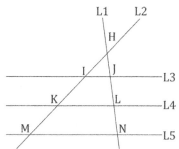

A. 3:16
B. 12:25
C. 24:50
D. 25:12
E. 16:3

10. In $\triangle ACD$ below, $\overline{EB} \parallel \overline{DC}$. The lengths are given in feet. What is CD?

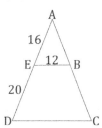

A. 6.7
B. 15
C. 9.6
D. 27
E. 60

11. In the figure below, $\overline{AB} \cong \overline{CD}$. Kayla wants to apply the Side-Angle-Side (SAS) congruence theorem to prove that $\triangle ABD \cong \triangle CBD$. Which of the following congruences, if established, is sufficient?

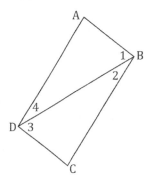

A. $\angle A \cong \angle C$
B. $\angle 1 \cong \angle 4$
C. $\angle 1 \cong \angle 3$
D. $\angle 2 \cong \angle 4$
E. $\angle 2 \cong \angle 3$

12. In the isosceles trapezoid $PQRS$ below, T is the intersection of the diagonals. What is the length of $\overline{SR}$ in terms of $\overline{PQ}$?

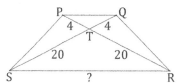

A. $5\overline{PQ}$
B. $10\overline{PQ}$
C. $16\overline{PQ}$
D. $20\overline{PQ}$
E. $24\overline{PQ}$

SIMILAR SHAPES — ANSWERS

ANSWER KEY
1. E 2. C 3. E 4. A 5. B 6. C 7. A 8. C 9. D 10. D 11. C 12. A

ANSWER EXPLANATIONS

1. **E.** Let the original volume be $V = \pi r^2 h$. The new volume is $\pi(3r)^2\left(\dfrac{h}{2}\right)$. Simplify by distributing the exponent $\dfrac{\pi 9 r^2 h}{2}$ or $\dfrac{9}{2}\pi r^2 h$ and then plug in "V" for $\pi r^2 h$ using group substitution: $\dfrac{9}{2}\pi r^2 h = \dfrac{9}{2}V$.

2. **C.** Tripling the measurement of each dimension turns our formula into:
$$2(3l)(3w) + 2(3l)(3h) + 2(3w)(3h) = 9(2lw) + 9(2lh) + 9(2lw) = 9(2lw + 2lh + 2wh) = 9S$$

3. **E.** $\triangle QTR$ is similar to triangle $\triangle PTS$. We know this because $PSQR$ is symmetric as an isosceles trapezoid and so these triangles within it, formed by its diagonals, must also be symmetric (AKA isosceles) and they share a vertex angle by vertical angles (two isosceles triangles are similar if their vertex angles are congruent) since their other angles can also be determined as $\left(\dfrac{180 - \text{vertex angle}}{2}\right)$. Because corresponding parts of similar triangles are similar. The ratio of $\overline{PT}$ to $\overline{QT}$ is $\dfrac{y}{z}$. The ratio of $\overline{TR}$ to $\overline{QT}$ is $\dfrac{z}{z}$. So, since $\overline{PR} = \overline{PT} + \overline{TR}$, $\dfrac{PR}{TQ} = \dfrac{PT + TR}{TQ} = \dfrac{y}{z} + \dfrac{z}{z} = \dfrac{y+z}{z}$.

4. **A.** The triangles formed by connecting the midpoints of the larger triangles are congruent. Thus, the center triangle is $\dfrac{1}{4}$ of the total area. Similarly, the three undotted triangles within the center triangle are $\dfrac{3}{4}$ of the center triangle's area. Thus, the undotted triangles are $\dfrac{3}{4} \times \dfrac{1}{4} = \dfrac{3}{16} = 18.75\%$ of the total triangle's area.

5. **B.** Draw a line straight down from U to the bottom of the rectangle. The two sets of triangles $\triangle QUT$ has been divided into are half the area of the respective smaller rectangles. So $\triangle QUT$ is half of the area of rectangle $QRST$.

6. **C.** Since the two shorter sides of the bottom rectangles together equal the longer side of the rectangle above them, we can conclude that the shorter side of any of the rectangles is half of the long side. Let s be the length of the short end and l be the length of the long end. $EF = l$. $AD = 2s + l = l + l = 2l$. Therefore $E:F = l:2l = 1:2$.

7. **A.** The ratio of the corresponding sides of similar shapes is equal to the ratio of their perimeters. In this case, the way to find the answer is to ensure that the sides we choose are, in fact, corresponding sides. The only choice of sides that are both across from congruent angles are $\overline{AB}$ and $\overline{BC}$, which are both across from 60° angles. Thus, since the ratio between the two is equal to the ratio of the perimeters, the ratio of the perimeters is equal to $AB:BC$.

8. **C.** The triangles are similar, but are mirror images to each other. The ratio of $\overline{TU}$ to $\overline{XY}$ is equal to the ratio of $\overline{UV}$ to $\overline{ZY}$. Thus $\dfrac{11}{10} = \dfrac{8.25}{ZY}$. We cross multiply to get $11ZY = 82.5$, and divide by 11 to get $ZY = 7.5$ or $7\dfrac{1}{2}$.

9. **D.** The ratio of $\triangle HKL$ to $\triangle HMN$, $5:8$, is equal to the ratio of $\overline{KL}$ to $\overline{MN}$. We express this as $\dfrac{KL}{MN} = \dfrac{5}{8}$. Since we are given that the ratio of $\overline{IJ}$ to $\overline{MN}$ is $3:10$, that is, $\dfrac{IJ}{MN} = \dfrac{3}{10}$, we can combine the fractions to get our desired ratio, $\dfrac{KL}{IJ}$. We first flip the $\overline{IJ}$ to $\overline{MN}$ ratio so that the $\overline{MN}$'s cancel out: $\dfrac{MN}{IJ} = \dfrac{10}{3}$. We then multiply $\dfrac{KL}{MN} \times \dfrac{MN}{IJ} = \dfrac{5}{8} \times \dfrac{10}{3} = \dfrac{50}{24} = \dfrac{25}{12}$. Since $\overline{KL}$ and 25 are the numerators, they correspond, and the same goes for $\overline{IJ}$ and 12. Thus, $KL:IJ = 25:12$.

10. **D.** The ratio of $\overline{DC}$ to $\overline{AD}$ is equal to the ratio between $\overline{EB}$ and $\overline{AE}$. $\dfrac{DC}{AD} = \dfrac{EB}{AE}$. We know that $AD = AE + ED = 16 + 20 = 36$. With other values given in the diagram, we can express our ratios as $\dfrac{DC}{36} = \dfrac{12}{16}$. Cross multiply to $16DC = 432$, and divide by 16 to get $DC = 27$.

11. **C.** Since $\overline{AB} \cong \overline{CD}$, and triangles $\triangle ABD$ and $\triangle CBD$ share $\overline{BD}$ as a side (and it is congruent to itself by definition), the only congruence necessary for the Side-Angle-Side congruence theorem is the angle between these congruent sides. The angle between $\overline{AB}$ and $\overline{BD}$ in $\triangle ABD$ is $\angle 1$, and in $\triangle CBD$, between $\overline{CD}$ and $\overline{BD}$ is $\angle 3$. Thus, proving a congruence between $\angle 1$ and $\angle 3$ is sufficient for the Side-Angle-Side congruence theorem.

12. **A.** $\triangle PQT$ and $\triangle SRT$ are similar because of AA (per the vertical angles theorem and parallel lines theorems.) Thus, the ratio of PQ to SR is equal to the ratio of PT to ST: $\dfrac{PQ}{SR} = \dfrac{4}{20}$. This simplifies to $\dfrac{PQ}{SR} = \dfrac{1}{5}$. Cross multiply to get $5PQ = SR$.

CHAPTER

16

SOLIDS

SKILLS TO KNOW

- Cubes
- Rectangular Solids
- Prisms
- Cylinders
- Cones
- Spheres
- Inscribed Solids

Instead of flat shapes like circles, squares, and triangles, solid geometry deals with spheres, cubes, and pyramids (along with any other three dimensional shapes). Solid geometry problems most often involve **surface area** and **volume**.

Because Solids build on knowledge of 2D shapes, I recommend you do this chapter after completing most other geometry topics except possibly similar shapes.

BASIC CONCEPTS:

Volume: Volume is a three-dimensional assessment of space filled. We always multiply three dimensions in some way to find volume, with units cubed (cm^3, ft^3, etc.).

Surface Area: Surface area is the sum of all the individual areas of each side of a solid shape. Surface area uses units squared (in^2, m^2, etc.).

NOTE: With tough solids, the equation you need is usually included as part of a question (though not always). You should know cube, prism, and rectangular solid formulas in any case.

CUBES

A cube's height, length, and width are all equal. The six faces of a cube are also all congruent.

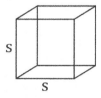

SURFACE AREA AND VOLUME OF A CUBE

Surface Area $= 6s^2$, where s is the side length
Volume $= s^3$, where s is the side length

The volume of a rectangular solid is $v = lwh$. For cubes, because l, w, and h are all equal to s, you can simplify the equation to s^3.

SOLIDS　　　　　　　　　　　　SKILLS

> The surface area of a cube is 216 sq.in.. What is the volume of the given solid?

Using algebra and formulas, we can back track and find out the side length:

$$\text{Cube Surface Area} = 6s^2$$
$$6s^2 = 216$$
$$s^2 = 36$$
$$s = 6$$

If each side of a cube is 6 inches in length, we can find the volume of the solid by cubing our side:

$$\text{Cube Volume} = s^3$$
$$6^3 = 216 \, in^3$$

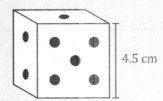

> Jil is planning on making a paper model of a 6-sided die. How much paper, in square centimeters, would Jil need to if he wanted to make the side lengths 4.5 cm?

This problem is asking us to find the surface area of this cube.

$$\text{Cube Surface Area} = 6s^2$$
$$6(4.5)^2 = 121.5 \, cm^2$$

RECTANGULAR SOLIDS (AKA RECTANGULAR PRISMS)

A rectangular solid (or rectangular prism) is essentially a box. It has three pairs of opposite sides that are congruent and parallel. A cube, as mentioned above, is a specific case of a rectangular solid.

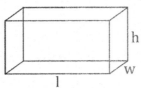

VOLUME OF A RECTANGULAR PRISM

Volume = lwh, where l is the length of the figure, w the width, and h the height.

This formula is the same as finding the area of the base of the rectangle (lw) times the height (h).

SURFACE AREA OF A RECTANGULAR PRISM

Surface Area = $2lw + 2lh + 2wh$

This formula finds the areas for all the flat rectangles on the surface of the figure (the faces) and adds those areas together.

In a rectangular solid, six faces cover the figure: three congruent pairs of opposite sides. To visualize this, I think of a book: the top and bottom, spine and length, and back/front cover. Find the individual areas of each of these three pairs, multiple each of those areas by 2, and sum.

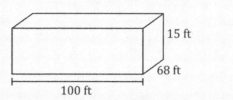

The local grocery store is planning on painting the inside of its newly renovated shop. The shape of the store is a rectangular prism that is 100 feet long, 68 feet wide, and has a ceiling height of 15 feet. Without counting the doors and windows of the store, how much paint would be needed, in square feet, to paint the ceiling and interior walls of the store?

Since we are not painting the floor, we only have to find the area of 5 of the 6 sides. First we know that the larger wall is $15 \times 100 = 1500 \, \text{ft}^2$ and the smaller wall is $15 \times 68 = 1020 \, \text{ft}^2$. The ceiling is $68 \times 100 = 6800 \, \text{ft}^2$.

Next, we multiply by two for each of the wall sizes and add the ceiling:

$$2(1500) + 2(1020) + 6800 = 11{,}840 \, \text{ft}^2$$

Answer: **11,840 ft^2**.

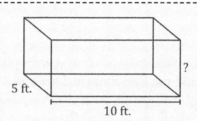

A large fish tank in the shape of rectangular prism has base measurements of 10 feet by 5 feet. If the fish tank must contain 350 cubic feet of water, how deep, in feet, is the tank at minimum?

The total volume of the rectangular prism is 350 cubic feet. We can use algebra to figure out the last missing side minimum.

$$5 \times 10 \times h = 350$$
$$50h = 350$$
$$h = 7 \, \text{ft.}$$

Answer: **7 ft**.

3D DIAGONAL LENGTH (AKA THE SUPER PYTHAGOREAN THEOREM)

The diagonal of a rectangular solid is the longest interior line of the solid. It touches from the corner of one side of the prism to the opposite corner on the other.

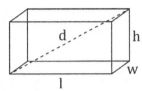

THE SUPER PYTHAGOREAN THEOREM

The diagonal of a rectangular solid with sides l, w, and h is: $d = \sqrt{l^2 + w^2 + h^2}$.

This formula is not given on the exam, but can come in handy on some problems. It's not that hard to remember, because it closely models the original Pythagorean theorem. However, you can also solve this problem without the formula **by breaking up the figure into two flat triangles and using the Pythagorean Theorem twice.**

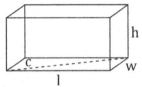

First, create an equation for the length of the diagonal (hypotenuse) of the base of the solid using the Pythagorean theorem.

$$c^2 = l^2 + w^2$$

Now, create an equation for the 3D-diagonal. Use h and c as pictured as the "legs" of the triangle shown in 3 dimensions, and again use the Pythagorean Theorem.

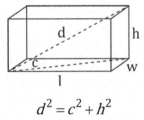

$$d^2 = c^2 + h^2$$

By substitution, we can now derive the Super Pythagorean Theorem, substituting in for c^2 in the 2nd equation, using the first equation:

$$d^2 = \left(l^2 + w^2\right) + h^2$$
$$d = \sqrt{l^2 + w^2 + h^2}$$

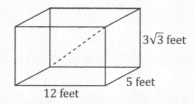

Engineers are looking to strengthen a portion of a bridge with a metal beam that will be a diagonal of a rectangular prism frame. How long does the beam have to be to fit the frame section shown below?

Using the Super Pythagorean Theorem:

$$\sqrt{12^2 + 5^2 + \left(3\sqrt{3}\right)^2} = \sqrt{144 + 25 + 27}$$
$$= \sqrt{196}$$
$$c = 14$$

We could also use the regular Pythagorean theorem to find the missing diagonal. First, identify the triangle created by the legs (5 and 12) and diagonal of the rectangular base. Seeing 5 and 12, we can solve for the hypotenuse quickly as we know this is a Pythagorean triple: 5-12-13. Thus, the diagonal of the base is 13 feet. Next, we must deal with the last triangle whose legs are the height of the figure ($3\sqrt{3}$) and the base's diagonal:

$$13^2 + 3\sqrt{3}^2 = c^2$$
$$169 + 27 = c^2$$
$$196 = c^2$$
$$14 = c$$

Answer: 14 ft.

PRISMS

A prism is a three-dimensional shape that has (at least) two congruent, parallel bases. Basically, you could pick up a prism and carry it with its opposite sides lying flat against your palms, or create one using a Play-Doh™ Fun Factory, chopping off what you extrude at 90° angles.

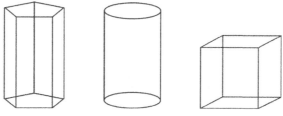

(A few of the many different kinds of prisms.)

VOLUME OF A PRISM

The **volume of a prism** (when at a right angle to the ground) = Bh, where B is the area of the base and h is the height.

This formula works for rectangles, as described earlier, and when the base is another shape (i.e. a circle, star, triangle, etc.). **Surface area** is simply the sum of all the areas of all the sides.

CHAPTER 16

The isosceles right triangular prism will be the top cover of a shading structure at the local park. With the given dimensions below, approximately how much material, to the nearest square foot, is needed to create the cover's sides, front, and back panels? (Note that the cover will not include a bottom floor piece.)

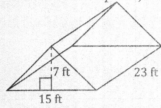

First, let's find the surface area of the individual shapes that make up the solid. The triangle bases each have an area of:

$$\frac{15 \times 7}{2} = 52.5 \text{ ft}^2$$

Two triangle bases thus have an area of $(52.5)2 = 105 \text{ft}^2$.

Next, we must use the Pythagorean theorem to find the missing sides of the triangle, which are also the missing lengths of the rectangles. Because the base triangle is isosceles, we know 7ft is a perpendicular bisector of the base, 15.

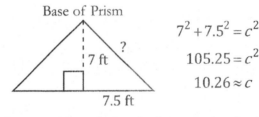

Now we find the area of the two sides of the shade cover by doubling the product of the two recatangle side lengths.

$$2(10.26 \times 23) = 471.96 \text{ ft}^2$$

When we add these areas we will have the surface area we are looking for; then round to the nearest square foot.

$$471.96 \text{ft}^2 + 105 \text{ft}^2 = 576.96 \text{ft}^2 \rightarrow 577 \text{ft}^2$$

Answer: 577ft^2.

The figure shown below is a pentagonal prism. The base of the prism is a regular pentagon with dimensions shown below. Find the volume of this solid in cubic inches.

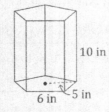

We must first find the area of the base pentagon. To do so we want the apothem: the distance from the center to the side at a 90-degree angle. We can draw this line on the pentagon base and see that it forms a right triangle with our given length 5.

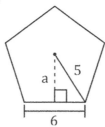

We also know it bisects the side that is 6 inches (knowing that the pentagon is regular, we know all lengths will be proportional).

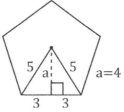

We now see that the triangle has a leg of 3 and hypotenuse of 5. Using our Pythagorean triples, we notice it is a 3-4-5 triangle, so the missing length, the apothem, must be 4. Next, we use this apothem in our formula for the area of a regular polygon:

$$\frac{1}{2}(apothem)(perimeter) = \frac{1}{2}(4)((5)(6)) = 60 \text{ in}^2$$

Next we multiply the base's area by the solid's height to find its volume:

$$60 \times 10 = 600 \text{ in}^3$$

Answer: 600 in^3.

CYLINDERS

A cylinder is a prism with two circular bases on its opposite sides:

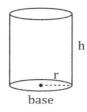

VOLUME OF A CYLINDER

Volume $= \pi r^2 h$,
where π is ≈ 3.14, r is the radius of the circular base, and h is the height of the cylinder (the straight line drawn connecting the two circular bases).

SURFACE AREA OF A CYLINDER

Surface Area $= 2(\pi r^2) + 2\pi r(h)$,
where π is ≈ 3.14, r is the radius of the base, and h is the height of the cylinder.

SOLIDS SKILLS

The surface area of a cylinder is equal to the areas of the two circular bases, $2(\pi r^2)$, added to the area of the side, which if unrolled, looks like a rectangle whose height (h) is the same as the cylinder's height, and whose base is the same as the circumference of the circle ($2\pi r$).

Imagine tearing apart a cylindrical oatmeal box at the seam. You can see how the side is derived:

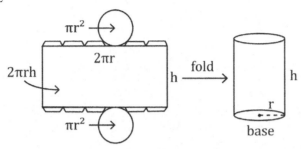

Total Area=$2\pi r^2 + 2\pi rh$

A company is shipping t-shirts inside tube packages. The dimensions of the cylindrical package are shown in the figure below. If each t-shirt requires 75 cubic inches of space, what is the maximum number of t-shirts that could fit inside the packaging?

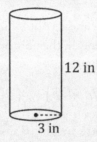

First, let's find the volume of the cylinder.

$$\text{Volume} = \pi r^2 h$$
$$\pi 3^2 \times 12 = 108\pi \approx 339.29 \text{ in}^2$$

Next, we divide our cylinder's volume by 75 to find out how many shirts can fit in our tube.

$$339.29 \div 75 \approx 4.52$$

Remember: be careful with rounding on word problems asking for "whole" amounts. Here we don't want to round up, because we can't fit 5 shirts if there is only room for 4.5!

We can fit a maximum of 4 whole shirts into our tube packaging.

Answer: 4.

CONES

A cone is similar to a cylinder, but has only one circular base instead of two. Its opposite end terminates in a point (called an apex), rather than a circle. There are two kinds of cones—right cones and oblique cones. For the purposes of the ACT, focus on right cones. Oblique cones are unlikely to appear on the ACT. In a right cone, when a height (h) is dropped from the apex to the center of the circle, it makes a right angle with the circular base.

SKILLS SOLIDS

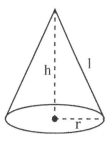

SURFACE AREA OF A RIGHT CIRCULAR CONE

***Surface Area** $= \pi r^2 + \pi r l$,
where l is the slant height from the apex to the circumference of the circular base. The surface area is the combination of the area of the circular base (πr^2) and the lateral surface area ($\pi r l$).

VOLUME OF A RIGHT CIRCULAR CONE

Volume $= \left(\dfrac{1}{3}\right)\pi r^2 h$,
where r is the radius of the base, and h is the height at a right angle to the base.

The volume of a cone is $\dfrac{1}{3}$ the volume of a cylinder. This makes sense logically, as a cone is basically a cylinder with one base collapsed into a point.

*(This is rarely tested on the ACT®, and the equation will likely be given to you should you need it.)

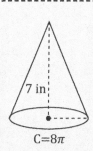

Raylene is creating party hats that are in the shape of right circular cones, but open along the bottom. She wants the height of the cone to be 7 inches tall with a circumference of the base circle 8π inches. With this information, approximately how much material, rounding up to the nearest square inch, would Raylene need to make 5 party hats of this size?

(Note: The total surface area of a right cone is equal to $\pi r^2 + \pi r l$, where π is a constant(≈ 3.14), r is the radius, and l is the slant height extending from the apex to the circumference of the circular base, and πr^2 is the area of the circular base.)

We need the lateral surface area that covers the side of the cone, but we are given the formula for the **total** surface area of a cone. Thus, we must subtract off the base (area of a circle) from this formula to isolate the lateral surface area.

Total surface area of a cone (including its base): $\pi r^2 + \pi r l$
Surface area of the "party hat" portion of a cone: $\pi r^2 + \pi r l - \pi r^2 = \pi r l$

First we must find the hypotenuse of the triangle represented by the dotted lines and cone side (a.k.a. l, the slant height of the cone). If the circumference of the circular base is $8\pi = 2\pi r$, then $r = 4$. Use the Pythagorean theorem to find l, using the radius, $r = 4$ in. and the height 7 in. as shown:

CHAPTER 16   259

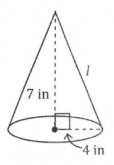

$$4^2 + 7^2 = l^2$$
$$65 = l^2$$
$$8.06 \approx l$$

Next, we use the formula $\pi r l$:

$$8.06 \times 4\pi = 32.34\pi \text{ in}^2 = 101.5 \text{ in}^2$$

Per the question, we need FIVE party hats, so we multiply this number by five: $5(101.5) = 507.5 \text{ in}^2$. Now, following directions in the question, we round up to the nearest square inch: **508**.
Answer: **508**.

> The city is erecting a statue and needs to know how big the base of the statue will be. The shape of the base is a right circular cone with its top removed with dimensions given below. What is the volume of the base of the statue in cubic feet? (Note: the formula for a cone with its top removed is $V = \frac{1}{3}\pi h \left(R^2 + r^2 + Rr \right)$ in which R is the larger radius and r is the shorter radius.)
>
>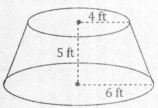

We have all the necessary dimensions, all we have to do is plug in our values:

$$\frac{1}{3}\pi(5)\left(6^2 + 4^2 + 24\right) = 126.67\pi \text{ ft}^3$$

Answer: $126.67\pi \text{ ft}^3$.

PYRAMIDS

Pyramids are geometric solids that are similar to cones, except that they have a polygon for a base and flat, triangular sides that meet at an apex.

There are many types of pyramids, defined by the shape of their base and the angle of their apex, but for the sake of the ACT®, square pyramids occur most often.

A right, square pyramid has a square base (each side has an equal length) and an apex directly above the center of the base. The height (h), drawn from the apex to the center of the base, makes a right angle with the base.

If you encounter a pyramid, you'll likely be given the formulas you need in the question.

VOLUME OF A PYRAMID

$$\text{Volume} = \frac{1}{3}(\text{area of base})h$$

To find the volume of a square pyramid, you could also say $\frac{1}{3}lwh$ or $\frac{1}{3}s^2h$, as the base is a square, so each side length is the same.

SURFACE AREA OF A PYRAMID

Surface Area = Area of Base + Area of Each Lateral Triangle

($\frac{1}{2}$(Slant length × Base Length))

A regular pyramid with a square base is shown below with a slant height of 15 inches and base side length of 18 inches. What is the volume of the pyramid, in cubic inches?

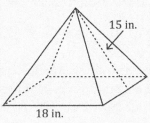

We must find the height of the pyramid. We can do this by drawing a line from the top vertex of the pyramid to the center of the square base. This creates a right triangle with a hypotenuse of 15 inches and one leg measurement of 9 inches. We use the Pythagorean theorem to determine the length of the other leg.

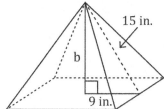

$$9^2 + b^2 = 15^2$$
$$81 + b^2 = 225$$
$$b^2 = 144$$
$$b = 12$$

Now that we know the height of the pyramid, we can find the volume of this solid.

$$\frac{1}{3}(18)(18)(12) = 1296 \text{ in}^3$$

Answer: 1296 in^3.

CHAPTER 16

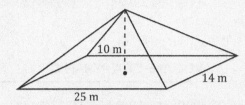

A homeowner plans to add solar panels onto his house's roof. The roof is shaped as a rectangular pyramid with the dimensions below. He plans to add solar panels on all four sides of his roof. What is the approximate surface area of the regions of the roof that will have solar panels on them?

Since the homeowner is only going to put solar panels on the triangular sides of his roof, we only need to find the surface area of the four triangles that make up the pyramid. First, we must find the two slant lengths of the pyramid. We do this by using the Pythagorean theorem and finding a triangle parallel to the sides of the base through the center:

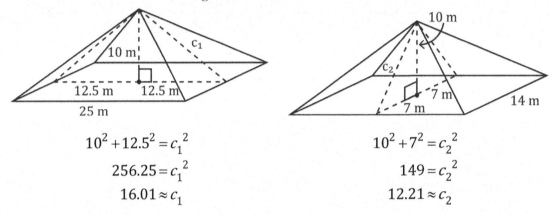

$$10^2 + 12.5^2 = c_1^2$$
$$256.25 = c_1^2$$
$$16.01 \approx c_1$$

$$10^2 + 7^2 = c_2^2$$
$$149 = c_2^2$$
$$12.21 \approx c_2$$

After we have the slant lengths, we know that these are the heights of our individual triangle surfaces. We can find the area of the triangles, and since we know this is a rectangular pyramid, there are two of the same triangles on the opposite side. So we double the area of each triangle face.

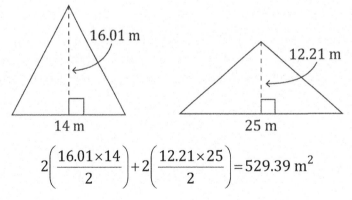

$$2\left(\frac{16.01 \times 14}{2}\right) + 2\left(\frac{12.21 \times 25}{2}\right) = 529.39 \text{ m}^2$$

Answer: 529.39 m^2 is approximately 530 m^2.

SPHERES

A sphere is essentially a 3D circle.

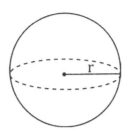

VOLUME AND SURFACE AREA OF A SPHERE

$$\text{Volume} = \frac{4}{3}\pi r^3$$
$$\text{Surface Area} = 4\pi (r)^2$$

> How much paint coverage, in square units, is needed to completely paint a spherical ball with a circumference of 36π units at its widest point? (Note: the Surface Area of a Sphere equals $4\pi(r)^2$.)

We can use the circumference to find out the radius of our sphere. If 36π is the circumference, $C = 2\pi r$, so we plug in:

$$36\pi = 2\pi r$$
$$18 = r$$

Now we just plug that into our given surface area formula:

$$4\pi (r)^2 = 4\pi (18)^2 = 1296\pi \text{ sq. units}$$

Answer: 1296π sq. units.

INSCRIBED SOLIDS

Sometimes, solid shapes are placed inside other solids, such as the sphere in the cube on the left, or the cube in the sphere on the right.

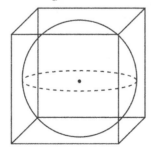 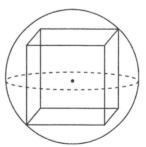

When dealing with inscribed shapes, draw on the diagram they give you, and **figure out what lengths are equal when surfaces touch.** For instance, on the left, **the diameter of the circle is the length of the cube side.** On the right, **the diagonal of the cube is the diameter of the sphere.** Always draw out inscribed shapes if no picture is given.

SOLIDS SKILLS

> In the figure below, a sphere is inscribed within a cube with side measurements of 5 inches. Find the total volume of the empty space between the sphere and cube in cubic inches.

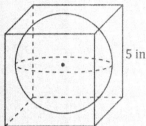

For this problem, we need to find the volume of the outer cube and the volume of the sphere and find the difference of the two to find out the volume of the void space. First we find the volume of the cube, s^3:

$$5^3 = 125 \text{ in}^3$$

Next, we must find the measurements of the sphere to find its volume. Since the sphere is inscribed, we know that its diameter is also the square's side length, 5.

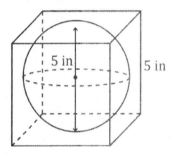

This means that the radius of the sphere is 2.5 inches, or half the diameter. Now we can find the volume using the formula:

$$V = \frac{4}{3}\pi(2.5)^3$$

$$V = \frac{62.5}{3} \times \pi$$

$$V \approx \frac{62.5}{3}\pi \text{ in}^3$$

Now just subtract the sphere's volume from that of the cube.

$$125 - \frac{62.5}{3}\pi = 125 - \frac{125}{6}\pi \text{ in}^3$$

Answer: $125 - \dfrac{125}{6}\pi \ in^3$.

1. The right triangle shown below with hypotenuse 25 inches long and vertical leg 7 inches long is rotated 360° around the vertical leg to form a right circular cone. What is the volume of this cone, to the nearest cubic inch? (Note: $V = \frac{1}{3}\pi r^2 h$).

   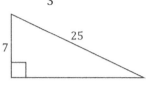

   A. 1322
   B. 1344
   C. 3878
   D. 4222
   E. 15080

2. The volume, V, of a right circular cylinder with radius r and height h is given by the formula $V = \pi r^2 h$. The right circular cylinder below has radius r, height h, and volume V. A second right circular cylinder has radius $3r$ and height $6h$. What is the volume of the second cylinder in terms of V?

   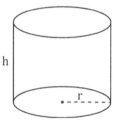

   A. $15V$
   B. $18V$
   C. $27V$
   D. $54V$
   E. $108V$

3. The radius and height of a right circular cylinder are given below in feet. What is the approximate volume of the cylinder?

   A. 188
   B. 240
   C. 754
   D. 815
   E. 900

4. The figure below is a cross section of a pool that is 20 ft wide and 65 ft long. The sides are vertical, and the bottom begins at an incline, and then becomes level. What is the volume of water, in cubic feet, it would take to fill the pool so that it is 10 ft deep at one end and 25 ft deep at the other end, as shown?

   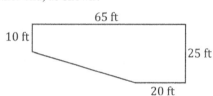

   A. 1137.5
   B. 1287.5
   C. 21250
   D. 22750
   E. 25750

5. A circle with radius 3 inches long is inscribed in a cube, as shown below. What is the length of a diagonal of the cube (i.e. the front top right corner to the back bottom left corner)?

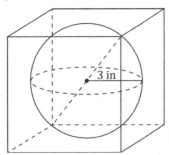

A. $3\sqrt{2}$
B. $3\sqrt{3}$
C. 6
D. $6\sqrt{2}$
E. $6\sqrt{3}$

6. A right triangular prism is 18 inches deep, 10 inches wide, and 6 inches tall. What is the volume of the prism in cubic inches?

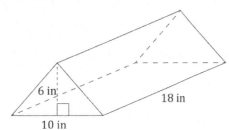

A. 300
B. 360
C. 540
D. 720
E. 1080

7. What is the total surface area, in inches, of the right triangular prism shown below?

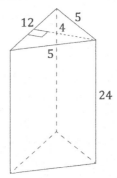

A. 500
B. 528
C. 552
D. 576
E. 624

8. A right rectangular pyramid is shown in the figure below. The slant height is $3\sqrt{3}$ meters and the length of the bases sides are 6 meters and 5 meters, respectively. What is the total length, in meters, of all 8 edges of the pyramid?

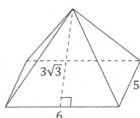

A. 22
B. 43
C. 46
D. 54
E. 64

9. The right rectangular prism shown below has a face that is 7 meters by 11 meters and a volume of 616 cubic meters. What is the total surface area of the prism in meters?

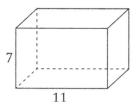

A. 221
B. 330
C. 365
D. 407
E. 442

10. Four cubes of equal volume can be rearranged in the two orientations shown below. To the nearest percent, the total surface area of the orientation on the left is what percent less than the total surface area of the orientation on the right?

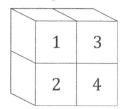 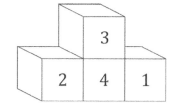

A. 11%
B. 20%
C. 27%
D. 89%
E. 94%

11. Which of the following is closest to the diameter, in meters, of a cylindrical container, shown below, with height 10 m and volume 1540 cubic meters? (Note: The volume of a cylinder with radius r and height h is $\pi r^2 h$.)

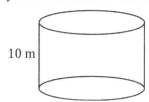

A. 4
B. 7
C. 10
D. 13
E. 14

12. A right circular cylinder has diameter 4 cm and height 8 cm. What is the total surface area of this cylinder, in square centimeters? (Note: The total surface area of a cylinder is given by $2\pi r^2 + 2\pi rh$ where r is the radius and h is the height.)

A. 10π
B. 20π
C. 30π
D. 40π
E. 96π

13. Steve decides to put away a standard gym mat (15 feet wide and 20 feet long) after practice. The mat is rolled up so that the two shorter edges just meet, forming a circular tube (cylinder) 15 feet tall. Given that there is no top or bottom to the cylinder, what is the cylinder's outer surface area?

A. $\dfrac{20}{2\pi}$
B. 40π
C. 225π
D. 300
E. 300π

14. Cubes each having a side length of 1.5 in are put together to form a rectangular solid with 4 layers. Each layer has 12 cubes. What is the volume, in cubic centimeters, of the rectangular solid?

A. 48
B. 72
C. 108
D. 162
E. 170

15. The volume of a cube is 216 cubic feet. What is the total surface area, in square feet, of the cube?

A. 24
B. 36
C. 72
D. 216
E. 1296

16. In order to build a new parking lot next to the beach, the city removes 2,000 cubic meters of sand from a lot. They decide to use the sand removed to create an indoor volleyball court of sand. If this sand was spread in an even layer over a volleyball court (which measures 18 meters by 9 meters), how many meters deep would the sand layer be?

   A. Less than 9
   B. Between 12 and 14
   C. Between 16 and 18
   D. Between 18 and 20
   E. More than 20

17. A solid right circular cone, picture below, is sliced horizontally, parallel to its base and perpendicular to its central line of height, somewhere between the base and the top of the cone. Which of the following best represents the possible plane section?

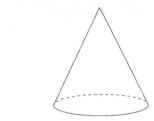

A.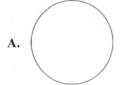

B.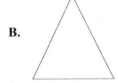

C.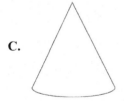

D.

E.

18. A chocolate company develops a chocolate bar that is in the shape of a trapezoidal prism. The length of the bottom side of the trapezoid base is 20 millimeters, and the length of the top side is 14 millimeters. What is the length, in millimeters, of the median of the trapezoid?

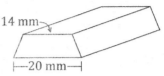

   A. 14
   B. 16
   C. 17
   D. 18
   E. 20

19. A solid beam of wood has 6 rectangular faces. The length, width, and height of the beam are h, j and k feet respectively. Which of the following represents the beam's surface area, in square feet?

   A. $2h+2j+2k$
   B. $2(hj+jk+hk)$
   C. $2hj+4jk$
   D. $(h+j+k)^2$
   E. $6hjk$

ANSWERS — SOLIDS

ANSWER KEY

1. D 2. D 3. C 4. E 5. E 6. C 7. D 8. C 9. E 10. A 11. E 12. D 13. D 14. D
15. D 16. B 17. A 18. C 19. B

ANSWER EXPLANATIONS

1. **D.** The radius of the cone is the bottom leg of the right triangle, which we can either recognize from the 7-24-25 Pythagorean triple, or as $\sqrt{25^2 - 7^2} = \sqrt{576} = 24$. The height of the cone will be the vertical leg of the triangle, 7. Plugging our values into the equation for volume of a cone gives us $\frac{1}{3}(\pi)(24^2)(7) = 1344\pi \approx 4222$.

2. **D.** The volume of the second cylinder will be $\pi(3r)^2(6h) = 54\pi r^2 h$. Since $\pi r^2 h = V$, the volume of the second cylinder can be found by plugging "V" for $\pi r^2 h$ to get $54V$.

3. **C.** The volume of a cylinder is $\pi r^2 h$. Plugging in our radius of 4 and height of 15 gives us $\pi(4^2)(15) = 240\pi \approx 754$.

4. **E.** We will first find the area of the polygon, and then multiply by the width of the pool to get the volume. Draw a vertical line that divides the shape into a trapezoid and a rectangle. The bases of the trapezoid are 10 and 25, and its height is $65 - 20 = 45$. The area of the trapezoid is $\frac{10 + 25}{2}(45) = 787.5$. The area of the rectangle is $(20)(25) = 500$. The total area of the figure is $787.5 + 500 = 1287.5$. We multiply by 20 to get the total volume: $(1287.5)(20) = 25750$.

5. **E.** The diagonal of the cube is going to be the hypotenuse of a right triangle with one leg, an edge of the cube, and the other, a diagonal across the face of the cube. Each edge of the cube is 6 inches since the radius of the sphere is 3 inches so its diameter is 6 inches, and that spans the entire cube. To solve fast, use the super Pythagorean theorem (pg 254): $\sqrt{6^2 + 6^2 + 6^2} = \sqrt{3(36)} = 6\sqrt{3}$. Or, draw a diagonal across a face of the cube forms a 45-45-90 triangle, so the diagonal equals $6\sqrt{2}$, according to the ratio of sides of a 45-45-90 triangle. Since the diagonal of the sphere (not the face) is the hypotenuse of a right triangle, and we now have the lengths of the legs, we can solve for it: $\sqrt{(6)^2 + (6\sqrt{2})^2} = \sqrt{36 + 72} = \sqrt{108} = \sqrt{36}\sqrt{3} = 6\sqrt{3}$.

6. **C.** The volume of a prism equals the area of the face that corresponds to the type of prism times the depth. In this case, the height is 18 and area of the triangle is $\frac{1}{2}(10)(6) = 30$. So the volume is $(30)(18) = 540$.

7. **D.** There are 5 faces of the prism: 2 triangles and 3 rectangles. The 2 triangles are congruent, and 2 of the rectangles are congruent, since they have the same side lengths. The area of each triangle is $\frac{1}{2}(12)(4) = 24$. The area of one of the rectangles is $(12)(24) = 288$ and the area of each of the other two rectangles is $(5)(24) = 120$. Adding together all of the faces' areas gives us $24 + 24 + 120 + 120 + 288 = 576$.

8. **C.** The perimeter of the base of the pyramid is $6 + 6 + 5 + 5 = 22$. The only remaining edges we need to find the total length of all 8 edges are the 4 edges from the vertices of the square to the apex of the pyramid. In a right triangle, these are all equal. We can solve for one of them as the hypotenuse with legs the length of the slant of height $3\sqrt{3}$ and of half the 6 unit base, which is 3. These form a 30-60-90 triangle with a hypotenuse of 6. Thus, we add $(4)(6) = 24$ to the perimeter of the base to get the total length of all 8 edges, which is $24 + 22 = 46$.

9. **E.** Let l be the height, 7. Let w be the width, 11. Let d be the depth. lwd equals the volume, 616. By substitution, $(d)(7)(11) = 616$. Using division, we find that $d = 8$. The surface area of a rectangular prism equals $2lw + 2ld + 2dw$. Plugging in our numbers, we get $2(7)(11) + 2(7)(8) + 2(8)(11) = 154 + 112 + 176 = 442$. See pg 252-253 for more surface area tips.

CHAPTER 16

SOLIDS — ANSWERS

10. A. Let f be the area of one face of a cube. The cubes are all congruent, since they have the same volume. Visualize and count out the number of exposed faces in the left figure to get 16; the surface area is thus $16f$. Count the number of exposed faces in the right figure, 18; the surface area is $18f$. Thus, the percent of surface area shown on the left out of the surface area on the right is $\frac{16f}{18f} = \frac{16}{18} = 88.9\%$. Subtract from 100% to find the percent not shown: $100\% - 88.9\% \approx 11\%$.

11. E. Plugging into the formula for the volume of a cylinder (pg 257), we can arrive at the equation $1540 = \pi r^2 (10)$. If we divide both sides by 10π, we can see that the square of the radius equals about 49 cubic meters. If we square root both sides, we find that the radius is 7 meters. However, we must carefully note that the problem asks for the number closest to the **diameter**—or twice the radius, so the answer is 14 meters.

12. D. The question tells us that the formula for surface area is $SA = 2\pi r^2 + 2\pi rh$. Although we are given that the diameter is 4 inches, we must solve for the radius (half the diameter), which is 2 in, in order to plug into the problem. Then, we can use the equation $SA = 2\pi(2^2) + 2\pi(2)(8) \to SA = 8\pi + 32\pi \to SA = 40\pi$ or answer (D).

13. D. The surface area is equal to the circumference of the cylinder times the height plus the top and bottom circles. However, since there is no top or bottom, we only need the lateral surface area portion, which equals circumference times the height (see top of pg 258 to visualize this). In this case, the gym mat is said to be 20 feet long. Once wrapped around in a cylinder, this would become the circumference of the circle. Therefore, we can find the surface area by multiplying 20 times the length of the mat, 15, to get 300. If it helps, draw out the "mat" as a rectangle whose width is the circumference and whose length is the length of the mat.

14. D. The volume of each individual cube is $(1.5)^3 = \left(\frac{3}{2}\right)^3 = \frac{27}{8}$. The volume of the solid is the sum of the volume of all the cubes that are in the solid. Because there are 4 layers of 12 cubes each, there is a total of $4(12) = 48$ cubes. Thus, the total volume is $\frac{27}{8}(48) = 162$.

15. D. If the volume is 216, then the length of one side s is equal to the cube root of 216: $s = \sqrt[3]{216} \to s = 6$. The surface area is the sum of the surface area of each side. The surface area of one side of this cube is $6(6) = 36$. Since a cube has 6 equal sides, the surface area is $36(6) = 216$.

16. B. We are essentially given a volume, a length, and a width, and are told to find the height, as height and depth fulfill the same function when finding volume of a rectangular solid. All measurements are in meters. Thus we have: $2000 = (18)(9)h$. Solve for h to get $h = 12.346$, which is between 12 and 14.

17. A. The cone is sliced horizontally. Because it is a circular cone, meaning it's base is a circle, any cross sections that are parallel with the base must be circles as well. Imagine chopping an ice cream cone horizontally.

18. C. The median is the average of the two bases, so the median of this trapezoid is $\frac{14+20}{2} = 17$ millimeters.

19. B. The beam's surface area is the sum of each individual side's area. This beam has 3 pairs of identical sides, one pair whose sides are h and j, giving it an area of hj, another whose sides are j and k, giving it an area of jk, and the final pair whose sides are h and k, giving it an area of hk. Thus the total surface area is $2hj + 2jk + 2hk = 2(hj + jk + hk)$.

CHAPTER 17

VECTORS

> ### SKILLS TO KNOW
> - Vector terminology: standard position, component form, magnitude, direction
> - Adding and subtracting vectors
> - Finding the magnitude of a vector
> - Scalar multiplication
> - Unit vector
> - Different forms of vectors

Vectors have appeared on the ACT® with slightly greater frequency over the past several years. Though many students learn these in physics class, not math class, for whatever reason, they're on the math portion of the ACT® exam. Still, vectors are far less important than many other areas on the test, so this chapter is brief. If you're aiming for a 32+ on the math section, consider reviewing this chapter.

VECTORS DEFINED:

A vector is defined by a magnitude (length) and a direction. We represent vectors visually as line segments at a particular angle:

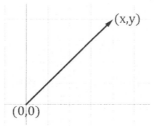

Direction is essentially the "angle" at which the vector is drawn, but we don't always measure it in degrees or radians. Rather, we think about how far over and how far up our vector extends, i.e. the x-direction movement and the y-direction movement. It's kind of like slope, but we keep the x and y parts separate.

We think of vectors much like we think of coordinate points. To represent a vector with symbols and variables, we write the vector in **component form**. Component form looks like a coordinate pair, but with funky dart shaped parentheses: $\langle x, y \rangle$. If we drew a directional line segment between $(0,0)$ and (x,y), for example, it could represent vector $\langle x, y \rangle$, showing the length of the vector (also called the **magnitude**) and the slope or angle at which that vector is drawn with respect to the axis.

A vector in **standard position** has a starting point at the origin such that the endpoint's coordinates are equal to the vector's component form. The above example, which starts at $(0,0)$ and ends at (x,y) is in **standard position**.

But the vector $\langle x, y \rangle$ does not have to start at $(0,0)$. We can move the vector around anywhere we want, but so long as it maintains the same **direction** (orientation "angle" wise), and has the same **magnitude** (i.e. length) it is the same vector, $\langle x, y \rangle$. I.e. placement on the graph does not define a vector: **length and direction alone define a vector**.

We can calculate the **component form** by finding the difference in x's and difference in y's: we take the endpoint's x and y values and subtract the starting point's x and y values, respectively.

COMPONENT FORM

$$\langle x_2 - x_1, y_2 - y_1 \rangle$$

I.e. If a vector began at point $(2,3)$ and ended at point $(5,7)$ we could calculate its component form as follows:

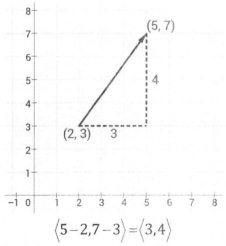

$$\langle 5-2, 7-3 \rangle = \langle 3, 4 \rangle$$

ADDING AND SUBTRACTING VECTORS:

Graphically, we add vectors by connecting them "tip to tail:" with one's ending point as the other's starting point.

We then play "connect the dots" and draw a vector line segment between the starting coordinate of our first vector and the endpoint of our second vector, creating a triangle. The vector that starts at the first vector's starting point and goes to the second vector's ending point is the **resultant vector**.

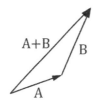

One way to subtract two vectors, visually speaking, is essentially to add the vectors, except first reverse the direction of the 2nd vector before adding (i.e. turn it around 180 degrees). For this problem, let's calculate Vector A minus Vector B. We first find $-B$ by **reversing our vector in the opposite direction**.

Then again we place them tip to tail:

Another way to think of subtraction is to put the lines "tail to tail":

To add or subtract vectors using the written representation is even easier: we just add or subtract the corresponding parts of the vectors. Don't forget to distribute the negative sign when subtracting!

Adding

$$\langle 3,5 \rangle + \langle 7,6 \rangle = \langle 3+7, 5+6 \rangle$$
$$= \langle 10, 11 \rangle$$

Subtracting

$$\langle 8,9 \rangle - \langle 3,6 \rangle = \langle 8-3, 9-6 \rangle$$
$$= \langle 5, 3 \rangle$$

If $\vec{a}=\langle -3,5 \rangle$ and $\vec{b}=\langle 4,7 \rangle$, what is $\vec{a}-\vec{b}$?

$\vec{a}-\vec{b}=\langle -3,5 \rangle - \langle 4,7 \rangle$ First, substitute in the vectors.

$\qquad = \langle -3-4, 5-7 \rangle$ Next, combine the corresponding terms.

$\qquad = \langle -7, -2 \rangle$ Finally, simplify.

Answer: $\langle -7, -2 \rangle$.

FINDING THE MAGNITUDE OF A VECTOR:

Again the **magnitude** (or **norm**) of a vector is the length of its arrow. Remember a vector has an x and y value, an amount of horizontal movement and vertical movement. These movements happen at a 90 degree angle to each other. Recognize that a vector is graphically the hypotenuse of this movement, and you'll realize finding the magnitude of vector $\langle x, y \rangle$ simply requires finding the hypotenuse of a right triangle that has legs x and y.

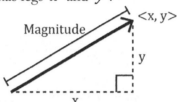

We can find the magnitude of the vector using the Pythagorean theorem on the values in the vector's component form.

What is the magnitude of vector $n = \langle 3, 4 \rangle$?

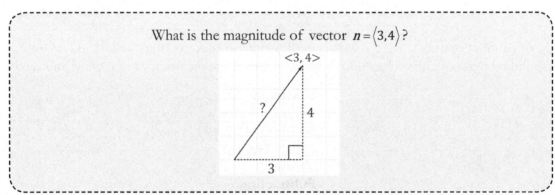

We take the values 3 and 4 from the component form and plug them into the Pythagorean theorem:

$$\sqrt{3^2 + 4^2} = \sqrt{9+16} = \sqrt{25} = 5$$

Answer: 5.

TIP: Sometimes magnitude is denoted by what looks like absolute value signs, don't get confused!

For example: $|a|$ or $|\vec{a}|$ or $\|a\|$ or $\|\vec{a}\|$ **all mean the same thing**: the magnitude of vector a. Also, the **magnitude** is also sometimes called the **norm** of a vector or the **Euclidian norm**.

> If a vector is defined such that $t = \langle 5, 12 \rangle$, what is its **norm**, $\|t\|$?

All we need to do is find the hypotenuse of a triangle with legs 5 and 12, that's 13. I know this because I have memorized this Pythagorean triple. I can also solve using the Pythagorean theorem:

$$5^2 + 12^2 = \|t\|^2$$
$$25 + 144 = \|t\|^2$$
$$169 = \|t\|^2$$
$$13 = \|t\|$$

Answer: 13.

SCALAR MULTIPLICATION

It sounds intimidating, but scalar multiplication is actually very simple. It *scales* the vector to a different magnitude, keeping the same direction, but changing the length.

For example, $2\langle 5, 12 \rangle$ will make a vector with a magnitude twice as large as the original. The component form of this vector will be the original vector's components times the scalar.

$$2\langle 5, 12 \rangle = \langle 2(5), 2(12) \rangle = \langle 10, 24 \rangle$$

Easy, just distribute the scalar!

> If $n = \langle 2, 4 \rangle$, what is $3n$?

$$3n = 3\langle 2, 4 \rangle$$
$$= \langle 3 \cdot 2, 3 \cdot 4 \rangle$$
$$= \langle 6, 12 \rangle$$

Answer: $\langle 6, 12 \rangle$.

UNIT VECTOR

A unit vector is a vector with a magnitude of 1, similar to how the unit circle has a radius of 1. We can find the corresponding unit vector for any vector by scaling that vector down such that the magnitude is one.

CHAPTER 17

To find the corresponding unit vector of any vector, you take two steps:
1. Find the magnitude
2. Divide each element in component form by the magnitude.

Why does this work? Essentially the "scalar" we multiply our component form by to scale it to the unit vector is the reciprocal of the magnitude. Remember the component form can form a triangle that shows the magnitude and the two sides.

Take for example this $3, 4, 5$ triangle formed by the vector $\langle -4, 3 \rangle$:

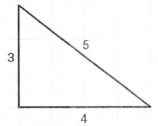

If we want the corresponding unit vector, we need to shrink that hypotenuse to 1. What can we multiply it by to do so? $\frac{1}{5}$ (the reciprocal of the magnitude). To create a similar triangle that defines the unit vector, we then multiply both legs by $\frac{1}{5}$ as well:

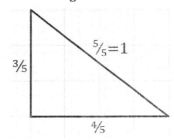

The vector defined by the "legs" of this triangle is now $\langle -\frac{4}{5}, \frac{3}{5} \rangle$. This is our unit vector. (Notice I keep the negative sign from the x-value in the original vector pair).

MORE VECTOR NOTATION

These don't come up very often on the ACT®. Nonetheless, for overachievers, the information below might be good for a quick read through just in case. You never know what curve balls the ACT® has for you, especially in subjects like vectors.

Magnitude and Direction Form

In addition to representing a vector using component form, we can also represent a vector using its magnitude and direction, where the magnitude is the geometric length of the arrow, and theta is the angle at which the vector is drawn in relation to the horizontal axis:

$$\|u\|, \theta$$

For example, I can define the vector below as 110 degrees, with a magnitude of 12:

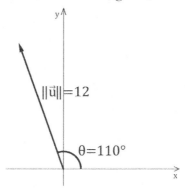

Unit Vector $(i+j)$ Form

Sometimes we express component form not with dart shaped parentheses, but **with the letters i and j** and with an addition sign in between them. Sometimes these letters have a little "hat" over them: $\hat{i}, \hat{j}$.

For example, $\langle 3, 4 \rangle$ can be written as $3i + 4j$ or $3\hat{i} + 4\hat{j}$. This is called **Unit Vector Form**.

We can also graph vectors in unit vector form on a coordinate plane such that the "i" part is graphed on an axis labeled i instead of x and the j part is graphed on an axis labeled j instead of y.

VECTORS QUESTIONS

1. What is the result of adding the vectors $\langle 8,5 \rangle$ and $\langle 2,8 \rangle$?

   A. $\langle 10,13 \rangle$
   B. $\langle 10,64 \rangle$
   C. $\langle 10,40 \rangle$
   D. $\langle 40,16 \rangle$
   E. $\langle 64,10 \rangle$

2. The component forms of vectors u and v are given by $u = \langle 6,3 \rangle$ and $v = \langle 4,-1 \rangle$. Given that $u - 2v + w = 0$, what is the component form of w?

   A. $\langle -2,5 \rangle$
   B. $\langle 2,-5 \rangle$
   C. $\langle -2,-5 \rangle$
   D. $\langle 2,5 \rangle$
   E. $\langle 4,-10 \rangle$

3. What vector is in the same direction as $\langle 9,-12 \rangle$ with a length of 1?

   A. $\langle 8,-11 \rangle$
   B. $\langle 1,-\frac{12}{9} \rangle$
   C. $\langle \frac{3}{5},-\frac{4}{5} \rangle$
   D. $\langle \frac{3}{5},\frac{4}{5} \rangle$
   E. $\langle \frac{3}{4},1 \rangle$

4. What is the magnitude of the vector formed from the addition of $\langle 4,6 \rangle$ and $\langle -3,9 \rangle$?

   A. $\langle 1,15 \rangle$
   B. $2\sqrt{13} + 3\sqrt{10}$
   C. $2\sqrt{13} - 3\sqrt{10}$
   D. $\sqrt{226}$
   E. 15

5. If $\vec{V_1} = \langle -3,7 \rangle$ and $\vec{V_2} = \langle 11,4 \rangle$, then what is $\langle 2\vec{V_1} + \vec{V_2} \rangle$?

   A. $\sqrt{349}$
   B. 26.94
   C. $\langle 5,18 \rangle$
   D. $\langle 5,11 \rangle$
   E. $\left\langle \frac{5\sqrt{349}}{349}, \frac{18\sqrt{349}}{349} \right\rangle$

6. If $|\vec{m}| = 23$ and $|\vec{n}| = 20$, which of the following could NOT be $\langle m+n \rangle$?

   A. 43
   B. 40
   C. 15
   D. 3
   E. 2

7. If vectors $\vec{t} = \langle 3,-4 \rangle$ and $\vec{u} = \langle 8,8 \rangle$, then $\langle \vec{t} - \vec{u} \rangle = ?$

   A. $\langle 11,4 \rangle$
   B. $\langle -5,-4 \rangle$
   C. $\langle -5,-12 \rangle$
   D. -6.31
   E. 16.31

8. A vector perpendicular to vector $\vec{V} = \langle 2,-5 \rangle$ is:

   A. $\langle -5,2 \rangle$
   B. $\langle -2,5 \rangle$
   C. $\langle -\frac{1}{2},\frac{1}{5} \rangle$
   D. $\langle 5,-2 \rangle$
   E. $\langle 5,2 \rangle$

9. If $\vec{a} = \langle 4, -1 \rangle$ and $\vec{b} = \langle 5, -7 \rangle$, what is the magnitude of $\vec{a} + \vec{b}$?

   A. $\langle 9, -8 \rangle$
   B. $\langle -1, 6 \rangle$
   C. $-\sqrt{145}$
   D. $\sqrt{145}$
   E. $\sqrt{17}$

10. If $\vec{V_1} = 3i + 2j$ and $\vec{V_2} = 2i - j$, the resultant vector of $3\vec{V_1} - \vec{V_2}$ equals:

    A. $\langle 3i + 2j, 2i - j \rangle$
    B. $11i + 5j$
    C. $7i + 7j$
    D. $5i + j$
    E. $7\sqrt{2}$

VECTORS ANSWERS

ANSWER KEY

1. A   2. B   3. C   4. D   5. C   6. E   7. C   8. E   9. D   10. C

ANSWER EXPLANATIONS

1. **A.** To add vectors, simply add the corresponding parts of the vector. So, $\langle 8,5 \rangle + \langle 2,8 \rangle = \langle (8+2),(5+8) \rangle = \langle 10,13 \rangle$.

2. **B.** When we perform simple operations (such as addition and scalar multipication) on vectors in component form, the answer will also be in component form. Any vector whose magnitude is zero has a direction of zero; given the right side of the equation is zero (though this is oddly not in component form), we can assume it equals $\langle 0,0 \rangle$. We now "plug in" this "zero" in component form and the component form for each variable into our equation, where $u = \langle 6,3 \rangle$, $v = \langle 4,-1 \rangle$ and w (our unknown) $= \langle w_1, w_2 \rangle$: $\langle 6,3 \rangle - 2\langle 4,-1 \rangle + \langle w_1, w_2 \rangle = \langle 0,0 \rangle$. We can now run the necessary calculation steps on the first element in each vector pair (i.e. the first number in each set of brackets) to solve for the first zero in the answer pair: $6 - 2(4) + w_1 = 0$ and on the second element in each vector pair to solve for the second zero in the answer pair: $3 - 2(-1) + w_2 = 0$. Simplify to find $w_1 = 2$ and $w_2 = -5$. Thus, w's component form is $\langle 2,-5 \rangle$.

3. **C.** We can find the vector with equal direction but a magnitude of 1 by dividing the entire vector by its current magnitude. Formally, we call this the "unit vector." We are essentially scaling down the triangle formed when we add the vectors down to a smaller triangle with a hypotenuse of 1. The current magnitude of the triangle is $\sqrt{9^2 + (-12)^2} = \sqrt{81 + 144} = \sqrt{225} = 15$. We divide the vector by 15: $\frac{\langle 9,-12 \rangle}{15} = \langle \frac{9}{15}, -\frac{12}{15} \rangle = \langle \frac{3}{5}, -\frac{4}{5} \rangle$.

4. **D.** First we add the vectors: $\langle 4,6 \rangle + \langle -3,9 \rangle = \langle 4-3, 6+9 \rangle = \langle 1,15 \rangle$. We now find the magnitude of this vector: $\sqrt{1^2 + 15^2} = \sqrt{1 + 225} = \sqrt{226}$.

5. **C.** First, plug the vectors into the equation in component form: $2\langle -3,7 \rangle + \langle 11,4 \rangle$. The scalar multiplier, 2, doubles the values of both components of the first vector; just distribute the 2 as you would with normal parentheses and then add the components that share the same position: $\langle -6,14 \rangle + \langle 11,4 \rangle = \langle 11-6, 14+4 \rangle \rightarrow \langle 5,18 \rangle$.

6. **E.** Remember that the sum of two vectors can be represented visually as arrows forming a straight line or arrows forming a triangle. If these two vectors share the same direction, we would just add 20 to 23 and get 43, Choice A. If they have opposite directions, we would subtract 20 from 23 and get 3, Choice D. Otherwise, their sum will visually look like a triangle, and logically, the possible numbers should fall within this range, 3-43 as these two cases were extremes. Now let's imagine the vectors as the legs of a triangle. We know from geometry that 2 legs of a triangle must have a sum greater than or equal to the remaining leg. Besides the choices we first eliminated (A and D), the only answer that contradicts this property is E. If the third leg of the triangle had a length of 2, then it and the leg of length 20 couldn't reach as far as the leg of length 23 even if they were lying flat against it. Thus choice E offers an impossible sum.

7. **C.** Plug the vectors into the equation in component form: $\langle 3,-4 \rangle - \langle 8,8 \rangle = \langle 3-8, -4-8 \rangle = \langle -5,-12 \rangle$.

8. **E.** A perpendicular vector will have a slope of the negative reciprocal of the original vector's slope. The original vector's slope is its y component (rise) divided by its x component (run) $\frac{-5}{2}$. So the slope of the answer will be $\frac{2}{5}$. The x component (run, on the bottom) of this slope is 5 and the y component (rise, on the top) is 2, so one vector with this slope will be $\langle 5,2 \rangle$.

9. **D.** The magnitude of $\vec{a} + \vec{b}$ is the magnitude of $\langle 4,-1 \rangle + \langle 5,-7 \rangle$, which is $\langle 4+5, -1-7 \rangle = \langle 9,-8 \rangle$. We find the magnitude using the Pythagorean theorem: $\sqrt{9^2 + (-8)^2} = \sqrt{81 + 64} = \sqrt{145}$.

10. **C.** $3\vec{V_1} - \vec{V_2} = 3(3i + 2j) - (2i - j) = 9i + 6j - 2i + j = 7i + 7j$.

PART FOUR: TRIGONOMETRY

CHAPTER 18: SOHCAHTOA

> ## SKILLS TO KNOW
> - SOHCAHTOA Basics
> - How to solve for a missing side or angle using SOHCAHTOA
> - Right Triangle Word problems
> - Choosing which trig function to use

This chapter is all about basic trigonometry involving right triangles. Most problems of this type can be solved by applying that wonderful mnemonic, SOHCAHTOA!

SOHCAHTOA BASICS

So what is "SOHCAHTOA"? It's an acronym to remember the three respective ratios described by Sine, Cosine, and Tangent in a right triangle:

Trigonometric function	Abbreviation	Acronym	Is equal to
Sine	sin	SOH	**O**pposite leg divided by **H**ypotenuse
Cosine	cos	CAH	**A**djacent leg divided by **H**ypotenuse
Tangent	tan	TOA	**O**pposite leg divided by **A**djacent leg

The diagram below shows how these sides are oriented. As you can see:

- The Hypotenuse is always the longest side of the triangle
- The Opposite is the side farthest from the angle of interest (here labeled θ)
- The Adjacent is the leg closest to the angle of interest (i.e. NOT the hypotenuse)

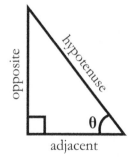

$$\sin(\theta) = \frac{opposite}{hypotenuse}$$

$$\cos(\theta) = \frac{adjacent}{hypotenuse}$$

$$\tan(\theta) = \frac{opposite}{adjacent}$$

 WARNING: SOHCAHTOA only works for RIGHT triangles. If you must solve for missing lengths on triangles without right angles, you'll need to use the **law of sines** or the **law of cosines** (covered in Chapter 20).

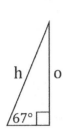

So, for example, when a question asks for $\sin(67°)$, it is asking for the ratio of the length of the opposite side from an angle of $67°$ to the length of the hypotenuse in a right triangle with those particular angles. Essentially, when we know two angles in a triangle ($67°$ and $90°$), we actually know all three angles, as all measures in a triangle sum to $180°$, and we can subtract to calculate the last angle. When we know all the angles, we also know that all triangles with those angles are similar. All sides of all similar triangles have the same ratios to each other, so it makes sense that we could store in our calculators all possible values for these ratios for the specific case of right triangles.

For right triangle QRS below, which of the following expressions is equal to $\tan Q$?

A. $\dfrac{q}{r}$ **B.** $\dfrac{q}{s}$ **C.** $\dfrac{r}{s}$ **D.** $\dfrac{s}{r}$ **E.** $\dfrac{r}{q}$

Here, we first think **tangent** is **TOA**. We need the Opposite over the Adjacent. We find the side opposite point Q: q is our numerator. Then we find the adjacent side (the leg that touches Q), which is r, our denominator.

$\dfrac{q}{r}$, or answer choice A, is correct.

Answer: **A**.

Right triangle XYZ below has angle measures μ, θ, and γ. Which of the following is true about the quotient of $\cos\mu$ and $\cos\gamma$?

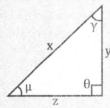

A. It equals 1.   **B.** It equals $\dfrac{y}{z}$.   **C.** It equals $\dfrac{z}{y}$.   **D.** It equals $\dfrac{y^2}{x^2}$.

E. The value cannot be determined from the given information.

For this problem, it's important you don't get overwhelmed by the weird Greek letters. Remember we can use Greek letters for angles and they are simply variables. Also remember quotient means divide. Finally, BE CAREFUL, the Greek letter γ looks a lot like y, and there are two other labels that say "y" that are different! This is written this way to try to fool you!

Now let's think of what $\cos\mu$ is: CAH tells us to find adjacent over hypotenuse, or $\dfrac{z}{x}$.

Now let's find $\cos\gamma$: adjacent over hypotenuse is $\dfrac{y}{x}$.

When we divide, order matters, so we must take $\dfrac{z}{x}$ and multiply by the reciprocal of $\dfrac{y}{x}$:

$$\dfrac{z}{x} \times \dfrac{x}{y} = \dfrac{z}{y}$$

Answer: **C**.

FINDING A MISSING SIDE

Often we encounter problems that ask us to solve for a missing side in a right triangle. Luckily, we can make use of our calculators, knowledge of right triangle anatomy, and SOHCAHTOA to find any missing sides or angles.

On the ACT®, though, you won't need your calculator's built in values for $\sin/\cos/\tan$. Typically, answer choices for these problems will NOT be simplified; i.e. the ACT® will only ask you to set up the problem, leaving expressions such as "$\sin$" or "$\arctan$" in the answer choices. Other times, the ACT® may give you a value for the $\sin$, $\cos$, or $\tan$, and ask you to approximate an answer.

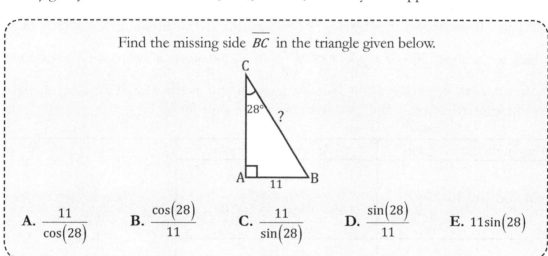

Find the missing side $\overline{BC}$ in the triangle given below.

A. $\dfrac{11}{\cos(28)}$ B. $\dfrac{\cos(28)}{11}$ C. $\dfrac{11}{\sin(28)}$ D. $\dfrac{\sin(28)}{11}$ E. $11\sin(28)$

When I see this problem, I first mark side $\overline{BC}$ with a question mark. Then I look at what I have: angle of 28 degrees, its OPPOSITE (O) side, 11, and I need $\overline{BC}$, the HYPOTENUSE (H). By first marking what I need, I can then figure out which trig function to use, whether it's $\sin$, $\cos$, or $\tan$. Given O and H, I need $\sin$ because SOH calls for O and H. Now let's set up our equation:

$$\sin(x°) = \dfrac{\text{opposite}}{\text{hypotenuse}}$$

$$\sin(28°) = \dfrac{11}{\overline{BC}}$$

CHAPTER 18

We're not done yet, as all the answer choices solve for an expression *equal* to $\overline{BC}$. We need to solve for $\overline{BC}$, and that takes a bit of algebraic manipulation. First, I multiply both sides by $\overline{BC}$:

$$\sin(28°)(\overline{BC}) = 11$$

And then I divide by $\sin(28°)$:

$$\overline{BC} = \frac{11}{\sin(28°)}$$

Answer: **C**.

FINDING A MISSING ANGLE

Sometimes in a right triangle, the ACT® will give you two or more side lengths of a right triangle and ask you to solve for an angle. As with missing side problems, the ACT® will ask you to simplify everything while leaving answers in trig notation (**arcsin**, $\tan^{-1}$, etc.), or will give you select values necessary to solve even if you don't have a calculator.

Let's talk **inverse trigonometric functions**.

INVERSE TRIG FUNCTIONS

Inverse trig functions help us turn a sin, cos, or tan value into the angle it represents. For instance, $\sin(30°) = \frac{1}{2}$. If we know that $\sin x = \frac{1}{2}$, we can use the inverse of sin (called **arcsin** or $\sin^{-1}$) to get back to the angle, $30°$. If I type $\arcsin\left(\frac{1}{2}\right)$ in my calculator, it will return the answer: 30.*

Inverse functions are denoted by a "-1" superscript or an "arc" prefix. The ACT® and your calculator may use these interchangeably. Both notations means the same thing.

Trigonometric Function	Inverse Function
sin (ex. $\sin(30°) = \frac{1}{2}$)	$\sin^{-1}$ or arcsin (ex. $\arcsin\left(\frac{1}{2}\right) = 30°$)
cos (ex. $\cos(60°) = \frac{1}{2}$)	$\cos^{-1}$ or arccos (ex. $\cos^{-1}\left(\frac{1}{2}\right) = 60°$)
tan (ex. $\tan(60°) = \frac{\sqrt{3}}{1}$)	$\tan^{-1}$ or arctan (ex. $\arctan(\sqrt{3}) = 60°$)

***TIP:** If using a calculator for inverse trig calculations, always be sure your calculator is in degree mode for degree problems!

> Find ∠B in the triangle given below. 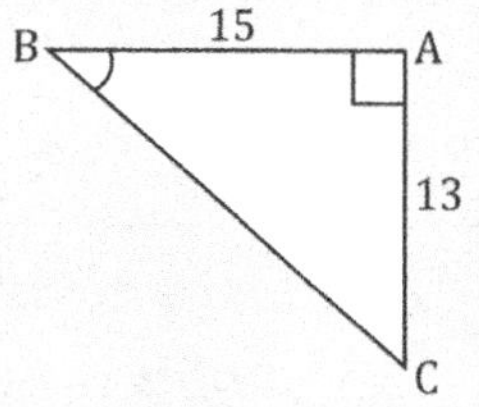
>
> [Triangle with B at top-left, A at top-right (right angle), C at bottom-right. BA = 15, AC = 13.]
>
> **A.** $\arcsin\left(\dfrac{13}{15}\right) = B$ **B.** $\arccos\left(\dfrac{13}{15}\right) = B$ **C.** $\arccos\left(\dfrac{15}{13}\right) = B$
>
> **D.** $\arctan\left(\dfrac{15}{13}\right) = B$ **E.** $\arctan\left(\dfrac{13}{15}\right) = B$

First, we know the two legs of the triangle. These are the OPPOSITE (O) and ADJACENT (A) sides to angle B (what we need). O and A are part of TOA: thus, we'll use the tangent function.

$$\tan(B) = \frac{opposite}{adjacent}$$
$$= \frac{13}{15}$$

We just solved for the tangent of the angle. But we need to solve for an expression that equals the angle itself. To translate a tangent into the angle it corresponds to, we use the inverse function, arctan or $\tan^{-1}$.

The above expression can be rewritten as:

$$\arctan\left(\frac{13}{15}\right) = B \text{ or } \tan^{-1}\left(\frac{13}{15}\right) = B$$

We now simply scour the answer choices for one of these two choices.

Answer: **E**.

COSECANT, SECANT AND COTANGENT

In addition to using sine, cosine and tangent to describe the ratios in a triangle, we can also write trigonometric ratios using cosecant, secant, and cotangent, all of which are the reciprocals of sine, cosine and tangent, respectively.

cosecant	$\csc(\theta) = \dfrac{hypotenuse}{opposite}$	$\dfrac{1}{\sin(\theta)}$
secant	$\sec(\theta) = \dfrac{hypotenuse}{adjacent}$	$\dfrac{1}{\cos(\theta)}$
cotangent	$\cot(\theta) = \dfrac{adjacent}{opposite}$	$\dfrac{1}{\tan(\theta)}$

Cotangent sounds like tangent, so that one is easy to remember. To remember secant, I see that the "s" and the "c" have flipped positions so it sort of looks like $\cos$ backwards, thus I remember that it is $\dfrac{1}{\cos}$ or $\cos$ flipped around. $\csc$ has "s" in the second position not the first, so I imagine that means it represents $\sin$ but flipped around also, or $\dfrac{1}{\sin}$.

We use these ratios in the same way we use $\sin$, $\cos$ and $\tan$. Know these appear rarely on the ACT®.

RIGHT TRIANGLE WORD PROBLEMS

TIP: Trigonometric word problems tend to be much easier to solve if you draw it out so we don't get any numbers mixed up. Always draw a picture if there isn't one already! Always label what you need if it isn't already labeled!

At a time when the sun's rays are striking the ground at 75°, a skyscraper casts a shadow that is 300 feet long. To the nearest foot, how tall is the skyscraper?

Let's draw this out first:

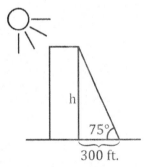

Now that we can visualize everything, we know we're trying to find height "h" of the building. The angle given, 75°, is OPPOSITE (O) to "h" and ADJACENT (A) to the given shadow length, 300 ft. Thus, observing O and A of TOA, we can solve for the opposite side "h" by using tangent.

Remember, tan is opposite over adjacent:

$$\tan(75°) = \frac{h}{300}$$

Isolating the h, we get $h = 300\tan(75°)$.

Answer: $h = 300\tan(75°)$.

Again, for most ACT® problems, this is as far as you need to go! If you look at the answers and need to solve, look for a table of values that includes $\tan(75)$ OR use your calculator. Be sure to round according to instructions and ensure your calculator is in DEGREE mode!

Scientists are attempting to measure the width of a mountain, represented in the figure below by $\overline{AB}$, using trigonometry. The distance from A to C is 5 km. The measure of $\angle CBA$ is 63°. What is the width of the mountain to the nearest tenth, if $\sin 63 \approx 0.891$, $\cos 63 \approx 0.454$, and $\tan 63 \approx 1.963$?

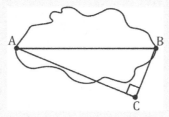

Here we first need to sort through what the diagram shows. Mark up your drawing.

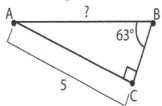

Length $\overline{AB}$ is what we need, the width of the mountain. Length $\overline{AC}$ is what we know. From our reference angle of 63° these are the **o**pposite side ($\overline{AC}$) and the **h**ypoteneuse ($\overline{AB}$). Thus we can use SOH, sin, to solve by setting the opposite over hypoteneuse:

$$\sin(63°) = \frac{AC}{AB}$$

$$\sin(63°) = \frac{5 \text{ km}}{AB}$$

$$AB = \frac{5 \text{ km}}{\sin(63°)}$$

Plugging in from our values, and using our calculator, we get:

$$\frac{5}{0.891} = 5.6 \text{ rounded to the nearest tenth}$$

Answer: **5.6**.

1. For right triangle $\triangle PQR$ shown below, what is $\cos(P)$?

   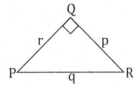

   A. $\dfrac{p}{r}$

   B. $\dfrac{p}{q}$

   C. $\dfrac{q}{r}$

   D. $\dfrac{r}{p}$

   E. $\dfrac{r}{q}$

2. The hypotenuse of the right triangle $\triangle QRS$ shown below is 18 feet long. The cosine of $\angle Q$ is $\dfrac{2}{3}$. About how many feet long is $\overline{QR}$?

   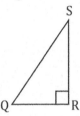

   A. 9
   B. 12
   C. 13.4
   D. 17.3
   E. 27

3. A zipline is to be built so that a taut cable is at a 12° incline relative to the level ground. If the zipline is 11 meters above the ground at its starting point and it ends at ground-level, how far horizontally will the endpoint be from the starting point?

   A. $11\sin 12°$
   B. $11\cos 12°$
   C. $11\tan 12°$
   D. $\dfrac{11}{\tan 12°}$
   E. $11\csc 12°$

4. A door is propped open 25° with a bar as shown in the diagram below. If the door is 40 inches long, how long does the bar need to be?
   (Note: $\sin 25° \approx 0.423$
   $\cos 25° \approx 0.906$
   $\tan 25° \approx 0.466$)

   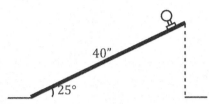

   A. 17
   B. 19
   C. 36
   D. 44
   E. 95

5. The sides of a right triangle measure 20 inches, 21 inches, and 29 inches. What is the cosine of the angle opposite the side that measures 20 inches?

   A. $\dfrac{20}{29}$

   B. $\dfrac{29}{20}$

   C. $\dfrac{21}{29}$

   D. $\dfrac{29}{21}$

   E. $\dfrac{21}{20}$

6. What is the sine of angle F in right triangle $\triangle DEF$ below?

   A. $\dfrac{2}{3}$

   B. $\dfrac{3}{2}$

   C. $\dfrac{3}{\sqrt{5}}$

   D. $\dfrac{\sqrt{5}}{2}$

   E. $\dfrac{\sqrt{5}}{3}$

7. A flashlight emits a cone of light as shown below. If the flashlight is projecting a circle of light on the wall that is 80 cm in diameter, how far is the flashlight from the wall in centimeters?

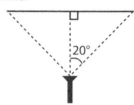

A. $40\tan 20°$
B. $\dfrac{\tan 20°}{40}$
C. $\dfrac{40}{\tan 20°}$
D. $\dfrac{80}{\tan 20°}$
E. $80\tan 20°$

8. In the figure below, $\cos S = \dfrac{3}{5}$. What is the approximate value of T using the given information and the diagram below?

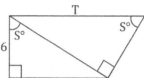

A. 10
B. 13.5
C. 0.60
D. 12.5
E. 53.1

9. In the right triangle shown below, the length of $\overline{WX}$ is 7 feet and the length of $\overline{XY}$ is 24 feet. For $\angle Y$, the value of which of the following trigonometric expressions is $\dfrac{7}{25}$?

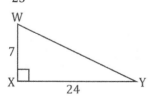

A. $\cos Y$
B. $\sin Y$
C. $\tan Y$
D. $\sec Y$
E. $\csc Y$

10. The dimensions of the right triangle shown below are given in meters. What is $\cos\theta$?

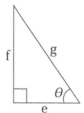

A. $\dfrac{e}{g}$
B. $\dfrac{f}{g}$
C. $\dfrac{g}{e}$
D. $\dfrac{g}{f}$
E. $\dfrac{f}{e}$

11. Which of the following trigonometric equations is valid for the measurements indicated below for the two sides and angle formed by a rectangle and its diagonal, k?

A. $\tan j° = \dfrac{l}{k}$
B. $\cos j° = \dfrac{k}{l}$
C. $\cos j° = \dfrac{l}{k}$
D. $\cot j° = \dfrac{k}{l}$
E. $\sec j° = \dfrac{l}{k}$

12. In the figure below, A, B, C, and D are on $\overline{QE}$. Which of the following angles has the smallest cosine?

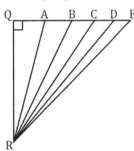

A. $\angle QRA$
B. $\angle QRB$
C. $\angle QRC$
D. $\angle QRD$
E. $\angle QRE$

13. Which of the following trigonometric equations is valid for the side measurement a nanometers, diagonal measurement c nanometers, and angle measurement $b°$ in the rectangle shown below?

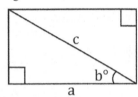

A. $\sin b° = \dfrac{a}{c}$
B. $\cos b° = \dfrac{a}{c}$
C. $\cos b° = \dfrac{c}{a}$
D. $\tan b° = \dfrac{a}{c}$
E. $\sec b° = \dfrac{a}{c}$

14. In right triangle $\triangle TUV$ below, $\overline{UV}$ is 125 yards long and $\angle T = 65°$. What is the measure of $\overline{TV}$?

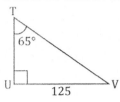

A. $125 \sin 65°$
B. $\dfrac{125}{\sin 65°}$
C. $\dfrac{\sin 65°}{125}$
D. $125 \cos 65°$
E. $\dfrac{125}{\cos 65°}$

15. For any right triangle as shown, $(\cos X)(\tan Y) = ?$ (All angle measurements are given in degrees)

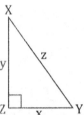

A. $\dfrac{y^2}{xz}$
B. $\dfrac{x^2}{yz}$
C. $\dfrac{x}{z}$
D. $\dfrac{y}{z}$
E. $\dfrac{z}{x}$

16. An arborist (a scientist who studies trees) comes across a tree that juts out of the ground at $12°$. When the sun is directly overhead, so that its rays are perpendicular to the ground, the tree's shadow is 8 meters long. If it can be determined, what is the length of the tree?

A. $\dfrac{8}{\cos 12°}$

B. $8\cos 12°$

C. $\dfrac{8}{\tan 12°}$

D. $8\tan 12°$

E. $\dfrac{8}{\sin 12°}$

17. In $\triangle HIJ$, $\angle J$ is a right angle and the measure of $\angle I$ is $63°$. Another triangle, $\triangle QRS$ is being constructed such that $\angle R$ is a right angle and the measure of $\angle S$ is one third the measure of $\angle I$. What is the measure of $\angle Q$?

A. $21°$
B. $22.5°$
C. $45°$
D. $69°$
E. $159°$

18. In the figure below, what is $\cos(x)$?

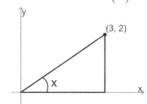

A. $\dfrac{2}{3}$

B. $\dfrac{3}{2}$

C. $\dfrac{3}{\sqrt{13}}$

D. $\dfrac{2}{\sqrt{13}}$

E. $\dfrac{4}{9}$

19. Roman is standing 55 feet from a cell-phone tower, on level ground. He can see the tip of the tower at an angle of inclination of $38°$. How many feet is the tip of the cell-phone tower from his eyes?

A. $\dfrac{55}{\sin 38°}$

B. $\dfrac{55}{\sin 142°}$

C. $\dfrac{55}{\sin 52°}$

D. $\dfrac{\sin 52°}{55}$

E. $\dfrac{55}{\cos 52°}$

ANSWER KEY

1. E   2. B   3. D   4. A   5. C   6. E   7. C   8. D   9. B   10. A   11. C   12. E   13. B   14. B
15. A   16. A   17. D   18. C   19. C

ANSWER EXPLANATIONS

1. **E.** Cosine equals the adjacent side divided by the hypotenuse. Side r is adjacent to $\angle P$, and the hypotenuse is q. Thus, $\cos(P) = \dfrac{r}{q}$.

2. **B.** The cosine equals the adjacent side divided by the hypotenuse. The cosine of $\angle Q = \dfrac{\overline{QR}}{18} = \dfrac{2}{3}$. Multiply both sides by 18 to get $\overline{QR} = 12$.

3. **D.** First, sketch the triangle formed by the starting point, the ground below it, and the ending point. The zipline forms the hypotenuse, and the height of the starting point and the horizontal distance to the ending point form the two legs. The distance we are given is the height of the starting point, and we want to find the horizontal distance. The height of the starting point divided by the horizontal distance, d, will be equal to $\tan 12°$. Thus $\dfrac{11}{d} = \tan 12°$, so $d = \dfrac{11}{\tan 12°}$.

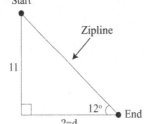

$$\tan = \dfrac{opposite}{adjacent} \qquad \tan 12° = \dfrac{11}{d}$$

$$d \tan 12° = 11$$

$$d = \dfrac{11}{\tan 12°}$$

4. **A.** The relation between the length of the bar, b, and the length of the door is the relation between the side opposite the 25° angle and the hypotenuse. $\dfrac{b}{40} = \sin 25° \approx 0.423$. Thus, $b \approx 40(0.423) = 16.92 \approx 17$.

5. **C.** First draw a picture. The cosine of the angle opposite the side that measures 20 inches is the adjacent side divided by the hypotenuse. We know that the hypotenuse must be the 29 because the hypotenuse is always the longest side of a right triangle. That leaves the adjacent side as 21. Thus, the cosine is $\dfrac{21}{29}$.

6. **E.** First find $\overline{DE}$ using the Pythagorean theorem: $\overline{DE}^2 + 4^2 = 6^2 \to \overline{DE} = \sqrt{20} = 2\sqrt{5}$. By SOCAHTOA, $\sin \angle F = \dfrac{opposite}{adjacent} = \dfrac{\overline{DE}}{6} = \dfrac{2\sqrt{5}}{6} = \dfrac{\sqrt{5}}{3}$.

7. **C.** Recognize that the side of the right triangle **opposite** 20° is the radius of the circle of light, not the diameter. The diameter is 80 so the radius is 40. Next, see that we need the **adjacent** distance (let's call it d) that starts in the flashlight's center and ends with the right angle at the wall (the solid line). Given the opposite and adjacent sides, we'll use TOA or tangent. Thus, $\tan 20° = \dfrac{40}{d}$, which can be rearranged with algebra to get $d(\tan 20°) = 40$ or $d = \dfrac{40}{\tan 20°}$.

8. **D.** If $\cos S = \dfrac{3}{5}$, since cosine is adjacent over hypotenuse, we can set up a ratio with side length 6: $\dfrac{3}{5} = \dfrac{6}{h}$, with h as the hypotenuse length of the bottom triangle. Cross multiplying we get $3h = 30$ so $h = 10$, and now we can use Pythagorean triples to find the third side of the lower triangle to be 8. By AA, the lower triangle is similar to the upper triangle. Using angle S as a reference point, I know opposite over hypotenuse ($\sin S$) of the upper triangle equals opposite over hypotenuse ($\sin S$) of the lower triangle. I set up a proportion so side 10 in the upper triangle corresponds to the side 8 in the lower triangle, so $\dfrac{10}{T} = \dfrac{8}{10}$. Cross multiplying I get $8T = 100$ or $T = 12.5$. Alternatively, given $\cos S = \dfrac{3}{5}$, we can solve for angle S: arccos of 0.6 produces $S = 53.1°$. Now, find T by setting up this equation: $\sin 53.1 = \dfrac{10}{T}$. Solving this, we get $T \approx 12.5$. See Chapter 15 for more problems with similar triangles.

9. **B.** This triangle is a $7-24-25$ right triangle (or use Pythagorean theorem to find the hypotenuse). For $\angle Y$, the 7 is opposite, and the 25 foot side is the hypotenuse. So sin (SOH)= $\dfrac{7}{25}$ (opposite divided by hypotenuse).

10. **A.** The cosine of an angle is the measure of the adjacent side divided by the measure of the hypotenuse. The side adjacent to $\angle \theta$ is e, and the hypotenuse is g. By plugging in these values we find that $\cos \theta = \dfrac{e}{g}$.

11. **C.** First label the diagonal, k. The side of measure l is adjacent to the angle of measure j, and the side of measure k is a hypotenuse. The trigonometric function for the relationship between an adjacent side and the hypotenuse is the cosine. The cosine is the adjacent side, l, divided by the hypotenuse, k. This is expressed as $\cos j° = \dfrac{l}{k}$.

12. **E.** The cosine is equal to the adjacent side divided by the hypotenuse. The smallest cosine will be the one with the smallest measure for the adjacent and largest measure for hypotenuse. All of the angles we can choose as answers have the same adjacent side, QR, so the only difference will be in the length of the hypotenuse. The largest hypotenuse will have the smallest cosine, and we can tell just by looking at the diagram that $\angle QRE$ has the largest hypotenuse, so it is our answer.

13. **B.** The two sides we are given are the adjacent side and hypotenuse relative to the angle of measure $b°$. The relationship between these two sides is expressed in the cosine. The cosine is the adjacent side divided by the hypotenuse, which in our case is $\dfrac{a}{c}$. Our answer is $\cos b° = \dfrac{a}{c}$.

14. **B.** We are given an angle and the measure of its opposite side, and we are looking for the measure of the hypotenuse. The relationship between an angle's opposite side and hypotenuse is the sine (SOH) of that angle. The sine of 65° is $\dfrac{125}{n}$ where n is the measure we are looking for. Since $\sin 65° = \dfrac{125}{n}$, $n(\sin 65) = 125$ and $n = \dfrac{125}{\sin 65°}$.

15. **A.** The cosine (CAH) of X is the adjacent side divided by the hypotenuse: $\dfrac{y}{z}$. The tangent (TOA) of Y is the opposite side divided by the adjacent side, $\dfrac{y}{x}$. Plugging these expressions in for the trigonometric functions, $\left(\dfrac{y}{z}\right)\left(\dfrac{y}{x}\right) = \dfrac{y^2}{xz}$.

16. **A.** The length of the tree is the hypotenuse, not the side. We are given the angle made by the tree and the ground and the length of its shadow, which relative to the angle is the adjacent side. We now have enough information to solve the problem. The cosine is the adjacent side (what we are given) divided by the hypotenuse (what we are looking for). $\cos 12° = \dfrac{8}{n}$, where n is the length of the tree. Rearranging this, we get $n(\cos 12°) = 8$ or $n = \dfrac{8}{\cos 12°}$.

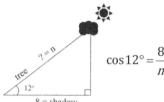

17. **D.** The measure of $\angle S = \dfrac{1}{3}\angle I = \dfrac{1}{3}(63°) = 21°$. We are given that $\angle R = 90°$. We know that $\angle Q + \angle R + \angle S = 180°$, since the interior angles of a triangle on a plane sum to 180°. Substitute in the values we have: $\angle Q + 90° + 21° = 180° \rightarrow \angle Q + 111° = 180° \rightarrow \angle Q = 69°$.

18. **C.** If we draw a triangle with a side on the x-axis, a side on the line $x = 3$, and the hypotenuse going from the origin to point $(3,2)$, we get a right triangle with $2 =$ the side opposite x and $3 =$ the side adjacent angle x. Since $\cos(x) = \dfrac{adjacent}{hypotenuse}$, we need to find the length of the hypotenuse first by using the Pythagorean theorem $a^2 + b^2 = c^2$. Plugging in $a = 3$ and $b = 2$ and solving for c we get $3^2 + 2^2 = c^2 \rightarrow 9 + 4 = c^2 \rightarrow 13 = c^2 \rightarrow c = \sqrt{13}$. Now, we can plug in $adjacent = 3$ and $hypotenuse = \sqrt{13}$ to get $\cos(x) = \dfrac{3}{\sqrt{13}}$.

19. **C.** We can draw a triangle $\triangle XYZ$ using the points from his eyes (X) to the tip of the tower (Y) to the point on the tower that matches his eye-level (Z). The angle at Z is equal to 90, because the tower is perpendicular to the ground and $\overline{XZ}$ is parallel to the ground. It would be easier to use cosine, but because none of the answer choices include a cosine, we must use sine. In order to use sine, we must find the other angle in the triangle, which we find by subtracting 90°-38°=52°: $\angle Y = 52°$ (in a right triangle, non-right angles are complementary). Now take the sin of Y, $\sin \angle Y = \sin 52° = \dfrac{opposite}{hypotenuse} = \dfrac{55}{\overline{XY}}$, so $\sin 52°(\overline{XY}) = 55$ and $\overline{XY}$, the distance we want to find, is equal to $\dfrac{55}{\sin 52}$.

CHAPTER 18   295

CHAPTER 19
TRIGONOMETRY

> ### SKILLS TO KNOW
> - Radian and Degree Measure
> - Unit Circle
> - ASTC (All Students Take Calculus) positive and negative values of trigonometric functions on a graph
> - Word problems involving Sine/Cosine/Angles
> - Reference angles
> - Trig Identities

 NOTE: This chapter is a continuation of skills covered in **SOHCAHTOA** (right triangle trigonometry). Trig concepts are additionally covered in Trig Graphs and Laws of Sines and Cosines. Trig questions (including SOHCAHTOA) typically occur 1-5 times on any given ACT© exam. They often take longer to drill and understand than many other areas of the exam and tend to be varied in form, hence the length of this chapter and the multiple chapters on trigonometry topics. **I recommend that ALL students feel proficient in SOHCAHTOA,** as it occurs on many exams and tends to be easier, but this chapter and the following two are best suited for those aiming for top scores (32+). Some of the explanations in this chapter go deep. Don't freak out if some items are confusing; all that matters is that you can do the problems. Just focus on what you get and which methods work for you.

RADIAN AND DEGREE MEASURE

Just as we can measure short lengths in inches or centimeters, angles in math can be measured in two ways: using **radians** or using **degrees.**

Most of us are familiar with measuring angles in **degrees**. We know that a straight angle is $180°$, or that there are $360°$ in a circle. Sometimes, however, it makes sense to use **radians**. Radians, like the area of a circle and the circumference of a circle, are typically described in terms of π (though they don't need to have a π in them to be in radians). You can see below how an angle of one radian has an arc length equal to the radius. Thus, a radian measure is a calculation of how many "radius lengths" along you are on a circle at a given angle.

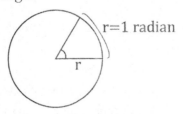

If you find that idea confusing, don't worry. Just know how to convert between degrees and radians:

There are π radians in 180 degrees, or 2π radians in 360 degrees:

$$\pi = 180°$$
$$2\pi = 360°$$

We can delineate radian measures, along with their corresponding degree measures, drawing angles centered at $(0,0)$ on a coordinate plane.

Angles in Quadrant I are between 0 and $\frac{\pi}{2}$, in Quadrant II they are between $\frac{\pi}{2}$ and π, in Quadrant III, between π and $\frac{3\pi}{2}$, and in Quadrant IV, between $\frac{3\pi}{2}$ and 2π.

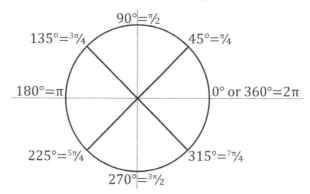

Converting between Radians and Degrees

In the chapter on **Ratios, Rates and Units** in the first book of this series, we learned how to convert between two equivalent measures using **dimensional analysis**. We use that same tactic to convert between radians and degrees.

Remember $\pi = 180°$, so our conversion factors are $\dfrac{\pi \text{ radians}}{180 \text{ degrees}}$ and $\dfrac{180 \text{ degrees}}{\pi \text{ radians}}$.

First, figure out what you **HAVE**. If you have **DEGREES**, multiply by the conversion factor with degrees **on the bottom so they cancel,** and **put what you need (radians) on top**. Here is how we set up that problem. Put the angle measure you know in the first blank.

$$\underline{} \, \cancel{\text{degrees}} \times \frac{\pi \text{ radians}}{180 \, \cancel{\text{degrees}}} = \underline{} \text{ radians}$$

If you have RADIANS, multiply by the conversion factor with radians on the bottom so they cancel, and put what you need (degrees) on top. Here is how we set up that problem. Put the angle measure you know in the first blank.

$$\underline{} \, \cancel{\text{radians}} \times \frac{180 \text{ degrees}}{\pi \, \cancel{\text{radians}}} = \underline{} \text{ degrees}$$

What is the measure, in degrees, of an angle of $-\frac{4}{5}\pi$ radians?

To solve, we set up our conversion as denoted above:

$$-\frac{4}{5}\pi \, \cancel{\text{radians}} \times \frac{180 \text{ degrees}}{\pi \, \cancel{\text{radians}}} = -\frac{4}{5} \times 180 \text{ degrees}$$

CHAPTER 19

Because the radians and π's cancel, we solve:

$$-\frac{4}{5} \times 180 = -144 \text{ degrees}$$

Answer: $-144°$

UNIT CIRCLE

The Unit Circle is a tool we use in trigonometry to solve and understand trigonometric problems. It is defined by the equation:

$$x^2 + y^2 = 1$$

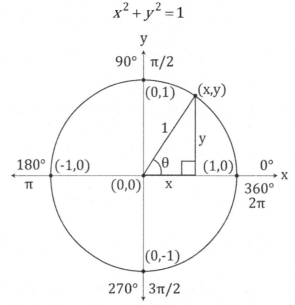

If we know the base form of a circle equation, or rather how that equation is derived, we realize that this is like the Pythagorean theorem: the x length and the y lengths are the sides of a little triangle we can draw that has a hypotenuse equal to the radius, 1 (1 for "unit" circle). Let's find **sin** and **cos** values on this little triangle.

Take the central angle in the picture θ. $\sin\theta = \frac{o}{h} = \frac{y}{1}$ right? So $\sin\theta = y$. $\cos\theta$ would equal $\frac{a}{h}$ or $\frac{x}{1}$ right? So $\cos\theta = x$. Because the radius is 1, we thus know in any unit circle:

$$x = \cos\theta \text{ and } y = \sin\theta$$

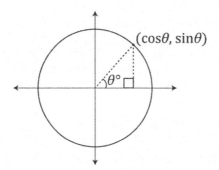

Thus, because $x^2 + y^2 = 1$, $\sin^2\theta + \cos^2\theta = 1$. We call this formula the **Pythagorean identity**.

Now that we realize this fact, we can use a unit circle to quickly jump between the **sin** of an angle and a **cos** of an angle by plugging into the Pythagorean identity ($\sin^2 x + \cos^2 x = 1$), or by using memorized values we visualize on the unit circle which adhere to this pattern.

Many students memorize the unit circle (x, y) values at angle measures such as $\frac{\pi}{3}$, $\frac{\pi}{2}$, or $\frac{\pi}{4}$ for their trig coursework. For the ACT, this level of memorization is likely overkill. However, you should know a few things that relate to the idea of a unit circle, and it's good to generally know what it is.

> If $1 - \sin^2 x = \frac{5}{12}$, what is $\cos^2 x$?

Using the Pythagorean Identity, we can quickly solve this problem. Knowing $\sin^2\theta + \cos^2\theta = 1$, I can plug in this expression where I see the **1** and solve:

$$1 - \sin^2 x = \frac{5}{12}$$

$$\left(\sin^2\theta + \cos^2\theta\right) - \sin^2 x = \frac{5}{12}$$

$$\cos^2\theta = \frac{5}{12}$$

Answer: $\frac{5}{12}$.

ASTC: ALL STUDENTS TAKE CALCULUS, ALL STUDENTS TAKE CLASSES, ADD SUGAR TO COFFEE, ALL SILVER TEA CUPS

Whatever phrase you use to remember it, ASTC is a way to remember when trig functions are positive or negative, based on the measure of the angle or the range of that measure.

To use this technique, first imagine a radius drawn on a unit circle.

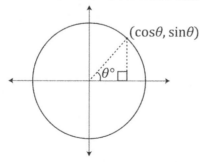

Imagine I know that my x value is $\frac{3}{5}$, or in other words, $\cos\theta = \frac{3}{5}$. I can solve:

$$\left(\frac{3}{5}\right)^2 + y^2 = 1$$
$$\frac{9}{25} + y^2 = 1$$
$$y^2 = \frac{16}{25}$$
$$y = \frac{4}{5}$$

But in this process, I always assume my trig value is positive. That works in this instance, but what if we were in Quadrant III? When we use the formula to solve for the trig identity we don't know, we can't use this formula alone to determine the correct sign. Because this formula squares each term, any negative signs vanish! Nor can I use the simple "triangle" method to track signs (we'll get to that soon)– triangle sides are always positive, never negative. We must thus track our signs in some other fashion. That's where ASTC comes in.

A=ALL

In the first quadrant ($0°$ to $90°$ or 0 to $\frac{\pi}{2}$), **A**ll trigonometric ratios are positive.

S=SINE

In the second quadrant ($90°$ to $180°$ or $\frac{\pi}{2}$ to π), **S**in is positive, and cos and tan are negative. (By extension, csc is also positive, and sec and cot are negative.)

T=TANGENT

In the third quadrant ($180°$ to $270°$ or π to $\frac{3\pi}{2}$), **T**an is positive, and sin and cos are negative. (By extension, cot is also positive, and csc and sec are negative.)

C=COSINE

In the fourth quadrant (in degrees, $270°$ to $360°$ or in radians, $\frac{3\pi}{2}$ to 2π), **C**os is positive, and tan and sin are negative. (By extension, sec is positive, and csc and cot are negative.)

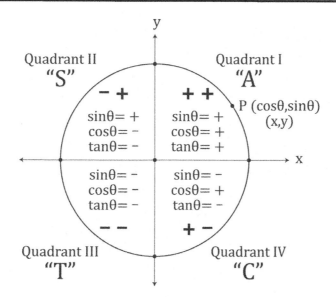

The idea is, if you measure an angle starting at the x-axis (horizontal line between Quadrant I and IV) as the base of the angle and draw a radius segment on the unit circle to form the other side of that angle in one of the four quadrants, the quadrant in which that radius (terminal side of the angle) falls will have trig function values according to this pattern.

An Alternative to ASTC: Think About X and Y

Remember on the unit circle, x corresponds to $\cos$ and y corresponds to $\sin$. Thus when x is positive on our quadrant, so is our cosine. When y is positive on the coordinate plane, so is our sine.

For instance if $\cos\theta = -\frac{1}{2}$ in quadrant II, when asked to solve for sine, we could draw out where the (x, y) end point would be on the circle and realize the x-value is negative while the y-value is positive. Visualizing this on our unit circle, we could then assume the sine value is positive becuase the y-value is positive. The tangent would also be negative (remember tangent equals $\frac{\sin}{\cos}$), because a negative cosine and positive sine make a negative tangent.

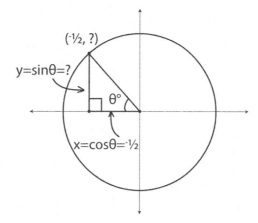

Much of this may confuse you, but what matters most is that you can apply this knowledge, so let's get into the problem type this information will help you solve.

TRIGONOMETRY SKILLS

DERIVING THE VALUE OF SINE, COSINE, OR THETA (AND COT, SEC, CSC) FROM ONE ANOTHER

Now that we know the basic idea of a unit circle, the Pythagorean identity, and ASTC, we can use these tools, along with SOHCAHTOA, to solve for any trig function of an angle when given a different trig function.

Oftentimes, you will be given the value of one trigonometric function (say $\sin\theta$) and be asked to find another, like $\cos\theta$. My favorite method, below, I call the **"triangle method."**

If $\sin\alpha = \dfrac{3}{5}$ and α is between $\dfrac{\pi}{2}$ and π, what is $\cos\alpha$?

1. First, sketch a little right triangle.

2. Next, label an angle (not the right angle). Make up side lengths that adhere to the SOHCAHTOA ratio you know. The easiest way is to use the fraction's numerator and denominator. Here, for example, we assume $\dfrac{3}{5} = \dfrac{opp}{hyp}$ and assign 3 to the opposite side and 5 to the hypotenuse opposite our reference angle. Now label the triangle:

3. Now, we solve for the missing side using the Pythagorean theorem.

$$a^2 + b^2 = c^2$$
$$3^2 + b^2 = 5^2$$
$$9 + b^2 = 25$$
$$b^2 = 16$$
$$b = 4$$

   Write down this value for our missing side:

4. Now we use these sides to find the value of $\cos$, $\sin$, $\tan$ or other trig function you need of this same angle. Here, to find cosine, just divide the adjacent side by the hypotenuse:

$$\cos\theta = \dfrac{4}{5}$$

But we're not done yet! This is the cos of the angle theta we made up in our little triangle, a triangle that has all positive side lengths and trig values. We need the angle in between $\frac{\pi}{2}$ and π or quadrant II that holds this ratio but may have a different sign.

5. Adjust the sign according to ASTC if necessary. Because the $\cos\alpha$ is in the 2nd quadrant (between $\frac{\pi}{2}$ and π), we know that quadrant is "S" meaning ONLY sin is positive. Thus, cosine is negative, so $\cos\alpha = -\frac{4}{5}$.

Answer: $-\frac{4}{5}$.

> If $\tan n = \frac{a}{b}$ and n is in the first quadrant, what is $\csc n$?

Here we don't have numbers but letters. Still the approach is the same.
1. Draw a right triangle.
2. Label according to what we know:

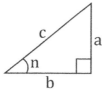

Angle n has an opposite side a and an adjacent side b.

3. Solve using the Pythagorean theorem. Since $a^2 + b^2 = c^2$, the third side, c, equals $\sqrt{a^2+b^2}$.

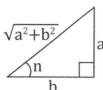

4. Find what you need, $\csc n$. Because csc is equal to $\frac{1}{\sin}$, we want $\frac{h}{o}$ instead of $\frac{o}{h}$. Our h is $\sqrt{a^2+b^2}$. And our o is a. That's $\frac{\sqrt{a^2+b^2}}{a}$.

5. Adjust the sign using ASTC if necessary. Our value is first quadrant, where sin and thus csc are positive. So, no adjustment is necessary.

Answer: $\frac{\sqrt{a^2+b^2}}{a}$.

REFERENCE ANGLES

Some students find trig values involving angles greater than 90° confusing. How do you do "SOHCAHTOA" if you can't create a little right triangle for reference that includes the angle you're analyzing? The answer lies in reference angles.

In each of the last two problems, we used a reference angle (the angle we "made up" in our little right triangle) to first find that angle's cosine and then to find the actual cosine of the angle in question. Here, I'll explain that concept in greater depth.

The reference angle is the angle that falls between 0 and 90° that expresses the positive (absolute) value of sine, cosine, and tangent that we are looking for in the other quadrants. It's the angle we focus on when we use the triangle method. When we combine the information we know about the reference angle and our ASTC, we can calculate the sin, cos, etc. of our larger angle.

With problems like the one previously, when we jump from sin to cos or cot to sec, we don't really have to think much of the reference angle's actual measure even though we're using it. We can simply use the triangle method and ASTC. But on problems in which we are given an angle measure that is negative or above 90 degrees, we need to understand more clearly what our reference angle is and how to calculate it.

Here, θ acts as a reference angle for any angle with terminal sides that match the following conditions:
Any angle with terminal side in Quadrant II at $\pi - \theta$ or $180° - \theta$.
Any angle with terminal side in Quadrant III as $\pi + \theta$ or $180° + \theta$.
Any angle with terminal side in Quadrant IV as $2\pi - \theta$ or $360° - \theta$.

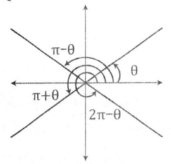

 To make this easier, **the reference angle is ALWAYS the acute angle formed by the horizontal x-axis and the terminal side of the angle we're evaluating.** Knowing this, we can now calculate trig values of angles that are greater than 90 degrees.

> What is the cosine of an angle that measures 225 degrees?

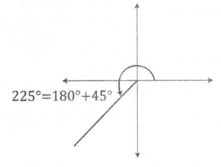

First, we need to calculate the reference angle. I know that 225 degrees is 45 more degrees than 180 degrees, so the acute angle I can draw using the terminal side of this angle and the horizontal x-axis is 45 degrees.

This is the 3rd quadrant, so following to the 3rd letter in AS**T**C. T (tangent) is the only positive value.

I know cosine must be negative.

Now I simply need the cosine of 45 degrees. I can use my calculator, the value I've memorized from the unit circle, or my knowledge of 45−45−90 triangles to answer this.

Let's sketch a quick 45−45−90 triangle.

We have x, x, $x\sqrt{2}$. Our adjacent side is x and our hypotenuse $x\sqrt{2}$, so $\dfrac{x}{x\sqrt{2}} = \dfrac{1}{\sqrt{2}}$. Rationalizing the denominator, we get:

$$\frac{1}{\sqrt{2}} \times \frac{\sqrt{2}}{\sqrt{2}} = \frac{\sqrt{2}}{2}$$

We apply the negative sign from ASTC (3rd quadrant, T only is positive so cosine is negative) to get $-\dfrac{\sqrt{2}}{2}$.

Answer: $-\dfrac{\sqrt{2}}{2}$

One more way to solve

One final way we can solve these sorts of problems is drawing the little triangle right on the coordinate plane itself, using the "reference" angle as the acute angle in our right triangle.

In this method, we now can accommodate for the differing signs as we have negative and positive values on the coordinate plane. In essence, we don't have to think about ASTC if we pay attention to the positive and negative values on the axis.

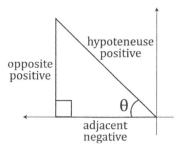

Here, you can see that angle is theta (θ).

When we take the $\cos\theta$ (CAH) here, for example, we divide the adjacent side, which is defined by a negative x-value coordinate, by the hypotenuse, which is always positive, so we get a negative value.

TRIG IDENTITIES & FORMULAS

(**NOTE**: formulas defining cot, csc, and sec are in the previous chapter **SOHCAHTOA**.)

TRIGONOMETRY IDENTITIES

$$\tan\theta = \frac{\sin\theta}{\cos\theta}$$
$$\sin^2\theta + \cos^2\theta = 1$$
$$\sec^2\theta - \tan^2\theta = 1$$
$$\csc^2\theta - \cot^2\theta = 1$$
$$\cos(2\theta) = \cos^2\theta - \sin^2\theta$$
$$\sin(2\theta) = 2\sin\theta\cos\theta$$

The first two of these, the equation for tangent and the Pythagorean identity, we've already discussed. It is vital that you memorize them. The others are not as important but still may be helpful on some ACT® problems. In general, if you need to know a trig identity or formula besides the first two, it will be given to you as part of the problem you're solving. As with the **law of sines** and **law of cosines** (covered in a separate chapter), however, there is a small chance memorizing the last four above may help you on the exam. **Doing so I would only recommend to those aiming for a perfect math score.** However, you will need to know how to manipulate expressions using these formulas and identities. Note these problems are pretty rare, occurring on maybe 10-20% of exams.

CALCULATOR TIP: For many trig identitity problems, you can make up numbers and use your calculator to back-solve the answer (i.e. using the answer choices and plugging in).

Whenever $\frac{\sin(2x)}{2\cos x}$ is defined it is equivalent to which of the following?

Note: $\sin(2\theta) = 2\sin\theta\cos\theta$

A. 1 **B.** $\sin x$ **C.** $\cos x$ **D.** $\tan x$ **E.** $\csc x$

Method 1: Plug in using the given equation.

$$\sin(2\theta) = 2\sin\theta\cos\theta$$

Thus, we know $\sin(2x) = 2\sin x \cos x$.

Then we "group substitute" (i.e. substitute in for a whole big, ugly piece rather than just a single variable) into the value of $\sin(2x)$, in our given expression, knowing that it will equal $2\sin x\cos x$.

$$\frac{\sin(2x)}{2\cos x} = \frac{2\sin x \cos x}{2\cos x}$$

Now we cancel the two instances of $\cos x$:

$$\frac{2\sin x \, \cancel{\cos x}}{2\cancel{\cos x}} = \sin x$$

Method 2: Use your calculator and make up numbers.

What do we do if we don't have the formula or it's not given by ACT? Make up numbers and use our calculator.

Here, I can make up $\dfrac{\sin(2x)}{2\cos x} = \dfrac{\sin(28)}{2(\cos 14)} \cong .24$

I plug into my calculator and then I start back-solving:

- A. 1—No!
- B. $\sin(14) \cong .24$ —Correct!

Depending on the amount of time I have, I may stop or may double check each additional answer choice. Know that to be 100% certain with this method I must check all possible answers.

Answer: **B**.

ADDITIONAL TIP: If you're still stuck, you can also use o's, a s, h s in place of $\sin$, $\cos$, $\tan$, etc. and sometimes simplify using those variables.

> The expression $\csc^2 x - \cot^2 x + 2$ is equal to:
>
> A. 1 B. 2 C. 3 D. 4 E. 5

Let's pretend I forgot the complex trig identities. What I can do instead is turn all into o's and h's.

$$\csc = \dfrac{1}{\sin} = \dfrac{h}{o}$$

$$\cot = \dfrac{1}{\tan} = \dfrac{a}{o}$$

We thus have:

$$\left(\dfrac{h}{o}\right)^2 - \left(\dfrac{a}{o}\right)^2 + 2$$

This simplifies to:

$$\dfrac{h^2 - a^2}{o^2} + 2$$

Now let's think. I know that everything with a variable must collapse or disappear because all my answers are integers. So, I know $\dfrac{h^2 - a^2}{o^2}$ will either have to equal 0 or an integer. 0 doesn't make sense. h is the hypotenuse, a greater value than the triangle leg, a. Hypotenuse, leg, two values squared...wait! That's it. This looks like the Pythagorean theorem! $a^2 + o^2 = h^2$ so $h^2 - a^2 = o^2$! Thus $\dfrac{h^2 - a^2}{o^2} = \dfrac{o^2}{o^2} = 1$.

Now we add $1 + 2$ and get 3, answer C.

Answer: **C**.

TRIGONOMETRY — SKILLS

SOLVING PROBLEMS WITH CSC, COT, AND SEC

These problems occur rarely on the ACT®, but when they do, many students miss them. For that reason, this chapter includes a good amount of practice on these.

> When μ is between π and $\frac{2\pi}{3}$, and $\csc\mu = \frac{13}{12}$, what is the value of $\cot\mu - \sec\mu$?

Start by remembering what each of these trig functions means:

$$\csc\theta = \frac{1}{\sin\theta} = \frac{hypotenuse}{opposite} \qquad \sec\theta = \frac{1}{\cos\theta} = \frac{hypotenuse}{adjacent} \qquad \cot\theta = \frac{1}{\tan\theta} = \frac{adjacent}{opposite}$$

Now I use the triangle method, and from $\csc\mu = \frac{13}{12}$, sketch the following:

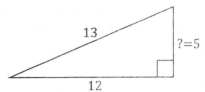

13 is the hypotenuse
12 is the opposite

By memorizing the $5-12-13$ Pythagorean triple, I know the other side (adjacent) must be 5.

Since our angle is located between π and $\frac{2\pi}{3}$, it is in the second quadrant (this is confusing though, as they are in reverse order from what you would expect, i.e. $\pi > \frac{2\pi}{3}$). Thus, **tan** and **cot** are negative and **cos** and **sec** are negative.

Now, **cot** equals adjacent/opposite: $-\frac{5}{12}$.

And **sec** equals hypotenuse over adjacent, $-\frac{13}{5}$.

Then I subtract $-\frac{5}{12} - \left(-\frac{13}{5}\right)$ using my TI-84 calculator, then I hit "MATH" ("FRAC" is highlighted once I do) and "ENTER" to turn my decimal into a fraction.

Answer: $\frac{131}{60}$.

QUESTIONS TRIGONOMETRY

1. If $\tan(x) = \dfrac{8}{15}$ and $\cos(x) = -\dfrac{15}{17}$, then $\sin(x) = ?$

   A. $\dfrac{8}{17}$
   B. $-\dfrac{7}{17}$
   C. $\dfrac{8}{15}$
   D. $-\dfrac{8}{17}$
   E. $-\dfrac{7}{15}$

2. If $90° < \theta < 180°$ and $\sin\theta = \dfrac{12}{20}$ then $\cos\theta = ?$

   A. $\dfrac{5}{3}$
   B. $\dfrac{3}{5}$
   C. $-\dfrac{4}{5}$
   D. $-\dfrac{5}{4}$
   E. $-\dfrac{5}{3}$

3. For right triangle $\triangle XYZ$, $\sin\angle X = \dfrac{5}{6}$. If $\angle Z$ is a right angle, what is $\tan\angle Y$?

   A. $\dfrac{\sqrt{11}}{5}$
   B. $\dfrac{5}{\sqrt{11}}$
   C. $\dfrac{6}{5}$
   D. $\dfrac{11}{5}$
   E. $\dfrac{\sqrt{61}}{5}$

4. If $\sin\theta = 0.6$, which of the following could be $\cot\theta$?

   A. $\dfrac{6}{\sqrt{164}}$
   B. $\dfrac{3}{5}$
   C. $\dfrac{3}{4}$
   D. $\dfrac{4}{3}$
   E. $\dfrac{4}{5}$

5. If $\sin\beta = \dfrac{4}{7}$ and $0 < \beta < 90°$ what is $\cos\beta$?

   A. $\dfrac{\sqrt{33}}{4}$
   B. $\dfrac{\sqrt{33}}{7}$
   C. $\dfrac{4}{\sqrt{33}}$
   D. $\dfrac{7}{\sqrt{33}}$
   E. $\dfrac{\sqrt{23}}{7}$

6. If θ is an angle measured in radians and $\dfrac{\pi}{2} < \theta < \pi$ and $\csc\theta = \dfrac{29}{21}$, what is $\cos\theta$?

   A. $\dfrac{-20}{29}$
   B. $\dfrac{20}{29}$
   C. $\dfrac{29}{20}$
   D. $\dfrac{21}{\sqrt{1282}}$
   E. $\dfrac{-21}{29}$

CHAPTER 19 309

7. Given that $\tan\theta = \dfrac{9}{40}$, what are all possible values of $\sec\theta$?

   A. $\dfrac{41}{9}$

   B. $\dfrac{41}{40}$

   C. $\dfrac{41}{40}$ & $\dfrac{-41}{40}$

   D. $\dfrac{41}{9}$ & $\dfrac{-41}{9}$

   E. $\dfrac{39}{40}$

8. In the standard (x,y) coordinate place below, an angle is shown whose vertex is the origin. One side of this angle with measure θ passes through $(-12,-5)$ and the other side includes the positive x-axis. What is the sine of θ?

   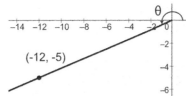

   A. $-\dfrac{5}{13}$

   B. $\dfrac{5}{13}$

   C. $\dfrac{5}{12}$

   D. $\dfrac{12}{13}$

   E. $-\dfrac{12}{13}$

9. An angle with measure γ such that $\cos\gamma = \dfrac{-15}{17}$ is in standard position with its terminal side extending into Quadrant III as shown in the standard (x,y) coordinate plane below. What is the value of $\sin\gamma$?

   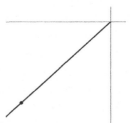

   A. $-\dfrac{17}{8}$

   B. $\dfrac{15}{17}$

   C. $-\dfrac{15}{17}$

   D. $\dfrac{8}{17}$

   E. $-\dfrac{8}{17}$

10. What are the values of θ in radians between 0 and 2π such that $\cot\theta = -1$?

    A. $\dfrac{\pi}{4}$ and $\dfrac{7\pi}{4}$

    B. $\dfrac{\pi}{4}$ and $\dfrac{3\pi}{4}$

    C. $\dfrac{3\pi}{4}$ and $\dfrac{7\pi}{4}$

    D. $\dfrac{\pi}{4}$ and $\dfrac{5\pi}{4}$

    E. $\dfrac{\pi}{4}, \dfrac{3\pi}{4}, \dfrac{5\pi}{4}$ and $\dfrac{7\pi}{4}$

11. For an angle with measure ϕ in a right triangle such that $\sin\phi = \dfrac{13}{85}$ and $\sec\phi = \dfrac{85}{84}$, what is the value of $\tan\phi$?

   A. $\dfrac{13}{84}$

   B. $\dfrac{84}{13}$

   C. $\dfrac{13}{\sqrt{6887}}$

   D. $\dfrac{\sqrt{6887}}{13}$

   E. $\dfrac{13}{\sqrt{7225}}$

12. If $\sin^2\theta = \dfrac{9}{16}$, then $\cos^2\theta = ?$

   A. $\dfrac{16}{25}$

   B. $\dfrac{3}{4}$

   C. 1

   D. $\dfrac{1}{4}$

   E. $\dfrac{7}{16}$

13. A man builds a tent (as shown in the figure below) in the shape of a triangular prism whose triangular front silhouette has three sides of length 6 feet. If the man would like to calculate the height of the tent, which of the following would be LEAST directly applicable?

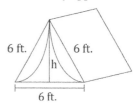

   A. The Law of Sines: For any $\triangle ABC$, where a is the length of the side opposite $\angle A$, b is the length of the side opposite $\angle B$, and c is the length of the side opposite $\angle C$, $\dfrac{\sin(\angle A)}{a} = \dfrac{\sin(\angle B)}{b} = \dfrac{\sin(\angle C)}{c}$.

   B. The Pythagorean Theorem.

   C. The ratios for the side lengths of 45–45–90 triangle.

   D. The ratios for the side lengths of a 30–60–90 triangle.

   E. The formula for the altitude of an equilateral triangle of side length x, where h is the altitude: $h = \dfrac{\sqrt{3}}{2}x$

14. If $\cot C = \dfrac{t}{p}$ and C is in the first quadrant of the unit circle, what is $\sec C$?

   A. $\dfrac{\sqrt{p^2 + t^2}}{t}$

   B. $\dfrac{\sqrt{p^2 - t^2}}{t}$

   C. $\dfrac{p^2 + t^2}{t}$

   D. $\dfrac{p}{\sqrt{p^2 + t^2}}$

   E. $\dfrac{\sqrt{p^2 + t^2}}{p}$

15. If a is in the first quadrant of the unit circle and $\cos a = \dfrac{5}{13}$, what is $\csc a$?

   A. $\dfrac{13}{12}$

   B. $\dfrac{12}{13}$

   C. $\dfrac{5}{12}$

   D. $\dfrac{5}{\sqrt{194}}$

   E. $\dfrac{13}{\sqrt{194}}$

16. The expression $\sec^2\theta - \tan^2\theta + 5$ is equal to?

   A. 5
   B. 6
   C. 4
   D. 9
   E. 3

17. The expression $8\cos^2\theta - 4$ is equivalent to? (Note: $\cos 2\theta = \cos^2\theta - \sin^2\theta$)

   A. $4\cos^2\theta$
   B. $4\cos 2\theta$
   C. 4
   D. 0
   E. $8\cos 2\theta$

18. Which of the following is equivalent to the function $h(x) = \csc x \tan x$? (Note: $\csc x = \dfrac{1}{\sin x}$)

   A. $\cot x$
   B. $\cos x$
   C. $\sin x$
   D. $\csc x$
   E. $\sec x$

19. For all values of θ where $\sin\theta$ and $\cos\theta$ are positive, which of the following is equal to $\dfrac{1}{8}\csc\theta \geq \sec\theta$?

   A. $\cot\theta \leq \dfrac{1}{8}$
   B. $\tan\theta \geq 8$
   C. $\cot\theta \geq 8$
   D. $\sin^2\theta\cos\theta \geq 8$
   E. $\cot\theta \leq -\dfrac{1}{8}$

20. For all θ such that $\sin\theta$ & $\cos\theta \neq 0$, which of the following is equivalent to $\dfrac{\cot^2\theta \sec\theta}{\csc^2\theta}$?

   A. $\dfrac{\sin^4\theta}{\cos^3\theta}$
   B. $\dfrac{\cos\theta}{\sin^4\theta}$
   C. $\cos\theta$
   D. $\cos^2\theta$
   E. $\dfrac{1}{\cos^3\theta}$

21. For all β such that $0° < \beta < 90°$ the expression $\dfrac{\sqrt{\sec^2\beta - 1}}{\tan\beta} - \dfrac{\sqrt{\csc^2\beta - 1}}{\cot\beta}$ is equal to?

   A. $\cot^2\beta - \tan^2\beta$
   B. $\csc\beta - \sec\beta$
   C. $\tan\beta - \cot\beta$
   D. 2
   E. 0

22. If x is in the first quadrant of the unit circle and $\cos x = \dfrac{5}{7}$, which of the following is equal to $\sin x \cot x$?

   A. $\dfrac{24}{35}$
   B. $\dfrac{5}{7}$
   C. $\dfrac{\sqrt{24}}{7}$
   D. $\dfrac{5\sqrt{24}}{49}$
   E. $\dfrac{\sqrt{24}}{5}$

23. When θ is between 0 and π and $\cot\theta = \dfrac{16}{5}$, what is the value of $\csc\theta - \sec\theta$?

   A. $\dfrac{11}{80}$

   B. $\dfrac{-11}{80}$

   C. $\dfrac{-11(281)}{80}$

   D. $\dfrac{11\sqrt{281}}{80}$

   E. $\dfrac{-11\sqrt{281}}{80}$

24. When $\sin x \neq 0$ and $\cos x \neq 0$, the expression $\csc^2 x - \cot^2 x$ is equal to?

   A. 1
   B. 0
   C. 2
   D. $\cos^2 x$
   E. $\tan^2 x$

25. Which of the following is equivalent to $\dfrac{\sec^2\alpha - 1}{\sin^2\alpha}$?

   A. $\sec^2\alpha$
   B. $\cos^2\alpha$
   C. $\dfrac{\cos^2\alpha}{\sin^4\alpha}$
   D. 1
   E. $\tan^2\alpha$

26. In trigonometry, an angle of $\dfrac{-11\pi}{4}$ has the same tangent as which of the following angle measures in degrees?

   A. $-225°$
   B. $-45°$
   C. $135°$
   D. $75°$
   E. $225°$

27. Two angles are coterminal if they have the same initial and terminal sides. The angles shown below with measures 30° and 390° are coterminal, for example.

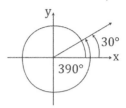

An angle with measure 45° is coterminal with a second angle. Which of the following measures could be that of the second angle?

   A. 225°
   B. 270°
   C. 315°
   D. 360°
   E. 405°

28. What are the values of θ, between 2π and 4π, when $\tan\theta = 1$?

   A. $\dfrac{9\pi}{4}$ and $\dfrac{11\pi}{4}$ only
   B. $\dfrac{9\pi}{4}$ and $\dfrac{13\pi}{4}$ only
   C. $\dfrac{11\pi}{4}$ and $\dfrac{13\pi}{4}$ only
   D. $\dfrac{11\pi}{4}$ and $\dfrac{15\pi}{4}$ only
   E. $\dfrac{13\pi}{4}$ and $\dfrac{15\pi}{4}$ only

29. If $\cos\theta = -\dfrac{1}{4}$, what is the value of $2\cos 2\theta$? (Note: $\cos^2\theta = \dfrac{1+\cos 2\theta}{2}$)

   A. $-\dfrac{9}{8}$
   B. $\dfrac{1}{8}$
   C. $-\dfrac{14}{8}$
   D. $-\dfrac{7}{8}$
   E. $\dfrac{1}{16}$

30. Tyler is lying on his back on a picnic blanket on the ground twenty-two feet away from the point where a tree touches the ground. If the tree is sixteen feet high, what is the measure of the angle of elevation of the Tyler's line of sight?

A. $\arcsin \dfrac{8}{11}$

B. $\arccos \dfrac{11}{8}$

C. $\arctan \dfrac{8}{11}$

D. $\text{arccot} \dfrac{8}{11}$

E. $\text{arccsc} \dfrac{11}{8}$

31. The vertex of $\angle A$ is the origin of the standard (x,y) coordinate plane shown below. One ray of $\angle A$ is the positive x axis. The other ray, AB, is positioned so that $\sin \angle A < 0$ and $\cos \angle A < 0$. In which quadrant, if it can be determined, is point B?

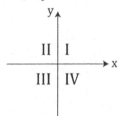

A. Quadrant IV
B. Quadrant III
C. Quadrant II
D. Quadrant I
E. Cannot be determined from the given information

32. In the standard (x,y) coordinate plane below, θ is the radian measure of an angle in standard position with the point (f,g) on the terminal side, where (f,g) falls in Quadrant IV of the coordinate plane. Which of the following points is on the terminal side of the angle in standard position having radian measure $\pi + \theta$?

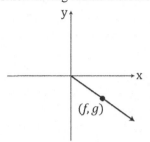

A. $(-f, g)$
B. $(f, -g)$
C. $(-f, -g)$
D. (g, f)
E. $(-g, -f)$

33. In the figure below, $\triangle ABC$ is a right triangle with all three sides of different lengths. What is the value of $\cos^2 A + \cos^2 B$?

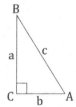

A. $\dfrac{1+\sqrt{3}}{2}$

B. $\sqrt{2}$

C. $\dfrac{1}{2}$

D. 1

E. Not enough information

34. The angle measures of $\triangle ABC$ are shown below. Point D of $\triangle DEC$ lies on $\overline{AC}$, and point C lies on $\overline{BE}$. What is the value of $\sin(\alpha + \beta)$?

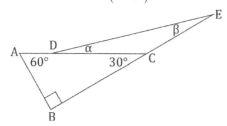

A. $\dfrac{\sqrt{2}}{2}$

B. $\dfrac{\sqrt{3}}{2}$

C. $\dfrac{1}{2}$

D. 1

E. 0

35. In $\triangle DEF$, the measure of $\angle D$ is $90°$, the measure of $\angle E$ is θ, $\overline{DF}$ is 24 units, and $\tan\theta = \dfrac{6}{5}$. What is the area of $\triangle DEF$ in square units?

A. 200

B. 240

C. $8\sqrt{30}$

D. 480

E. $24\sqrt{30}$

36. The measure of the sum of interior angles of a regular polygon with n sides is $[(n-2)180]$ degrees. What is the measure of the sum of the interior angles of a regular polygon with n sides in radians?

A. $(n-2)2\pi$

B. $\dfrac{(n-2)}{\pi}$

C. $(n-2)\pi$

D. $\dfrac{n-2}{2\pi}$

E. $\dfrac{(n-2)\pi}{2}$

37. What is the measure, in degrees, of an angle of $-\dfrac{2\pi}{9}$ radians?

A. $20°$

B. $-20°$

C. $-40°$

D. $-80°$

E. $90°$

38. If the value in radians to the nearest thousandth of $\cos\theta$ is $-.734$, which of the following could be true about θ?

A. $\dfrac{\pi}{2} < \theta < \dfrac{2\pi}{3}$

B. $\dfrac{5\pi}{6} < \theta < \pi$

C. $\dfrac{7\pi}{6} < \theta < \dfrac{4\pi}{3}$

D. $\dfrac{\pi}{6} < \theta < \dfrac{\pi}{3}$

E. $\dfrac{5\pi}{3} < \theta < \dfrac{11\pi}{6}$

39. Angle A measures $\dfrac{10\pi}{3}$ from its initial to its terminal side. Angle B has the same initial and terminal side as Angle A. Which of the following measures could be that of Angle B?

A. $54°$

B. $60°$

C. $108°$

D. $120°$

E. $240°$

40. A right triangle has legs of length $49\sin\theta$ units and $49\cos\theta$ units for some angle θ that satisfies $0° \le \theta \le 90°$. What is the length, in units, of the longest side of the triangle?

A. θ

B. 1

C. 7

D. 49θ

E. 49

TRIGONOMETRY QUESTIONS

41. An angle in a right triangle has measure β. If $\cos\beta = \dfrac{5}{13}$ and $\tan\beta = \dfrac{12}{5}$, then $\sin\beta = ?$

   A. $\dfrac{12}{5}$

   B. $\dfrac{12}{13}$

   C. $\dfrac{13}{12}$

   D. $\dfrac{12}{\sqrt{194}}$

   E. $\dfrac{12}{\sqrt{313}}$

42. If $\sin(A) = \dfrac{10}{26}$, which of the following values could $\cos A$ equal?

   A. $\dfrac{24}{26}$

   B. $\dfrac{16}{26}$

   C. $\dfrac{26}{24}$

   D. $\dfrac{10}{24}$

   E. 24

ANSWERS — TRIGONOMETRY

ANSWER KEY

1. D	2. C	3. A	4. D	5. B	6. A	7. C	8. A	9. E	10. C	11. A	12. E	13. C	14. A
15. A	16. B	17. B	18. E	19. C	20. C	21. E	22. B	23. D	24. A	25. A	26. E	27. E	28. B
29. C	30. C	31. B	32. C	33. D	34. C	35. B	36. C	37. C	38. C	39. E	40. E	41. B	42. A

ANSWER EXPLANATIONS

1. **D.** By SOHCAHTOA, we know that $\tan(x) = \frac{opposite}{adjacent}$ and $\cos(x) = \frac{adjacent}{hypotenuse}$ so we can draw a little right triangle and label it according to the pieces of these fractions, setting the opposite, adjacent, and hypotenuse sides as 8, 15, and 17 respectively. Because $\cos(x)$ is negative, and $\tan(x)$, which equals sine divided by cosine, is positive, the sine of x must also be negative, because a negative divided by a negative is a positive. Now find the opposite and divide it by the hypotenuse using values on the little triangle, but make the answer negative: $\sin(x) = \frac{opposite}{hypotenuse} = \frac{-8}{17}$.

2. **C.** Remembering SOH, we know $\sin(x) = \frac{opposite}{hypotenuse}$. Using the triangle method (pg 302), we'll find the cos of the reference angle first; draw a right triangle, label an angle (not the right angle!), and assign a measure of 12 to the opposite side and of 20 to the hypotenuse. Solve for the adjacent side, a, by seeing how the sides are multiples of a 3–4–5 triangle (fastest) or with the Pythagorean theorem ($a^2 + b^2 = c^2$) $a^2 + 12^2 = 20^2 \rightarrow a^2 + 144 = 400 \rightarrow a^2 = 256 \rightarrow a = 16$. Then $\cos(x) = \frac{adjacent}{hypotenuse} = \frac{16}{20}$ or simplified $\frac{4}{5}$. This is the cos of the reference angle. However, since $90° < \theta < 180°$, angle theta is in the 2nd quadrant (S in ASTC for "sin"): thus only sine is positive, and the cosine is negative. So, $\cos(x) = -\frac{4}{5}$.

3. **A.** If $\angle Z$ is the right angle, then the hypotenuse is opposite $\angle Z$. Since we know $\sin \angle X = \frac{5}{6}$, applying SOHCAHTOA, we can sketch the hypotenuse as equal to 6 and the side opposite of $\angle X$ as equal to 5. We now use the Pythagorean theorem $a^2 + b^2 = c^2$ to find the adjacent side. We plug in $a = 5$ and $c = 6$ to get $5^2 + b^2 = 6^2 \rightarrow 25 + b^2 = 36 \rightarrow b^2 = 11 \rightarrow b = \sqrt{11}$ So, the adjacent side of $\angle X$ is $\sqrt{11}$. Now, if we look at the graph from a different perspective, taking $\angle Y$ as our featured angle but keeping the same lengths on either side of it, the adjacent side becomes 5 and the opposite side becomes $\sqrt{11}$ So, $\tan \angle Y = \frac{opposite}{adjacent} = \frac{\sqrt{11}}{5}$. We don't need ASTC as the angles are part of a right triangle so all angles are less than 90.

4. **D.** $\sin \theta = 0.6 = \frac{6}{10}$ and by SOHCAHTOA we know that $\sin \theta = \frac{opposite}{hypotenuse}$, so we can draw a right triangle and make the opposite side equal to 6 and the hypotenuse equal to 10. Then, using Pythagorean triples, 6–8–10 is a just multiple of 3–4–5 so our other leg is 8. Alternatively, use the Pythagorean Theorem $a^2 + b^2 = c^2$, and plug in $a = 6$ and $c = 10$ to solve for the adjacent side b. $6^2 + b^2 = 10^2 \rightarrow 36 + b^2 = 100 \rightarrow b^2 = 64 \rightarrow b = 8$. So, the adjacent side is equal to 8. By SOHCAHTOA, we know that $\cot \theta = \frac{1}{\tan \theta} = \frac{1}{\frac{opposite}{adjacent}} = \frac{adjacent}{opposite} = \frac{8}{6} = \frac{4}{3}$.

5. **B.** As $\sin \beta = \frac{4}{7}$, use **SOH**CAHTOA: draw a right triangle and set the **o**pposite side equal to 4 and the **h**ypotenuse equal to 7. Then, using the Pythagorean theorem $a^2 + b^2 = c^2$, we plug in $a = 4$ and $c = 7$ to solve for the adjacent side b. We have $4^2 + b^2 = 7^2 \rightarrow 16 + b^2 = 49 \rightarrow b^2 = 33 \rightarrow b = \sqrt{33}$. So, the adjacent side is equal to $\sqrt{33}$. By SOHCAHTOA, we know that $\cos \beta = \frac{adjacent}{hypotenuse} = \frac{\sqrt{33}}{7}$. Because the angle is less than 90 degrees and in Quadrant I, we leave it positive.

TRIGONOMETRY ANSWERS

6. **A.** Since $\frac{\pi}{2} < \theta < \pi$, the angle is in the second quadrant (the S in ASTC), where sine is positive and cosine is negative. Since $\csc\theta = \frac{29}{21}$, we know by **SOH**CAHTOA that $\csc\theta = \frac{1}{\sin\theta} = \frac{1}{\frac{opposite}{hypotenuse}} = \frac{hypotenuse}{opposite}$. Draw a triangle and make the hypotenuse length 29 and the opposite side length 21. Use the Pythagorean theorem $a^2 + b^2 = c^2$ to solve for the adjacent side b. Plugging in $a = 21$ and $c = 29$, we get $21^2 + b^2 = 29^2 \to b^2 = 400 \to b = 20$. So, the adjacent side is 20. Now solve for cosine, being sure to make the answer negative given our ASTC figuring above: $\cos\theta = \frac{adjacent}{hypotenuse} = -\frac{20}{29}$.

7. **C.** Given $\tan\theta = \frac{9}{40}$, by SOHCA**HTOA**, we know that $\tan\theta = \frac{opposite}{adjacent}$, so we can draw a right triangle in which the opposite side is 9 and the adjacent side is 40. Using the Pythagorean theorem and plugging in $a = 9$ and $b = 40$ we solve for the hypotenuse c: $(9)^2 + (40)^2 = c^2 \to 1681 = c^2 \to c = 41$. Because we need all possible values, let's consider when tangent is positive as it is in our given original value: only in the first quadrant (A of ASTC) and third quadrant (T of ASTC). So we'll need to calculate secant for each of those quadrants. The sign of secant relates to the sign of cosine, as it is its inverse. Cosine (and by extension secant) is positive in quadrant 1 and negative in quadrant 3. Thus, the possible values of $\sec\theta = \frac{1}{\cos\theta} = \frac{1}{\frac{adjacent}{hypotenuse}} = \frac{hypotenuse}{adjacent}$ are $\frac{41}{40}$ or $\frac{41}{-40}$.

8. **A.** The sine of this angle is the negative of the sine of its reference angle (the angle the segment makes with the closest side of the x-axis, which in this case is the negative x-axis). It we draw a triangle with vertices at $(0,0)$, $(0,-12)$, and $(-12,-5)$, we can use the Pythagorean theorem to find the length of the hypotenuse, which is 13. Using the triangle, we see that the sine of the reference angle is $\frac{5}{13}$. However, because the actual angle lies in the third quadrant, we know it is negative (because sine is negative in the third quadrant, the T in ASTC). Thus, our answer is $-\frac{5}{13}$.

9. **E.** In Quadrant III, where the terminal side of this angle is, sine is negative because y is negative. We can find the length of the opposite side by first drawing it as a vertical line from the dot to the x-axis. Now label the sides we know using the lengths of the adjacent side (15) and hypotenuse (17) which constitutes the cosine (CAH or A over H). Now use Pythagorean triples or theorem to solve for the missing side: $a^2 + b^2 = c^2 \to a^2 + (-15)^2 = 17^2 \to a^2 + 225 = 289 \to a^2 = 64 \to a = 8$. Now we plug that value in atop the value of the hypotenuse to find the sine (opposite over hypotenuse). However, since we are in Quadrant III, where the y-value is negative, sine must be negative: $\sin\gamma = \frac{-8}{17}$.

10. **C.** $\cot\theta = \frac{1}{\tan\theta}$. If we want the cotangent to be equal to -1, then the tangent must be equal to -1. $\tan\theta = \frac{\sin\theta}{\cos\theta}$, so it will only equal -1 when sine and cosine have the same magnitude but opposite signs, which we can express as $\sin\theta = -\cos\theta$. The quadrants where sine and cosine have opposite signs are Quadrants II and IV. Sine and cosine have equal magnitude when both sides of a right triangle are equal, (this is because $\sin\theta = \cos\theta \to \frac{opposite}{hypotenuse} = \frac{adjacent}{hypotenuse} \to opposite = adjacent$). This makes it an isosceles triangle, and a right-isosceles triangle has angles of $45-45-90$. In terms of radians, the $45°$ angle is $\frac{\pi}{4}$. However, this is only the reference angle. We are looking for the angles that fall within Quadrants II and IV. The angle in Quadrant II will be $\pi - \frac{\pi}{4} = \frac{3\pi}{4}$. The angle in Quadrant IV will be $2\pi - \frac{\pi}{4} = \frac{7\pi}{4}$. Thus, our answer is C.

11. **A.** If sine is $\frac{13}{85}$, we can draw a right triangle and define the measure of the opposite side as 13 and the measure of the hypotenuse as 85. Since the secant is $\frac{85}{84}$, we can define the adjacent side as 84, since $\sec\theta = \frac{hypotenuse}{adjacent}$. We have everything we need to find the tangent now. $\tan\phi = \frac{opposite}{adjacent} = \frac{13}{84}$.

12. **E.** The fastest way to solve this is with the Pythagorean identity: $\sin^2\theta + \cos^2\theta = 1$. We simply plug into this using our

given value $\sin^2\theta = \frac{9}{16}$: $\frac{9}{16} + \cos^2\theta = 1$, $\cos^2\theta = 1 - \frac{9}{16}$, $\cos^2\theta = \frac{7}{16}$. Alternatively, we can take the root of our given value, $\sin^2\theta = \frac{9}{16}$, to find $\sin\theta = \frac{3}{4}$. Now by SOHCAHTOA we can draw a little triangle and assign opposite and hypotenuse sides as 3 and 4, respectively. We can solve for the adjacent side, a, by using the Pythagorean theorem: $a^2 + b^2 = c^2$: $a^2 + 3^2 = 4^2 \to a^2 + 9 = 16 \to a^2 = 7 \to a = \sqrt{7}$. Then, $\cos^2\theta = \left(\frac{opposite}{hypotenuse}\right)^2 = \left(\frac{\sqrt{7}}{4}\right)^2 = \frac{7}{16}$.

13. **C.** Since the front silhouette of the tent is an equilateral triangle, the triangle is split down the middle to make two 30–60–90 triangles. There are no 45–45–90 triangles, so the answer choice C is the least relevant to calculating the height of the tent. Answer choice A could be applied to solve for the height by plugging in $\frac{\sin 90°}{6} = \frac{\sin 60°}{h}$ and solving for h. Answer choice B can be applied to solve for the height by plugging in $h^2 + 3^2 = 6^2$ and solving for h. Answer choice D can be applied to find the height by using the ratio $x - x\sqrt{3} - 2x$ and plugging in $x = 3$ to solve for $h = x\sqrt{3}$. Answer choice E could be applied to solve for the height by plugging in $x = 6$ to get $h = \frac{\sqrt{3}}{2}(6)$.

14. **A.** If C is in the first quadrant, that means all its sides are positive. We know that $\cot C = \frac{1}{\tan C} = \frac{adjacent}{opposite} = \frac{t}{p}$. So, $t =$ adjacent side and $p =$ opposite side. We can then use the Pythagorean theorem $t^2 + p^2 = c^2$ and take the square root of both sides to write the hypotenuse side as $\sqrt{p^2 + t^2}$. We then find $\sec C = \frac{1}{\cos C} = \frac{1}{\frac{adjacent}{hypotenuse}} = \frac{hypotenuse}{adjacent} = \frac{\sqrt{p^2 + t^2}}{t}$.

15. **A.** Since a is in the first quadrant of the unit circle, all of its trig functions are positive. Using SOHCAHTOA, we know that $\cos a = \frac{adjacent}{hypotenuse} = \frac{5}{13}$, so the adjacent side is equal to 5 and the hypotenuse is 13. Using the Pythagorean Theorem $a^2 + b^2 = c^2$, we can plug in $a = 5$ and $c = 13$ to solve for the opposite side, b. Thus, we have $5^2 + b^2 = 13^2 \to 25 + b^2 = 169 \to b^2 = 144 \to b = 12$. We now know that $\csc a = \frac{1}{\sin a} = \frac{1}{\frac{opposite}{hypotenuse}} = \frac{hypotenuse}{opposite} = \frac{13}{12}$. Memorize the 5, 12, and 13 Pythagorean triple to do this faster!

16. **B.** This problem is easiest if you remember the identity: $\tan^2\theta = \sec^2\theta - 1$. You can rearrange this to get $1 = \sec^2\theta - \tan^2\theta$. We are told to find the value of the expression $\sec^2\theta - \tan^2\theta + 5$, so we substitute $\sec^2\theta - \tan^2\theta$ as 1 and get $1 + 5 = 6$. If you can't remember the identity or how to derive it, you have two options: make up any angle value for theta and plug this into your calculator (probably best/fastest) or make up o's, a's and h's in place of tan and sec and solve down with those letters. Here let's try the former method, using 37° for θ. I get out my calculator, remembering that sec is $\frac{1}{\cos}$, and plug in $\frac{1}{\cos^2(37°)} - \tan^2(37°)$ to get 1. Then I add 1 to 5 and get 6.

17. **B.** Although our answer is in terms of cosine, our expression in its current form cannot be manipulated to one of the answers provided. Thus, we'll need to either make up a number for theta and back solve using our calculator OR factor and substitute. Let's try the 2nd method (I give an example of the 1st method earlier in the chapter). We can factor 4 out of our expression to get $4(2\cos^2\theta - 1)$. Whenever I see a 1 and I know I need to substitute, I try group subbing in the entire left hand side of the Pythagorean identity for 1: $\sin^2 x + \cos^2 x = 1$. Here we have θ, not x, so I sub in: $4(2\cos^2\theta - (\sin^2\theta + \cos^2\theta)) = 4(2\cos^2\theta - \sin^2\theta - \cos^2\theta)$. Now I can simplify this to $4(\cos^2\theta - \sin^2\theta)$. At this point, I regroup and look at my given information and see that the elements I have in parenthesis are equal to $\cos 2\theta$. Now I substitute in again and get $4(\cos 2\theta)$. There is also an identity you can memorize that makes solving this problem much faster, but is very rarely tested on the ACT: $\cos 2\theta = 2\cos^2\theta - 1$. Using this identity, we substitute after factoring out the 4.

18. **E.** We can rewrite $\csc x$ as $\frac{1}{\sin x}$ and $\tan x$ as $\frac{\sin x}{\cos x}$. So, $\csc x \tan x = \frac{1}{\sin x}\left(\frac{\sin x}{\cos x}\right) = \frac{1}{\cos x} = \sec x$.

TRIGONOMETRY ANSWERS

19. C. We can rewrite $\csc\theta$ as $\frac{1}{\sin\theta}$ and $\sec\theta$ as $\frac{1}{\cos\theta}$. The inequality $\frac{1}{8}\csc\theta \geq \sec\theta$ can then be written as $\frac{1}{8}\left(\frac{1}{\sin\theta}\right) \geq \frac{1}{\cos\theta}$. Multiplying by $8\cos\theta$ on both sides we get $\frac{\cos\theta}{\sin\theta} \geq 8$. Since $\cot\theta = \frac{1}{\tan\theta} = \frac{1}{\frac{\sin\theta}{\cos\theta}} = \frac{\cos\theta}{\sin\theta}$, we can rewrite the inequality by substituting $\cot\theta$ in for $\frac{\cos\theta}{\sin\theta}$. We get $\cot\theta \geq 8$.

20. C. We can rewrite $\cot^2\theta$ as $\frac{1}{\tan^2\theta} = \frac{1}{\frac{\sin^2\theta}{\cos^2\theta}} = \frac{\cos^2\theta}{\sin^2\theta}$, $\sec\theta$ as $\frac{1}{\cos\theta}$, and $\csc^2\theta$ as $\frac{1}{\sin^2\theta}$. Substituting these expressions into the equation $\frac{\cot^2\theta \sec\theta}{\csc^2\theta}$, we get $\frac{\frac{\cos^2\theta}{\sin^2\theta}\left(\frac{1}{\cos\theta}\right)}{\frac{1}{\sin^2\theta}} = \frac{\cos^2\theta}{\sin^2\theta}\left(\frac{1}{\cos\theta}\right)(\sin^2\theta) = \frac{\cos^2\theta}{\sin^2\theta}\left(\frac{\sin^2\theta}{\cos\theta}\right) = \cos\theta$.

21. E. $0° < \beta < 90°$ means that all of the values are positive. Take the trig identity $\sin^2\beta + \cos^2\beta = 1$, and then divide both sides of the equation by $\cos^2\beta$ to get $\frac{\sin^2\beta}{\cos^2\beta} + 1 = \frac{1}{\cos^2\beta} \rightarrow \tan^2\beta + 1 = \sec^2\beta \rightarrow \tan^2\beta = \sec^2\beta - 1$ On the other hand, taking $\sin^2\beta + \cos^2\beta = 1$ and dividing both sides by $\sin^2\beta$, we get $1 + \frac{\cos^2\beta}{\sin^2\beta} = \frac{1}{\sin^2\beta} \rightarrow 1 + \cot^2\beta = \csc^2\beta \rightarrow \cot^2\beta = \csc^2\beta - 1$. We can now substitute $\tan^2\beta = \sec^2\beta - 1$ and $\cot^2\beta = \csc^2\beta - 1$ into the expression $\frac{\sqrt{\sec^2\beta - 1}}{\tan\beta} - \frac{\sqrt{\csc^2\beta - 1}}{\cot\beta}$ to get $\frac{\sqrt{\tan^2\beta}}{\tan\beta} - \frac{\sqrt{\cot^2\beta}}{\cot\beta} = \frac{\tan\beta}{\tan\beta} - \frac{\cot\beta}{\cot\beta} = 1 - 1 = 0$. Alternatively, make up an angle, plug into your calculator, and back solve using the answers.

22. B. If x is in the first quadrant, this means that all sides of the triangle formed by x are positive. $\sin x \cot x = \sin x\left(\frac{1}{\tan x}\right) = \sin x\left(\frac{1}{\frac{\sin x}{\cos x}}\right) = \sin x\left(\frac{\cos x}{\sin x}\right) = \cos x$. We are given that $\cos x = \frac{5}{7}$.

23. D. Because θ is between 0 and π, it must be in the first or second quadrant, and because we are dealing with cotangent, which is only positive in the first and third quadrant, we know that for this problem we are working only in the first quadrant. cosecant, secant, and cotangent are reciprocals of sine, cosine, and tangent, respectively, so $\cot\theta = \frac{adjacent}{opposite}$, $\csc\theta = \frac{hypotenuse}{opposite}$, and $\sec\theta = \frac{hypotenuse}{adjacent}$. We are given that $\cot\theta = \frac{16}{5}$, so our adjacent side must be 16 and our opposite side must be 5. Using the Pythagorean theorem, we solve to find our hypotenuse, which we find is $\sqrt{281}$. Now that we know our three side lengths, we can determine that $\csc\theta = \frac{\sqrt{281}}{5}$ and $\sec\theta = \frac{\sqrt{281}}{16}$. Subtracting the two, $\frac{\sqrt{281}}{5} - \frac{\sqrt{281}}{16} = \frac{16\sqrt{281} - 5\sqrt{281}}{5(16)} = \frac{11\sqrt{281}}{80}$.

24. A. Take the trigonometric identity $\sin^2 x + \cos^2 x = 1$, and divide both sides by $\sin^2 x$ to get: $\frac{\sin^2 x}{\sin^2 x} + \frac{\cos^2 x}{\sin^2 x} = \frac{1}{\sin^2 x} \rightarrow 1 + \cot^2 x = \csc^2 x$. Subtract $\cot^2 x$ on both sides, to get $1 = \csc^2 x - \cot^2 x$.

25. A. There are many different ways to solve this. One method: take the trigonometric identity $\sin^2\alpha + \cos^2\alpha = 1$ and divide both sides of it by $\cos^2\alpha$ to get $\frac{\sin^2\alpha}{\cos^2\alpha} + \frac{\cos^2\alpha}{\cos^2\alpha} = \frac{1}{\cos^2\alpha} \rightarrow \tan^2\alpha + 1 = \sec^2\alpha$ (remember sin over cos equals tan and sec is the reciprocal of cos). Subtracting 1 from both sides of the equation, we get $\sec^2\alpha - 1 = \tan^2\alpha$. Plugging this into the expression $\frac{\sec^2\alpha - 1}{\sin^2\alpha}$, we get $\frac{\tan^2\alpha}{\sin^2\alpha}$. Substituting in, $\frac{\sin^2\alpha}{\cos^2\alpha}$ for $\tan^2\alpha$ we get $\frac{\sin^2\alpha}{\cos^2\alpha}\left(\frac{1}{\sin^2\alpha}\right) = \frac{1}{\cos^2\alpha} = \sec^2\alpha$.

26. E. Because our answer is in terms of degrees, it's best to convert $\frac{-11\pi}{4}$ from radians to degrees first, by multiplying by $\frac{180}{\pi}$: $\frac{-11\pi}{4} \times \frac{180}{\pi} = -495°$. We don't know the tangent of $-495°$, but we at least know that $-495°$ has the same

320   CHAPTER 19

tangent as an angle it is coterminal with. $-495°$ is coterminal with every angle that is a full $360n°$ before or after it, where n is any integer, such as $-855°, -135°$, and $225°$. We get these numbers simply by subtracting or adding 360 multiple times. Luckily $225°$ is one of our answers, which we can visualize in quadrant III, so we can end there. But if we had not been so fortunate, we would have had to look at the reference angle of $225°$, which is $45°$, and find another answer whose reference angle is $45°$ and has a positive tangent. Alternatively we could add/subtract 360 over and over until we get an available answer.

27. **E**. An angle in a circle that is greater than $360°$ is congruent, or coterminal, to an angle with the same measure plus or minus $360°$, or any multiple thereof (e.g. $370°$ is coterminal with $10°$ and $750°$ is coterminal with $30°$). So to find an angle that is coterminal with $45°$, we can try adding $360°$. This gives us $405°$, which is one of the answers provided.

28. **B**. Because of the unit circle, we know that tangent is equal to one only when sin and cos values are equal at angles with the reference angle of 45 degrees or radians $\pi/4$ so $\tan\frac{\pi}{4}=1$ and $\tan\frac{5\pi}{4}=1$. But because we want an angle between 2π and 4π, we must find the coterminal angles. $\frac{\pi}{4}+2\pi=\frac{9\pi}{4}$ and $\frac{5\pi}{4}+2\pi=\frac{13\pi}{4}$, so our answer is $\frac{9\pi}{4}$ and $\frac{13\pi}{4}$.

29. **C**. If $\cos\theta=-\frac{1}{4}$, then $\cos^2\theta=\frac{1}{16}$. Because we are given an identity, we can set the two equal and manipulate the expression to get our answer: $\cos^2\theta=\frac{1}{16}=\frac{1+\cos 2\theta}{2}$. Multiply both sides by 2: $\frac{1}{8}=1+\cos 2\theta$. Subtract 1 from both sides: $\frac{1}{8}-1=\cos 2\theta=-\frac{7}{8}$. Since we're looking for $2\cos 2\theta$, we can multiply this by 2 to get $-\frac{14}{8}$.

30. **C**. Draw a diagram based on what we are told: a right triangle, where the distance between Tyler and the tree house, 22, is the adjacent side, and the height of the tree house, 16, is the opposite side. We could find the third side using the Pythagorean Theorem, but since all of the answer choices actually use multiples of the side lengths already given $\left(\frac{22}{2}=11 \text{ and } \frac{16}{2}=8\right)$, our answer must be the trigonometric function that only uses the opposite and adjacent sides, either arctan or arccot. $\arctan\frac{opposite}{adjacent}=\theta$, so our answer is $\arctan\frac{16}{22}=\arctan\frac{8}{11}$.

31. **B**. Per the unit circle, since $\sin\angle A<0$, then point B must be below the x-axis, and since $\cos\angle A<0$, then point B must also be to the left of the y-axis. Essentially, we want our sin and cos to both be negative. Because sin is equivalent to y on the unit circle, we need a negative y-value. Because cos is equivalent to x on the unit circle, we need a negative x-value. Quadrant III is the only quadrant with negative x and y values.

32. **C**. In this problem f and g are the x and y components of a point on a line. For simplicity's sake, imagine that it is the endpoint of a radius of the unit circle. The point we want to find is on the line whose angle measures $\pi+\theta$ radians, in other words, the line directly opposite of the original line on the unit circle (a $180°$ degree rotation from the original position to the new one). The new angle will, therefore, be in quadrant II, and this will change both the sign of the x and the sign of y components. Thus, our answer is $(-f,-g)$.

33. **D**. From SOHCAHTOA we know $\cos A=\frac{b}{c}$ and $\cos B=\frac{a}{c}$, so $\cos^2 A+\cos^2 B$ equals $\left(\frac{b}{c}\right)^2+\left(\frac{a}{c}\right)^2$. Distribute the exponent and combine the fractions to get $\frac{b^2+a^2}{c^2}$. By the Pythagorean theorem we know $a^2+b^2=c^2$, so we can substitute c^2 for b^2+a^2, giving us $\frac{c^2}{c^2}$, which equals 1. Alternatively, recognize how in a right triangle, the sin of one acute angle is the cos of the other acute angle. So $\cos A=\sin B$. Substituting we get $\sin^2 B+\cos^2 B$ which equals 1 by the Pythagorean identity.

34. **C**. Although there is a formula to find $\sin(\alpha+\beta)$ by using the sines and cosines of α and β, that actually doesn't matter here. This is actually more of a triangle angle question. Assuming we don't use the exterior angle theorem as a shortcut, we could find the measure of $\alpha+\beta$ as follows: $\overline{BE}$ is a straight line, which means $\angle BCA$ and $\angle DCE$ are supplementary. Thus, we can solve for $\angle DCE$: $\angle BCA+\angle DCE=180° \rightarrow 30°+\angle DCE=180° \rightarrow \angle DCE=150°$. Now, looking at $\angle DEC$, we know by the triangle sum theorem that $\angle DCE+\alpha+\beta=180°$. If we subtract $\angle DCE$, which we just found, from both sides, we get $\alpha+\beta=30°$. $\sin 30°=\frac{1}{2}$.

TRIGONOMETRY ANSWERS

35. **B.** We have a right triangle with one angle and one side given. We know that $\tan\theta = \tan\angle E = \frac{6}{5} = \frac{opposite}{adjacent}$. If we sketch the triangle, we see that $\overline{DF}$ is our "opposite" side, so we can plug that in to solve for the "adjacent" side, $\overline{DE}$. $\frac{6}{5} = \frac{24}{DE}$, so $\overline{DE} = 20$. Because this is a right triangle, the side lengths are the base and height of the triangle, so we can use them to find the area. $A = \frac{1}{2}bh = \frac{1}{2}(24)(20) = 240$. If you missed this, you may want to review SOHCAHTOA, Ch 18.

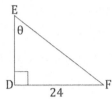

36. **C.** To convert from degrees to radians, simply multiply the degrees by the conversion ratio $\frac{\pi}{180°}$: $(n-2)180° \times \frac{\pi}{180°} = (n-2)\pi$. Put what you "need" on top of the conversion ratio (radians) and what you want to cancel (degrees) on the bottom. See Ch 9 in Book 1 for more on unit conversions.

37. **C.** Convert by multiplying by the conversion ratio $\frac{180°}{\pi}$: $-\frac{2\pi}{9} \times \frac{180°}{\pi} = -40$. Put what you "need" on top of the conversion ratio (degrees) and what you want to cancel (radians) on the bottom. See Ch 9 in Book 1 for more on unit conversions.

38. **C.** Cosine is negative in the II and III quadrants, so we know that $\frac{\pi}{2} < \theta < \frac{3\pi}{2}$ at least, so D is out. We know that the cosine of angles with a reference angle of $\frac{\pi}{6}$ is $\pm\frac{1}{2}$, and that the sine of angles with a reference angle of $\frac{\pi}{3}$ is $\pm\frac{\sqrt{3}}{2} \approx .866$, so our mystery angle must be between a pair like that. In the second quadrant, the angle with a reference angle of $\frac{\pi}{3}$ is $\frac{2\pi}{3}$ which the angle with a reference angle of $\frac{\pi}{6}$ is $\frac{5\pi}{6}$, so a possible answer is $\frac{2\pi}{3} < \theta < \frac{5\pi}{6}$. That answer is not provided but using the same logic in the third quadrant, we get $\frac{7\pi}{6} < \theta < \frac{4\pi}{3}$, which is provided. If you forget the unit circle values, use your calculator (in radian mode) to calculate the cosines of the appropriate reference angles for the answers given until you spot a range the given cos value is in.

39. **E.** First we convert $\frac{10\pi}{3}$ to degrees: $\frac{10\pi}{3} \times \frac{180°}{\pi} = 600°$. Angles that share the same initial and terminal sides are coterminal, and coterminal angles are all $\pm 2\pi$ (or $\pm 360°$) of each other. While $600°$ is not one of the answers, $600° - 360° = 240°$ is a given answer.

40. **E.** First draw a picture! The longest side of the right triangle is the hypotenuse, which we can find using the Pythagorean theorem. Plug in the measures given: $\sqrt{(49\sin\theta)^2 + (49\cos\theta)^2} = \sqrt{c^2}$, where c is the hypotenuse. Simplify the right side and factor out the 49's: $c = \sqrt{49^2(\sin\theta^2 + \cos\theta^2)}$. Since $\sin\theta^2 + \cos\theta^2 = 1$, this becomes $c = \sqrt{49^2(1)} = 49$.

41. **B.** The sine of an angle is the opposite side divided by the hypotenuse. Because these angles are in a right triangle, they correspond to Quadrant I (less than 90 degrees) and we don't need to worry about any negative signs. We can use the triangle method (pg 302), drawing a triangle and labelling the angle we want. Given the tangent $\frac{12}{5}$, which is the opposite side divided by the adjacent side, we assign 12 to the side opposite our angle and 5 to the side adjacent. We can find the hypotenuse using the Pythagorean triple $5-12-13$, or from the denominator of the cosine, $\frac{5}{13}$. If the adjacent side is 5, the hypotenuse is 13. The opposite side is 12, and the hypotenuse is 13, so the sine (opposite/hypotenuse) is $\frac{12}{13}$.

42. **A.** Since we know that $\sin(A) = \frac{10}{26}$, we know by SOHCAHTOA that the opposite and hypotenuse sides are 10 and 26, respectively. We can solve for the adjacent side, a, knowing it is a multiple of a $5-12-13$ triangle or using the Pythagorean Theorem $a^2 + b^2 = c^2$. $a^2 + 10^2 = 26^2 \rightarrow a^2 + 100 = 676 \rightarrow a^2 = 576 \rightarrow a = 24$. $\cos(A) = \frac{adjacent}{hypotenuse} = \frac{24}{26}$. Because we only need what cosine COULD equal, ASTC isn't necessary and we look for an equivalent answer with any sign.

CHAPTER 20
LAWS OF SINES AND COSINES

SKILLS TO KNOW
- Law of Sines
- Law of Cosines

The ACT® sometimes incorporates questions that require you to understand how to apply the **law of sines** and **law of cosines**. Though historically, the ACT® nearly always published these equations as part of the question itself whenever you needed to apply them, recent tests have diverged from this trend. In other words, ACT® has asked questions solvable with these rules that did not include the formulas for reference. I recommend that students **aiming for a 34+ on the math section memorize** these laws (or have a calculator program that performs them and adheres to **ACT's® strict rules**). Students aiming for a 30+ should simply know how to use them if given the equation. Students aiming for below a 30 can skip this chapter until other areas are mastered.

Given the calculator program rule, **you could potentially program a short program into your calculator that performs each of these functions.** We offer some programs for download at **supertutortv.com/bookowners** if you're interested in this option.

Why do we need these laws?

We use SOHCAHTOA to solve for missing pieces of right triangles, but what about triangles that have no 90 degree angles? That's when the laws of sines and cosines come in.

If you have enough pieces of a triangle, namely ASA, SAS, SSA, or SSS, you can solve for every angle and side length in the triangle using these formulas. It can be confusing to know which one to use when, but with practice that part gets easier.

LAW OF SINES

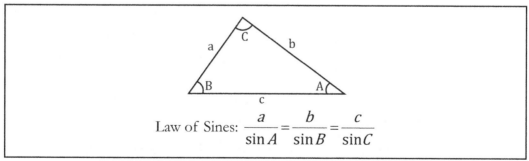

Law of Sines: $\dfrac{a}{\sin A} = \dfrac{b}{\sin B} = \dfrac{c}{\sin C}$

This rule states that the length of a side over the sine of the opposite angle equals another side over that respective opposite angle's sine. We use the **law of sines** when we know two sides and one of the corresponding angles (**SSA**) or two angles and one of the corresponding sides (**ASA** or **AAS**). When solving using Law of Sines, you'll end up with two angles and two corresponding sides. Feel free to come up with some silly way to remember all this.

LAWS OF SINES AND COSINES — SKILLS

LAW OF COSINES

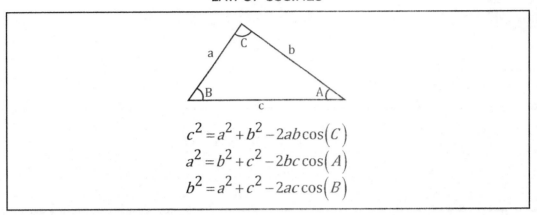

As you can see, **law of cosines** has three "versions"—but in fact they all are essentially the same idea. The side to the left of the equal sign corresponds to the angle we take the cos of on the right. Note that if the ACT® gives you the **law of cosines** for reference, **it will NOT give you all three versions**. Even more confusing, it may reuse letters that are arbitrarily already on your diagram, so don't rely on the "letters" as much as the concept. The biggest mistake students make is that they assign the longest side arbitrarily to C because they mix up this formula with the Pythagorean theorem, and think C must always be the long side. That's incorrect. With the law of cosines we know three sides and one angle by the end of the calculation, and again the variable to the left of the equals sign is always representative of the side opposite the angle in question (whether we know the angle or are solving for it.)

We use **law of cosines** when we know the lengths of three sides (**SSS**) and need an angle, or we know the length of two sides and the angle that joins them (**SAS**) and need the third side.

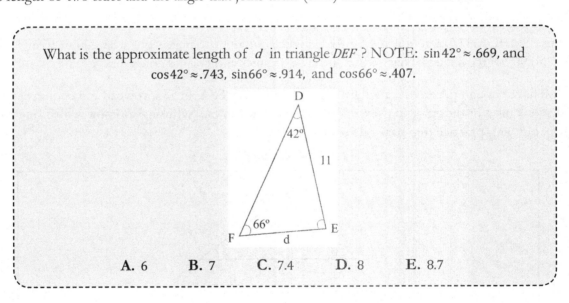

We know the angle of F, which is $66°$, the angle of D, which is $42°$, and side $\overline{DE}$, which is 11. This gives us AAS. Anytime we know **two angles** we are always using the **law of sines** not the **law of cosines**.

Using the **law of sines** we set up the proportion:

SKILLS　　　　　　　　　　　　LAWS OF SINES AND COSINES

$$\frac{d}{\sin 42°} = \frac{11}{\sin 66°}$$

Then we multiply both sides by $\sin 42°$ to solve for d.

$$d = \left(\frac{11}{\sin 66°}\right) * (\sin 42°)$$

$$d = 8.057$$

Considering the answer choices, we see the closest answer to this value is choice (D).
Answer: **D**.

WARNING: If you're going to use your calculator's values for sine instead of those given by ACT, be sure that your calculator is in degree mode (assuming all angles are given in degrees.)

Using the diagram below, find the approximate angle C.

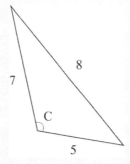

A. $\sin^{-1}\left(\frac{7}{8}\right)$　　**B.** $\sin^{-1}\left(\frac{1}{7}\right)$　　**C.** $\cos^{-1}\left(\frac{1}{7}\right)$　　**D.** $\cos^{-1}\left(\frac{7}{8}\right)$　　**E.** $\tan^{-1}\left(\frac{8}{5}\right)$

Using the law of cosines, $c^2 = a^2 + b^2 - 2ab\cos(C)$ we can plug in the numbers into the equation. Again, I am very careful to place the 8 on the LEFT side of the equation, as it is the side opposite the angle I need. **The corresponding side opposite the angle in question always goes on the left by itself.** This gives us:

$$8^2 = 5^2 + 7^2 - 2(5)(7)\cos(C)$$
$$64 = 25 + 49 - 70\cos(C)$$
$$-10 = -70\cos(C)$$

After combining like terms, 64, 25, and 49, by subtracting 25 and 49 from both sides, divide both sides by -70:

$$\frac{1}{7} = \cos(C)$$

Most of the time, unless the ACT provides you with the values for sine and cosine of particular angles, you won't be required to solve down for the exact angle, as all ACT questions can be solved without a calculator. Instead, the answer choices will denote trig function you would need to perform to get the answer. Here that is $\cos^{-1}$ (the inverse of $\cos$) of $\frac{1}{7}$.

CHAPTER 20

LAWS OF SINES AND COSINES — QUESTIONS

1. Two sailboats, at points R and T as shown in the figure below, are 3 miles and 2 miles away from a lighthouse at point S (respectively). If $\angle RST = 97°$, which of the following is equal to the distance between the two boats at R and T?
(Note: For any $\triangle ABC$, where a is the length of the side opposite $\angle A$, b is the length of the side opposite $\angle B$, and c is the length of the side opposite $\angle C$, $\dfrac{\sin(\angle A)}{a} = \dfrac{\sin(\angle B)}{b} = \dfrac{\sin(\angle C)}{c}$, and $c^2 = a^2 + b^2 - 2ab\cos\angle C$.

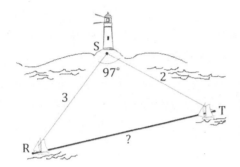

A. $2\left(\dfrac{\sin(97°)}{3}\right)$

B. $2^2 + 3^2 - (2)(2)(3)\cos(97°)$

C. $\sqrt{2^2 + 3^2 - (2)(3)\cos(97°)}$

D. $3\left(\dfrac{\sin(97°)}{2}\right)$

E. $\sqrt{2^2 + 3^2 - (2)^2(3)\cos(97°)}$

2. In $\triangle XYZ$, the measure of $\angle Z$ is $39°$, the measure of $\angle Y$ is $65°$, and the length of $\overline{ZY}$ is 13 inches. Which of the following is an expression for the length of $\overline{XZ}$? (Note: The law of sines states that for any triangle, the ratios of the lengths of the sides to the sines of the angles opposite those sides are equal.)

A. $\dfrac{13\sin 39°}{\sin 76°}$

B. $\dfrac{13\sin 39°}{\sin 65°}$

C. $\dfrac{13\sin 65°}{\sin 39°}$

D. $\dfrac{507}{76}$

E. $\dfrac{13\sin 65°}{\sin 76°}$

3. Red Ranch and Crimson Ranch lie on opposite sides of a river. The nearest ferry is located 800 meters from Crimson ranch. The ranchers estimate the angles between these locations to be as shown on the map below. Using these estimates, which of the follow expressions gives the distance, in meters, between Red Ranch and Crimson Ranch?

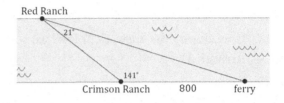

A. $\dfrac{800}{21}$

B. $\dfrac{800}{\sin 21°}$

C. $\dfrac{800\sin 18°}{\sin 21°}$

D. $1000\tan 18°$

E. $\dfrac{800}{\cos 141°}$

4. In the figure below, a radar screen shows 2 boats. Boat A is located at a distance of 45 nautical miles and bearing $110°$, and Boat B is located at a distance of 60 nautical miles and bearing $220°$. Which of the following is an expression for the straight line distance, in nautical miles, between the two boats?

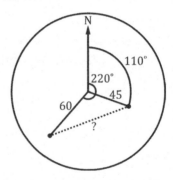

A. $\sqrt{45^2 + 60^2}$

B. $45^2 + 60^2 - 2(45)(60)(\cos 110°)$

C. $45^2 + 60^2 - 2(45)(60)(\sin 110°)$

D. $\sqrt{45^2 + 60^2 - 2(45)(60)(\cos 110°)}$

E. $\sqrt{45^2 + 60^2 - 2(45)(60)(\cos 220°)}$

5. For △XYZ below, the length of $\overline{XZ}$ is 30 yards. Which of the following equations, when solved, will give the length, in yards, of $\overline{XY}$?
(Note: The law of sines states that given △ABC, $\frac{\sin(\angle A)}{a} = \frac{\sin(\angle B)}{b} = \frac{\sin(\angle C)}{c}$.)

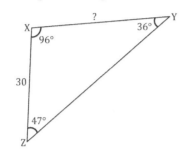

A. $\frac{\sin 36°}{30} = \frac{\sin 47°}{XY}$

B. $\frac{\sin 47°}{30} = \frac{\sin 36°}{XY}$

C. $\frac{\sin 97°}{30} = \frac{\sin 47°}{XY}$

D. $\frac{\sin 97°}{30} = \frac{\sin 36°}{XY}$

E. $\frac{\sin 36°}{30} = \frac{\sin 97°}{XY}$

6. In △ABC below, ∠A measures 71°, ∠C measures 52°, and the length of $\overline{BC}$ is 42 feet. To the nearest foot, what is the length of $\overline{AB}$?
(Note: The law of sines states that the lengths of the sides of a triangle are proportional to the singes of the opposite angles. Note also that $\sin 71° \approx .946$ and $\sin 52° \approx .788$.)

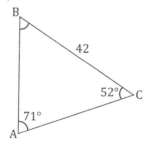

A. 50
B. 45
C. 37
D. 35
E. 57

7. A warehouse is 6 miles from a factory, and a retail store is 7.1 miles from the factory, as shown below. The angle between straight lines from the warehouse to the factory and store is 11°. The approximate distance, in miles, from the warehouse to the retail store is given by which of the following expressions?
(Note: The law of cosines states that for any triangle with vertices A, B, and C and sides opposite those vertices with lengths a, b, and c, respectively, $c^2 = a^2 + b^2 - 2ab\cos C$.)

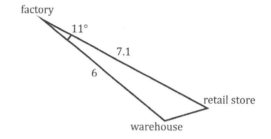

A. $\sqrt{6^2 + 7.1^2 + 2(6)(7.1)(\cos 11°)}$

B. $\sqrt{6^2 + 7.1^2 - 2(6)(7.1)(\cos 79°)}$

C. $\sqrt{6^2 + 7.1^2 - 2(6)(7.1)(\cos 11°)}$

D. $\sqrt{7.1^2 + 6^2}$

E. $\sqrt{7.1^2 - 6^2}$

8. Triangle △ABC is shown in the figure below. The measure of ∠A = 70°. AB = 15, and AC = 9. BC = ?

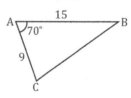

A. $15\sin 70°$
B. $9\cos 70°$
C. $\sqrt{15^2 - 9^2}$
D. $\sqrt{9^2 + 15^2}$
E. $\sqrt{9^2 + 15^2 - 2(9)(15)\cos 70°}$

9. In △ABC, shown below, angle measures are as marked. Which of the following is an expression for the length of AB?

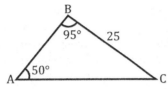

A. $\dfrac{25\sin 50°}{\sin 95°}$

B. $\dfrac{25\sin 35°}{\sin 50°}$

C. $\dfrac{25\sin 35°}{\sin 95°}$

D. $\dfrac{25\sin 95°}{\sin 50°}$

E. $\dfrac{25\sin 95°}{\sin 35°}$

10. Austin is pitching a tent for a camping trip. He wants the entrance to the tent to be an isosceles triangle with a base of 5 feet and base angles measuring 70°, as shown below. Which of the following expressions gives the perimeter of the entrance in feet?

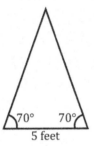

A. $5+3\left(\dfrac{5\sin 70°}{\sin 40°}\right)$

B. $5+2\left(\dfrac{5\sin 70°}{\sin 40°}\right)$

C. $5+2(5\sin 70°)$

D. $5+2(5\tan 70°)$

E. $3\left(\dfrac{5\sin 70°}{\sin 40°}\right)$

11. In △FGH below, FH = 12 meters. To the nearest tenth of a meter, how many meters long is FG?
(Note: $\sin 75° \approx 0.966, \sin 65° \approx 0.906, \sin 40° \approx 0.643$)

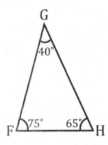

A. 8.0
B. 8.5
C. 11.3
D. 16.9
E. 18.0

12. An isosceles triangle has legs of equal length 12 furlongs, and a third side of x furlongs. The degree measure between the sides that are 12 furlongs long is θ. In terms of x, $\theta = ?$

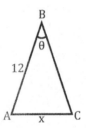

A. $\cos^{-1}\left(\dfrac{x^2+288}{288}\right)$

B. $\dfrac{-x^2+288}{288}$

C. $\dfrac{x^2-24}{288}$

D. $\cos^{-1}\left(\dfrac{288-x^2}{288}\right)$

E. $\dfrac{24-x^2}{288}$

ANSWER KEY

1. E   2. E   3. C   4. D   5. A   6. D   7. C   8. E   9. B   10. B   11. D   12. D

ANSWER EXPLANATIONS

1. **E.** We are given two side lengths and the angle opposite the mystery side, so we should use the law of cosines. Plug in $a = 3, b = 2$, and $\angle C = 97°$ to get the equation $c^2 = 3^2 + 2^2 - 2(3)(2)\cos°\angle C$. Taking the square root of both sides, we get $c = \sqrt{3^2 + 2^2 - 2(3)(2)\cos(97°)} = \sqrt{2^2 + 3^2 - 2^2(3)\cos(97°)}$.

2. **E.** We know that the angles in a triangle always add up to 180°, so we can find $\angle X$ by subtracting $\angle Z$ and $\angle Y$ from 180°. We get $\angle X = 180° - 39° - 65° = 76°$. Using the law of sines, we can solve for the length of $\overline{XZ}$ by plugging in $\angle Y = 65°$ as the opposite angle of $\overline{XZ}$, and $\angle X = 76°$ as the opposite angle of $\overline{ZY}$, which is given to be a length of 13. Since the law of sines is a ratio, as long as the numerators and denominators correspond consistently, we can arrange the proportion any way (although traditionally the opposite angles are the numerators and their corresponding sides the denominators). In this case, however, we will put the side length on top to make simplifying easier, giving us $\dfrac{13}{\sin 76°} = \dfrac{\overline{XY}}{\sin 65°}$. Multiplying both sides by $\sin 65°$, we get $\overline{XY} = \dfrac{13(\sin 65°)}{\sin 76°}$.

3. **C.** Because this is not a right triangle and we are told to solve for a distance, we must use either the law of sines or the law of cosines. We have all three angles of a triangle (since two are given, we can solve for the third with the triangle sum theorem) and one side length, which means we cannot use the law of cosines (as that requires 2 side lengths). We must use the law of sines, and the only answer choice that even somewhat resembles what our answer would be if we did is C. Solve for the third angle by subtracting the other two from 180. $180 - 21 - 141 = 18°$. Then apply the law of sines: $\dfrac{\sin 21°}{800} = \dfrac{\sin 18°}{d}$, cross multiplying we get $d(\sin 21°) = 800(\sin 18°)$ and then can isolate $d = \dfrac{800 \sin 18°}{\sin 21°}$, C.

4. **D.** In this problem, again it is not specified that we need to use the law of sines or the law of cosines, but in considering the form of the answer choices, and knowing we have SAS and need the third side, we can determine we should use the law of cosines: $c^2 = a^2 + b^2 - 2ab\cos \angle C$. Use law of cosines whenever you only have one angle involved in the problem at any time. The angle between the two boats is found by subtracting the smaller angle from the larger angle: $220° - 110° = 110°$. Looking at the answer choices, we don't have to have the law of cosines memorized, but we do have to be familiar enough to recognize it. Only D and E have the "root" around the whole calculation as would be necessary to solve for the side. In this case, in order to know that D is the right answer, we recognize we need $\cos 110°$ in the answer as it is the angle opposite our missing side in the "triangle." Answer D is the only possible answer. See page 324 to review the law of cosines.

5. **A.** Using the law of sines, where the ratio of the sides and the angles opposite to each side are equal, we see that $\dfrac{\sin(\angle X)}{YZ} = \dfrac{\sin(\angle Y)}{XZ} = \dfrac{\sin(\angle Z)}{XY}$, which means $\dfrac{\sin 97°}{YZ} = \dfrac{\sin 36°}{30} = \dfrac{\sin 47°}{XY}$. Thus, our answer is A, $\dfrac{\sin 36°}{30} = \dfrac{\sin 47°}{XY}$.

6. **D.** Using the law of sines, which states that given $\triangle ABC$, $\dfrac{\sin(\angle A)}{a} = \dfrac{\sin(\angle B)}{b} = \dfrac{\sin(\angle C)}{c}$, we get $\dfrac{\sin 71°}{42} = \dfrac{\sin \angle B}{AC} = \dfrac{\sin 52°}{AB}$. If we only look at part of this equation that is relevant to us, the part with the angles and side length plugged in, and plug in the sine values given to us in the note, we get $\dfrac{.946}{42} = \dfrac{.788}{AB}$. Solving for $\overline{AB}$, we get $\overline{AB} = 34.9852... \approx 35$.

7. **C.** Because it is not a right triangle and we are given the formula for the law of cosines, we simply need to plug in the appropriate values. Let the retail store be vertex A, the warehouse be vertex B, and the factory be vertex C; then a is the distance between the factory and the warehouse, b is the distance between the factory and the store, and c is the distance between the warehouse and retail store, which is what we are solving for. Plug in the given values into the law of cosines: $c^2 = 6^2 + 7.1^2 - 2(6)(7.1)\cos 11°$. Taking the square root of both sides, we get $c = \sqrt{6^2 + 7.1^2 - 2(6)(7.1)\cos 11°}$.

LAWS OF SINES AND COSINES   ANSWERS

8. **E.** The law of cosines states that $c^2 = a^2 + b^2 - 2ab\cos\angle C$. In our case, we want to find the measure of $\overline{BC}$. We plug in the other legs of the triangle for b and c, and the angle opposite from $\overline{BC}$ for $\angle C$. Our formula is $(BC)^2 = 9^2 + 15^2 - 2(9)(15)\cos 70°$. Square root both sides to isolate our answer: $BC = \sqrt{9^2 + 15^2 - 2(9)(15)\cos 70°}$.

9. **B.** The law of sines states that for a triangle with lengths $a, b,$ and c opposite angles $\angle A, \angle B$, and $\angle C$ respectively, $\frac{\sin(\angle A)}{a} = \frac{\sin(\angle B)}{b} = \frac{\sin(\angle C)}{c}$. We want to find AB, so we will use the measure of the angle opposite it and a given opposite angle-side pair to plug into our formula: $\frac{\sin 35°}{AB} = \frac{\sin 50°}{25}$. Cross multiply and divide by $\sin 50°$ to get $AB = \frac{25\sin 35°}{\sin 50°}$.

10. **B.** The law of sines states that for a triangle with lengths $a, b,$ and c and with opposite angles $\angle A, \angle B$, and $\angle C$ respectively, $\frac{\sin(\angle A)}{a} = \frac{\sin(\angle B)}{b} = \frac{\sin(\angle C)}{c}$. We know that the missing angle of the isosceles triangle is $40°$ because all of the angles must sum to $180°$. To find the measure of the other two legs of the isosceles triangle, n, we plug into the law of sines: $\frac{\sin 70°}{n} = \frac{\sin 40°}{5}$. Rearranging the equation yields $n = \frac{5\sin 70°}{\sin 40°}$. The perimeter is equal to 5, the bottom leg, plus $2n$, where n equals the length of one of the other two legs: $5 + 2\left(\frac{5\sin 70°}{\sin 40°}\right)$.

11. **D.** The law of sines states that for a triangle with lengths $a, b,$ and c and with opposite angles $\angle A, \angle B$, and $\angle C$ respectively, $\frac{\sin(\angle A)}{a} = \frac{\sin(\angle B)}{b} = \frac{\sin(\angle C)}{c}$. Plugging in our values, including the approximations of the sines, we get $\frac{0.906}{FG} = \frac{0.643}{12}$. Rearranging gives us $FG = \frac{12(0.906)}{0.643} \approx 16.9$.

12. **D.** The law of cosines states that for a triangle with lengths $a, b,$ and c and with opposite angles $\angle A, \angle B$ and $\angle C$ respectively, $a^2 = b^2 + c^2 - 2bc\cos\angle A$. Plugging in the values we have, where $a = x$, b and c are identical sides of length 12 and $\angle A = \theta$, $x^2 = 12^2 + 12^2 - 2(12)(12)\cos\theta$. Simplify to isolate the $\cos\theta$ term: $(-288)\cos\theta = x^2 - 288$. $\cos\theta = \frac{(x^2 - 288)}{-288} = \frac{288 - x^2}{288}$. Now to find θ, we take the arccos: $\cos^{-1}\left(\frac{288 - x^2}{288}\right)$.

CHAPTER

21 TRIGONOMETRY GRAPHS

> ## SKILLS TO KNOW
>
> - The shape and orientation of basic trigonometric function graphs
> - Standard equations for trig graphs and their basic transformations (scalars, coefficients, and constants
> - Graphs of inverse trigonometric functions (i.e. cosecant, secant, cotangent)
> - Finding minimum and maximum values

NOTE: This chapter covers problems that mostly appear in the last ten to fifteen or so problems of the exam. If you're not aiming for a math score of 32+, your time may be better spent on other chapters first. I also recommend studying this chapter close to your exam date. Remembering the intricacies of amplitude, period, etc. can get you fast and easy points, but requires memorizing details that can be easily forgotten with time.

GRAPHS OF BASIC TRIGONOMETRIC FUNCTIONS

If you're aiming for a 32+, I recommend you become familiar with the basic graph forms of trig functions. If you have them memorized, these questions are much easier.

Before we get too far, let's define some terms:

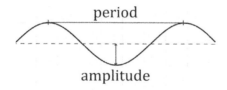

Period: The period of a trig graph is **the horizontal distance it takes the graph to complete one full cycle**. You can measure the distance between two maximum points (shown above) or two minimum points. Alternatively, for sin and cos: measuring from the midline, trace one wave up to the maximum and one wave down to the minimum and then back to the midline. For tan, the period is the portion of the graph between two adjacent asymptotes or the width of one single "squiggly" line that approaches positive and negative infinity. (More on tangent on the next page.)

Amplitude: The amplitude is the vertical height from the midline you could draw horizontally through a sin or cos function and the maximum or minimum of the graph. You can also think of it as **the half height of the wave**. Note: the graph of tan has no amplitude.

TRIGONOMETRY GRAPHS — SKILLS

Sine Function

The sine function, which is the ratio of the opposite to the hypotenuse, starts at **0** and moves between y-values of **1** and **−1** in a sinusoidal pattern (in other words, it looks like a wave).

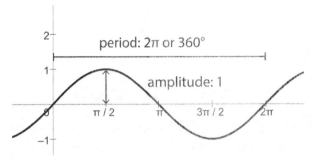

$\sin(0) = 0$ (passes through the origin)

Cosine Function:

The cosine function, which is the ratio of the adjacent to the hypotenuse, starts at **1** and moves between y-values of **−1** and **1** in a sinusoidal pattern (aka in a wave).

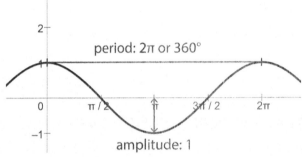

$\cos(0) = 1$, passes through $(0,1)$, not the origin.

Tangent Function

The tangent function, which is the ratio of the opposite to the adjacent, has a range of $-\infty$ to ∞.

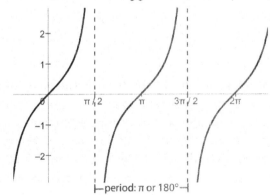

Amplitude: N/A
Passes through origin $(0,0)$ with a complete cycle centered at the origin.

LESS COMMON TRIG GRAPHS

If you don't want a 34-36, skip these!

Cosecant Function $\left(\dfrac{1}{\sin x}\right)$

The cosecant function is the reciprocal of sine, and therefore the ratio of the hypotenuse over the opposite. It starts at ∞ and has a range of $-\infty < y \leq -1$ and $1 \leq y < \infty$, with a period of 2π radians or $360°$. It has no amplitude.

Secant Function $\left(\dfrac{1}{\cos x}\right)$

The secant function is the reciprocal of cosine, and therefore the ratio of the hypotenuse over the adjacent. It starts at 1 and has a range of $-\infty < y \leq -1$ and $1 \leq y < \infty$, with a period of 2π radians or $360°$. It has no amplitude.

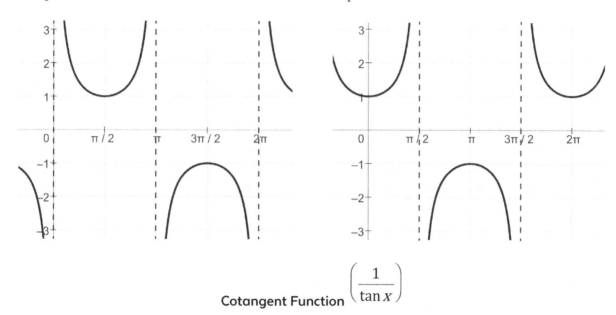

Cotangent Function $\left(\dfrac{1}{\tan x}\right)$

The cotangent function is the reciprocal of tangent, and therefore the ratio of the adjacent over the opposite. It starts at ∞ and has a range of $-\infty$ to ∞ with a period of π radians or $180°$. Like tangent, cotangent has no amplitude (a).

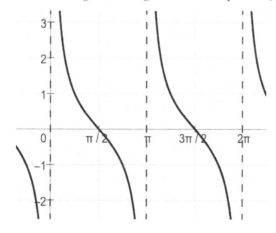

TRIGONOMETRY GRAPHS — SKILLS

BASIC TRIGONOMETRIC TRANSFORMATIONS

Most trigonometric graph problems require you to know the parent equation of the graph and how each constant affects the graph. As with all translations, remember you can always shift a trig graph to the right replacing every x with $(x-h)$ or to the left with $(x+h)$. Similarly, you can shift it up by replacing y with $(y-k)$ and down by replacing y with $(y+k)$. In the standard forms below, the sign on the k will be positive for upward movements and negative for downward movements, as it is on the opposite side of the equation as y. Here's an overview of the variables and constants in the standard equation forms:

	sin	cos	tan
Parent Equation	$y = a\sin b(x-h)+k$	$y = a\cos b(x-h)+k$	$y = a\tan b(x-h)+k$
Amplitude	a	a	undefined
Period	$\dfrac{2\pi}{b}$	$\dfrac{2\pi}{b}$	$\dfrac{\pi}{b}$
Phase Shift (horizontal)*	h to the right	h to the right	h to the right
Vertical Shift*	k up	k up	k up

*If h or k are negative numbers, then the shift is in the opposite direction. **The value of k is the y-value of the "midline" of the graph,** halfway vertically between the maximimum and minimum.

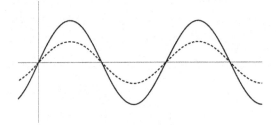

You should be able to look at graphs of different trig functions and see how one is a transformation of the other. Here, for example, we have two sin waves. The solid line is taller, so it has a greater amplitude. Thus its equation should have a larger number in the "a" position of its equation, $y = a\sin b(x-h)+k$. Conversely, the dotted line should have a smaller number in the "a" position of its equation. Since the graphs appear to have the same period, the rest of the values in their equations are likely identical to each other.

 CALCULATOR TIP: When in doubt on trig graph translations, use your graphing calculator! Make up numbers that adhere to the parameters of the question and you can gauge how these graphs behave.

Translations with Period

To affect the period of a trig graph, you'll need to change the value at "b." **Remember, the coefficient b and the value of the period are INVERSELY proportional.** The bigger the value of "b" the smaller the period. The smaller the "b," the larger the period.

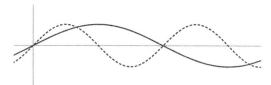

Notice how the solid line graph is "wider"? This graph would have a smaller value of b than the dotted line because it has a larger period.

Here's another example: $\tan(x)$ (dashed line) has a period of π, while $\tan(2x)$ (solid line) has a period of $\dfrac{\pi}{2}$.

As you can see, the one with the smaller "b" value of 1 ($\tan(x)$) has a larger period.

Let's do a transformation from $y = \sin(x)$ to $y = \sin(4x)$.

Because $b = 4$, we know that there will be 4 complete cycles within the span of 2π. Let's take a point that is easily visible and divisible by 4 to do our transformation: $(2\pi, 0)$

To be accurate, we can plot the points.

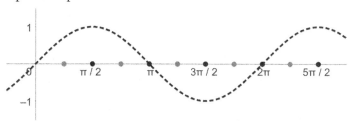

The dashed line is our original equation, $y = \sin(x)$. We should dot $\dfrac{\pi}{2}$, π, $\dfrac{3\pi}{2}$, and 2π first. Each of these is a "stopping" point of a full cycle in the graph. So the new graph will pass through each of these points on the x-axis. The graph will also pass through the points half way between these points. So I dot halfway between each dark dot to label the remaining times the trig function will hit $x = 0$. Then, I can start drawing the graph:

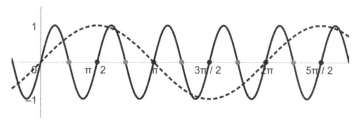

Notice how at every dark dot, the sine graph makes a full wave and restarts? And that there are 4 complete cycles by the time the function hits 2π?

TRIGONOMETRY GRAPHS — SKILLS

The graph below shows $y = \sin(x)$. Which of the following choice is a graph of $y = \sin\left(x - \dfrac{\pi}{2}\right)$?

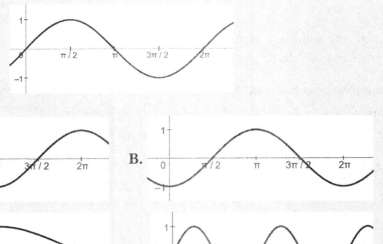

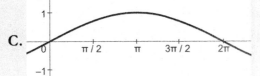

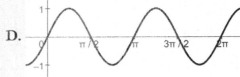

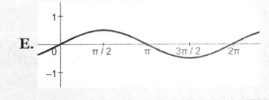

To solve, we know this has to be a phase shift (horizontal x movement), as the transformation is limited to the $(x - h)$ position. Because h equals $\dfrac{\pi}{2}$, we simply move the original graph $\dfrac{\pi}{2}$ to the right. Thus the answer is B.

Answer: **B.**

SKILLS TRIGONOMETRY GRAPHS

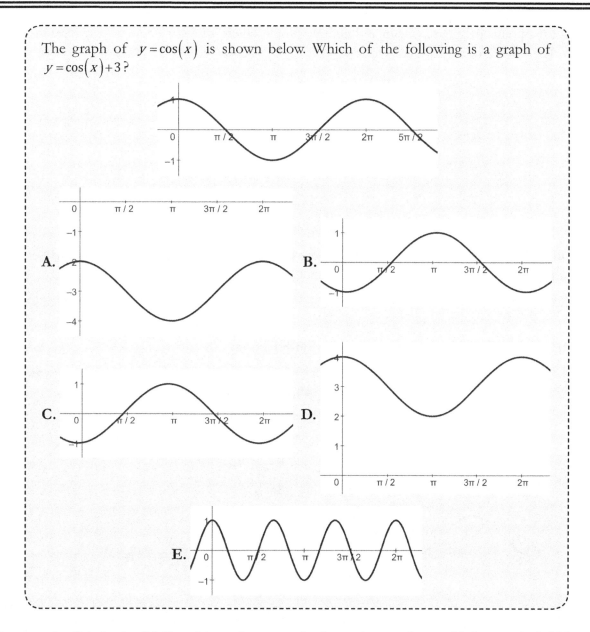

The graph of $y = \cos(x)$ is shown below. Which of the following is a graph of $y = \cos(x) + 3$?

Because the 3 is in the "k" position of our standard equation, we know this is a vertical shift and the midline of our graph is at 3.

A safe way to translate a trig function vertically or horizontally is to pluck points and move them up or sideways as reference points. To illustrate, you can draw the points of all the maximums and minimums of each crest before drawing the entire line. Move each up 3. Here we draw:

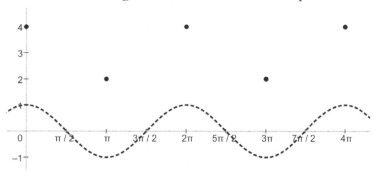

We plot the points of the maximum and minimum of each crest to make sure we will have the right graph. Then we can draw the actual line:

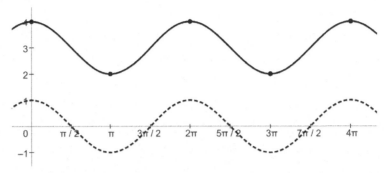

The original equation, $y = \cos(x)$, is seen in the dashed line. So now we can pick the right answer from the graphs given. Clearly the solid line matches answer choice D.

Answer: **D**.

COMBINING VERTICAL AND LATERAL SHIFTS

Which of the following is the graph of $\cos(x+2) - 5$?

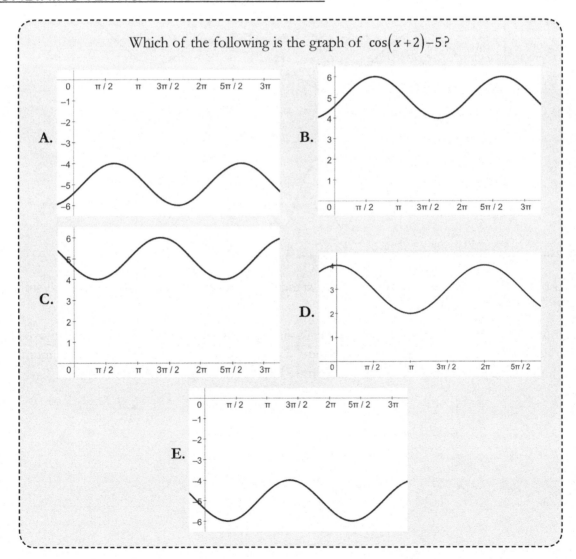

Here we not only need to know the cosine graph, we also have to modify it. This is harder than most trig problems on the ACT®, but it could happen!

First I'll sketch cosine and then shift the max/min points left by 2 and down by 5 (using h and k of $h=-2$ and $k=-5$—remember when h and k are negative, we move left and down).

For h, I'll take the minimum at π (3.14) and move it to ~1.14, or about $\frac{1}{3}$ the distance from the origin towards π. For k, I'll take the mid-line at 0 and move it to -5.

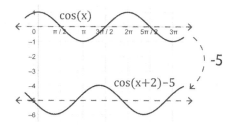

For the ACT®, if all this overwhelms you, you can always use your calculator. If the problem you're solving has variables not numbers as coefficients, make up numbers that adhere to any parameters given.

HOW TO GRAPH TRIG FUNCTIONS ON YOUR TI-84 CALCULATOR

1. Hit MODE. Now, put your calculator in "FUNCTION" and "RADIAN" mode.
2. Use $y=$, enter your equation. Hit GRAPH.
3. Press "ZOOM" then 7.

If you have a different calculator, search online for instructions.

Let's try an example with every possible transformation:

> What is the transformation from $y=\sin(x)$ to $y=0.5\sin(3x-1)+2$?

Simply graph both:

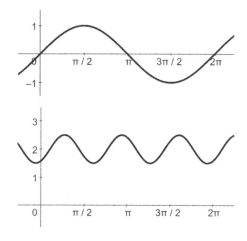

CHAPTER 21

TRIGONOMETRY GRAPHS SKILLS

CALCULATOR TIP: Be sure to make the most efficient use of your calculator so you don't run out of time! Graphs take time. Move on while your calculator is rendering the picture and come back once it's done!

FINDING MAXIMUMS AND MINIMUMS (35+ ONLY)

Like full transformations, the (local) maximum or minimum can be found using your graphing calculator.

On the TI-series, after you graph your function, press "2ND" and then "TRACE" (to get "CALC"), to reveal a drop down menu. You can choose "MINIMUM" or "MAXIMUM" and then you'll be brought back to the graph screen. It will ask for "LEFT BOUND," so move your marker to the left side of the local max/min and press "ENTER." Then do the same but on the right for "RIGHT BOUND". After you press enter, the calculator will automatically give you the local min/max (depending on what you chose to look for). For example, look at the screencaps below:

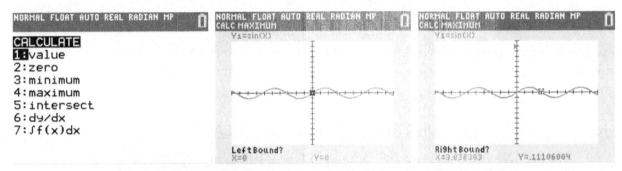

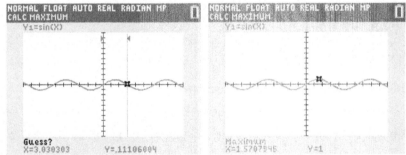

340 CHAPTER 21

For example, if our equation is $y = \csc(0.5x)$, and we're looking for the local minimum between 0 and 2π, we can first graph the equation on our calculator, knowing **csc** is **1/sin** (graph y=(1/(sin(.5x)):

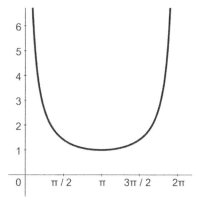

Then, following the instructions above, we can move the "left bound" line to 0 and the "right bound" line to 2π. Then we can let the calculator work for us and see that the local minimum (i.e. y-value that is smallest) for all values on the x-axis between 0 and 2π occurs at $(\pi, 1)$. The minimum, remember is always a value of y, so the answer here would be 1.

You can also use the trace function to more quickly estimate a minimum or maximum value.

Some of the problems in the set that follow may be more complex than these examples. Whenever they are, however, you can almost always make it simply by just using your calculator to graph!

TRIGONOMETRY GRAPHS — QUESTIONS

1. The following graph of $y = \sin(x)$ is shown in the standard (x, y) coordinate plane below. What is the period of $\sin(x)$?

   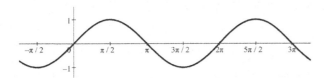

   A. $\dfrac{\pi}{4}$

   B. $\dfrac{\pi}{2}$

   C. π

   D. $\dfrac{3\pi}{2}$

   E. 2π

2. If x, y, z, t, and w represent positive real numbers what is the minimum value of the function $f(x) = w\,[\sin t(x - y)] + z$?

   A. $wx - wy + z$
   B. $w\sin(tx) - w\sin(ty) + z$
   C. z
   D. $w + z$
   E. $z - w$

3. In the standard (x, y) coordinate plane, what is the range of the function $f(a) = -7\cos 3(a + 4) + 5$?

   A. $-12 \le f(a) \le 2$
   B. $-2 \le f(a) \le 12$
   C. $5 - 14\pi \le f(a) \le 5$
   D. $0 \le f(a) \le 2\pi$
   E. $-2 \le f(a) \le 5$

4. What is the amplitude of the graph of the equation $y - 4 = \dfrac{1}{2}\cos\dfrac{7\alpha}{9}$?

   A. $\dfrac{7a}{9}$

   B. 2

   C. 1

   D. $\dfrac{1}{2}$

   E. 4

5. A trigonometric function with equation $y = \cos(bx + c)$ where b and c are real numbers, is graphed in the standard (x, y) coordinate plane below. The period of this function $f(x)$ is the smallest possible number p such that $f(x + p) = f(x)$ for every real number x. One of the following is the period of the function. Which is it?

   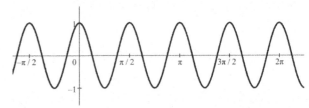

   A. $\dfrac{\pi}{2}$

   B. π

   C. 2π

   D. 1

   E. 2

6. The graph of $y = -a\sin(bx) + c$ is shown below for certain positive values of a, b, and c. One of the following values is equal to a. Which one?

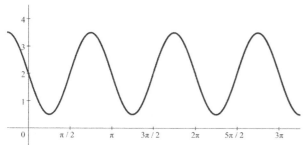

A. $\dfrac{1}{2}$

B. $\dfrac{\pi}{2}$

C. $\dfrac{3}{2}$

D. 2

E. $\dfrac{7}{2}$

7. Which of the following trigonometric functions has an amplitude of $\dfrac{1}{3}$?

A. $f(x) = 3\tan\left(x + \dfrac{1}{3}\right)$

B. $f(x) = \dfrac{1}{3}\tan(x)$

C. $f(x) = \dfrac{1}{3}\cos(x)$

D. $f(x) = 3\sin(x)$

E. $f(x) = \sin\left(\dfrac{1}{3}x\right)$

8. The graph of $f(x) = \sin x$ and $g(x) = \sin\left(x - \dfrac{3\pi}{2}\right) - 2$ are shown in the standard (x, y) coordinate plane below. After one of the following pairs of transformations is applied to the graph of $f(x)$, the image of the graph of $f(x)$ is the graph of $g(x)$. Which pair is it?

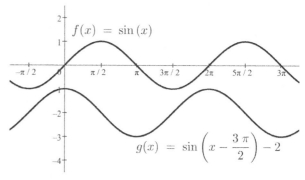

A. Phase shift $\dfrac{3\pi}{2}$ units to the right, and vertical translation 2 units down.

B. Phase shift $\dfrac{3\pi}{2}$ units to the left, and vertical translation 2 units up.

C. Phase shift 2 units to the right, and vertical translation $\dfrac{3\pi}{2}$ units down.

D. Phase shift 2 units to the left, and vertical translation $\dfrac{3\pi}{2}$ units down.

E. Phase shift $\dfrac{3\pi}{2}$ units to the right, and vertical translation 2 units up.

9. For the function graphed below, the x-axis can be partitioned into intervals, each of length p radians, and the curve over any one interval is a repetition of the curve over each of the other intervals. What is the least possible value for p, the period of the function?

A. $\dfrac{\pi}{4}$

B. $\dfrac{\pi}{2}$

C. π

D. 2π

E. 4π

10. If a, b, c, d, and x represent positive real numbers, what is the maximum value of the function $f(x) = a\left[\sin(b(x+c))\right] + d$?

A. $a - d$
B. $d - a$
C. $a + d$
D. d
E. a

11. The domain of function $y(x) = \dfrac{\cos(3x+1)}{2} + 4$ is all real numbers. Which of the following is the range of function $y(x)$?

A. $0 < y < 4$
B. $\dfrac{7}{2} < y < \dfrac{9}{2}$
C. $2 < y < 4$
D. $-\dfrac{1}{2} < y < \dfrac{1}{2}$
E. $2 < y < 6$

12. The graph of $y = 2\tan x$ is shown in the standard coordinate plane below. What is the period of $2\tan x$?

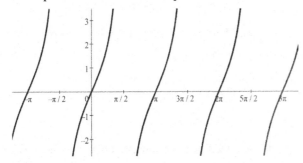

A. $\dfrac{1}{2}$
B. $\dfrac{\pi}{2}$
C. π
D. 2
E. 2π

13. A trigonometric function with equation $y = a\cos bx + c$, where a, b, and c are real numbers, is graphed in the standard (x, y) coordinate plane below. The period of this function is the smallest positive number p such that $f(x + p) = f(x)$ for all real numbers x. What is the period of the following function?

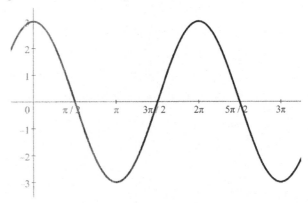

A. 1
B. $\dfrac{\pi}{2}$
C. π
D. 2π
E. 2

14. The graph below could represent which of the following equations?

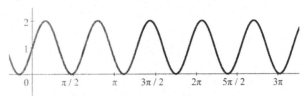

A. $\sin(\pi x + 1)$
B. $\pi \sin(x + 1)$
C. $\sin(x) + 1$
D. $\sin(2\pi x) + 1$
E. $\sin(\pi x) + 1$

15. If $0 \leq x \leq \pi$, what is the maximum value of the function $f(x) = -2\sin x$?

A. -2π
B. -2
C. 0
D. 2
E. 2π

ANSWER KEY

1. E 2. E 3. B 4. D 5. A 6. C 7. C 8. A 9. C 10. C 11. B 12. C 13. D 14. E
15. C

ANSWER EXPLANATIONS

1. **E.** The period of a graph is the horizontal length of one cycle of the curve. Since $\sin 0 = 0$ and $\sin 2\pi = 0$, the cycle starts at 0 and ends at 2π. Thus the period is $2\pi - 0 = 2\pi$.

2. **E.** We know that the range of $\sin x$ is from -1 to 1. In order to minimize the value of the entire function in the question, we want the sine term to be its minimum possible value. Let's isolate a portion of the function that doesn't affect the y-values or range of the function: $\sin t(x-y)$ at its very minimum is equal to -1 because the y and the t do not affect the vertical height of the function. We can thus substitute in -1 for this portion of the more complex given function as that is the minimum value that portion of the equation will generate. Substituting, this leaves us with $f(x) = w(-1) + z$. So at its minimum, $f(x) = z - w$. If this was tough, ignore everything that doesn't affect the vertical range and apply the amplitude and z value only to the equation to create w(sinx)+z. Then make up numbers for w and z and plug this simpler function into your graphing calculator.

3. **B.** We know that the range of $\cos x$ is from -1 to 1. So, we can find the range of the function by plugging in the maximum and minimum possible values of $\cos 3(a+4)$, in order to simplify our calculations and ignore the effect that 3 and 4 have on our answer (as these only affect horizontal movements). We know the max and min of this part of the function are 1 and -1, so we plug those in: Plugging in the value negative one for the portion of the expression equal to it ($\cos 3(a+4) = -1$), we get $-7(-1) + 5 = 12$. Plugging in the value positive one since at a maximum $\cos 3(a+4) = 1$, we get $-7(1) + 5 = -2$. So the range is $-2 \leq f(a) \leq 12$. Alternatively, just graph the function in your graphing calculator! It's fast and it works!

4. **D.** A function represented in the form $y = a\cos(bx - c) + d$ has an amplitude of a. In this problem, the function $y - 4 = \frac{1}{2}\cos\frac{7\alpha}{9}$ has amplitude $\frac{1}{2}$ because it is the value in the "a" position.

5. **A.** The question background is complicated and honestly not worth untangling. What you do know is that you have a graph and can analyze it visually. If you trace one full cycle of the graph, you see that it completes a cycle in $\frac{\pi}{2}$ units, so that is the answer.

6. **C.** The negative sign does not affect the answer, because the question is asking for the amplitude. The amplitude is the difference between the maximum and the minimum point the graph reaches, divided by two. It can also be though of as how far above/below the graph reaches from the central axis. In this case, the graph reaches a maximum at what looks like $\frac{7}{2}$ and a minimum at $\frac{1}{2}$, so the amplitude is $\frac{\frac{7}{2} - \frac{1}{2}}{2} = \frac{\frac{6}{2}}{2} = \frac{3}{2}$. You can also eyeball this value on the graph as the distance between the midline and the maximum (around 1.5).

7. **C.** In a trigonometric equation, the amplitude is the coefficient of the cosine or sine function. Only C has $\frac{1}{3}$ in front of a cosine or sine function.

8. **A.** Let's look at the equations given rather than the graph. Trigonometric translations are actually the same as regular graph translations. The difference between $f(x)$ and $g(x)$ is that $\frac{3\pi}{2}$ is subtracted from the x term of $g(x)$, so we get a horizontal translation (or phase shift) $\frac{3\pi}{2}$ to the right, and in $g(x)$, 2 is subtracted from the entire equation, which yields a vertical translation 2 units down.

TRIGONOMETRY GRAPHS   ANSWERS

9. **C.** Starting from the relative maximum (highest y-point) where $x = 0$, we want to find the first point at which the curve is back at this maximum again. Visually, we can see this occurs at $x = \pi$. Because a period is defined as the horizontal distance of one complete cycle of a graph, the distance between pi and 0, or pi, is our period. Choice C is correct.

10. **C.** a is the amplitude constant, and d is essentially our central axis, so the maximum is found by adding the amplitude to the central axis. This gives $d + a$, which is answer C.

11. **B.** Rewrite the function like this: $y(x) = \frac{1}{2}\cos(3x+1) + 4$. The minimum value that a sine or cosine function can equal is the y value of the central axis (which in this graph is 4 or the "k" value in the above form of the equation as explained on pg 334) minus the amplitude (which in this graph is $\frac{1}{2}$, or the a value as explained on pg 334) and the maximum of a sine or cosine function is the sum of the central axis y-value and the amplitude. The range is $minimum < y < maximum$ so for this function the range is $4 - \frac{1}{2} < y < 4 + \frac{1}{2}$, which is $\frac{7}{2} < y < \frac{9}{2}$. Or just graph the function in your graphing calculator and trace or eyeball it!

12. **C.** The period is only affected by a coefficient to the x term (or "b" as described in the equations on page 334), not the entire equation, so because the x coefficient is one, this graph has the same period as the parent tangent graph, which equals π. Alternatively, eyeball the distance between two repeating points on the graph horizontally to see the period is π.

13. **D.** The period of a function is how far a graph travels along the y-axis before it repeats itself. In this case, the cosine graph reaches its peak first at x=0 and then every 2π units in a recurring pattern. Thus, its period is also 2π.

14. **E.** The most obvious change to the graph is that it has been shifted upwards. We can tell it has been shifted up by 1 since its new maximum is 2, while the maximum of the parent function, $\sin x$, is 1. Likewise the midline is not at zero, but at 1. We thus want an equation with a "k" value of 1. Only C, D, and E have a $k = 1$, so eliminate A and B. The midline is equal to "k" in the form listed on page 334. However, note that the period of this function appears to be around 2, not 2π and thus C is incorrect. The period of $\sin(bx)$ is equal to $\frac{2\pi}{b}$. Since here $\frac{2\pi}{b} = 2$, $b = \pi$. Thus, our function is $\sin(\pi x) + 1$. If we are uncertain of the exact value of the period from eyeballing it, we could also backsolve using choice D but would find the period equals: $\frac{2\pi}{b} \to \frac{2\pi}{2\pi} \to 1$. We know the period is greater than 1 as it occurs after $\frac{\pi}{2}$ (which equals around 1.57), so D cannot be correct.

15. **C.** One way to do this is to graph in a calculator and trace from 0 to pi to find the max y-value on that range, which is zero. Alternatively we can consider what we know about the graph given the form described on page 334: parent graph $\sin x$ is flipped upside down by the negative sign in front of -2, the -2 stretches the graph vertically but does not affect its horizontal movement, and we only include values horizontally from 0 through $x = \pi$, which means because sign usually goes positive for its first curve and negative for its second that this pattern is inverted: we start off dipping vertically downward and then climb back up around x= π or half way through a single cycle (2π or the period of sine) of the graph. We can thus determine that we stop before the graph would go above the x-axis. Thus, our maximum value is 0.

CRUSH THE SAT®!

Head to SUPERTUTORTV.COM

Video-Based Online Prep by a Perfect Scoring Tutor*!

>Over 100 hours of video content
>Know-how for every section: Evidence Based Reading, Writing and Language, Math and Essay
>Includes a FREE copy of the Official SAT® Study Guide**, with eight practice tests
>Downloadable Writing & Language, Math, and Reading content drills
>Explanations for 8 Official SAT practice tests**
>Private tutoring experience & results at a fraction of the cost!

Learn the private tutoring secrets that have helped Brooke's students score in the 99+ percentile on the new SAT and get perfect scores on individual sections!

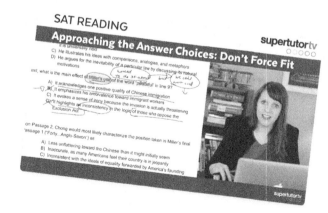

supertutortv
○○○○○

Or get BUNDLE SAVINGS with our SAT & ACT Course Bundle!

*As an adult, Brooke has scored perfectly on the SAT. In high school, she scored perfectly on the math section.
**SAT is a registered trademark of The College Board. This course is not endorsed by or associated with The College Board or ETS.

About the Author:

As a teenager, Brooke scored perfectly on the PSAT, SAT, and ACT math sections, and won awards at regional and state math competitions. As an adult, she also scored perfectly on the GRE math section and perfectly overall on the SAT and ACT. She has been tutoring for over a decade working with classes, small groups, and individual students to increase their opportunity in the college admissions marketplace, whether that means going to Harvard or surviving community college. She's worked for or contracted with over 10 education companies, and has also developed curriculum for three different education firms in everything from test prep to reading comprehension and debate. Her YouTube channel, SupertutorTV, has over 135,000 subscribers and 10,000,000 views. She also has designed a complete video based prep course for the ACT (The Best ACT Prep Course Ever) and for the SAT (The Best SAT Prep Course Ever) available at SupertutorTV.com.

In addition to her education work, Brooke is also a filmmaker. Before launching SupertutorTV, Brooke worked for two seasons as a showrunner for a short form series on the Yahoo! Screen platform with Emmy nominated host Cat Deeley (So You Think You Can Dance); has co-produced, written, and edited a documentary, Dear Albania, for public television; and also has field produced EPK and digital content for Stand Up to Cancer, featuring top talent from Tom Hanks to Dave Matthews Band.

Brooke graduated with honors from Stanford University with a BA in American Studies and also holds an MFA in Cinematic Arts Production from the University of Southern California. She lives in Santa Monica, California with her husband and energetic toddler.

Made in the USA
Monee, IL
05 February 2021